软件项目开发全程实录

Java Web 项目开发全程实录
（第 2 版）

明日科技　编著

清华大学出版社
北京

内 容 简 介

本书精选 8 个热门项目，涉及 Servlet、SSM 框架和 Spring Boot 框架三大 Java Web 重点应用方向，实用性非常强。具体项目包含：明日科技门户网、购好物网络商城、员工信息管理系统、好生活个人账本、嗨乐影评平台、电瓶车品牌信息管理系统、寻物启事网站和明日之星物业管理系统。本书从软件工程的角度出发，按照项目开发的顺序，系统、全面地讲解每一个项目的开发实现过程。在体例上，每章一个项目，统一采用"开发背景→系统设计→技术准备→数据库设计/公共模块实现/各功能模块实现→项目运行→源码下载"的形式完整呈现项目，给读者明确的成就感，可以让读者快速积累实际项目经验与技巧，早日实现就业目标。

另外，本书配备丰富的 Java 在线开发资源库和电子课件，主要内容如下：

- ☑ 技术资源库：426 个核心技术点
- ☑ 实例资源库：707 个应用实例
- ☑ 源码资源库：747 套项目与案例源码
- ☑ PPT 电子课件
- ☑ 技巧资源库：583 个开发技巧
- ☑ 项目资源库：40 个精选项目
- ☑ 视频资源库：644 集学习视频

本书可为 Java 和 Java Web 入门自学者提供更广泛的项目实战场景，可为计算机专业学生进行项目实训、毕业设计提供项目参考，可供计算机专业教师、IT 培训讲师用作教学参考资料，还可作为软件工程师、IT 求职者、编程爱好者进行项目开发时的参考书。

本书封面贴有清华大学出版社防伪标签，无标签者不得销售。

版权所有，侵权必究。举报：010-62782989，beiqinquan@tup.tsinghua.edu.cn。

图书在版编目（CIP）数据

Java Web 项目开发全程实录 / 明日科技编著. -- 2 版. -- 北京：清华大学出版社, 2025.1. -- (软件项目开发全程实录). -- ISBN 978-7-302-67568-6

Ⅰ. TP312.8

中国国家版本馆 CIP 数据核字第 2024AA2037 号

责任编辑：贾小红
封面设计：秦 丽
版式设计：楠竹文化
责任校对：范文芳
责任印制：杨 艳

出版发行：清华大学出版社
网　　址：https://www.tup.com.cn, https://www.wqxuetang.com
地　　址：北京清华大学学研大厦 A 座
邮　　编：100084
社 总 机：010-83470000
邮　　购：010-62786544
投稿与读者服务：010-62776969, c-service@tup.tsinghua.edu.cn
质量反馈：010-62772015, zhiliang@tup.tsinghua.edu.cn

印 装 者：定州启航印刷有限公司
经　　销：全国新华书店
开　　本：203mm×260mm　　印　张：20　　字　数：648 千字
版　　次：2019 年 1 月第 1 版　　2025 年 1 月第 2 版　　印　次：2025 年 1 月第 1 次印刷
定　　价：89.80 元

产品编号：107427-01

如何使用本书开发资源库

本书赠送价值 999 元的"Java 在线开发资源库"一年的免费使用权限,结合图书和开发资源库,读者可快速提升编程水平和解决实际问题的能力。

1. VIP 会员注册

刮开并扫描图书封底的防盗码,按提示绑定手机微信,然后扫描右侧二维码,打开明日科技账号注册页面,填写注册信息后将自动获取一年(自注册之日起)的 Java 在线开发资源库的 VIP 使用权限。

读者在注册、使用开发资源库时有任何问题,均可咨询明日科技官网页面上的客服电话。

2. 开发资源库简介

Java 开发资源库中提供了技术资源库(426 个核心技术点)、技巧资源库(583 个开发技巧)、实例资源库(707 个应用实例)、项目资源库(40 个精选项目)、源码资源库(747 套项目与案例源码)、视频资源库(644 集学习视频),共计六大类、3147 项学习资源。学会、练熟、用好这些资源,读者可在最短的时间内快速提升自己,从一名新手晋升为一名软件工程师。

3. 开发资源库的使用方法

在学习本书的各项目时,可以通过 Java 开发资源库提供的大量技术点、技巧、热点实例等快速回顾或了解相关的 Java 编程知识和技巧,提升学习效率。

除此之外,开发资源库还配备了更多的大型实战项目,供读者进一步扩展学习,以提升编程兴趣和信心,积累项目经验。

另外,利用页面上方的搜索栏,还可以对技术、技巧、实例、项目、源码、视频等资源进行快速查阅。

万事俱备后,读者该到软件开发的主战场上接受洗礼了。本书资源包中提供了 Java 各方向的面试真题,是求职面试的绝佳指南。读者可扫描封底的"文泉云盘"二维码获取。

前言
Preface

丛书说明："软件项目开发全程实录"丛书第 1 版于 2008 年 6 月出版，因其定位于项目开发案例、面向实际开发应用，并解决了社会需求和高校课程设置相对脱节的痛点，在软件项目开发类图书市场上产生了很大的反响，在全国软件项目开发零售图书排行榜中名列前茅。

"软件项目开发全程实录"丛书第 2 版于 2011 年 1 月出版，第 3 版于 2013 年 10 月出版，第 4 版于 2018 年 5 月出版。经过十六年的锤炼打造，不仅深受广大程序员的喜爱，还被百余所高校选为计算机科学、软件工程等相关专业的教材及教学参考用书，更被广大高校学子用作毕业设计和工作实习的必备参考用书。

"软件项目开发全程实录"丛书第 5 版在继承前 4 版所有优点的基础上，进行了大幅度的改版升级。首先，结合当前技术发展的最新趋势与市场需求，增加了程序员求职急需的新图书品种；其次，对图书内容进行了深度更新、优化，新增了当前热门的流行项目，优化了原有经典项目，将开发环境和工具更新为目前的新版本等，使之更与时代接轨，更适合读者学习；最后，录制了全新的项目精讲视频，并配备了更加丰富的学习资源与服务，可以给读者带来更好的项目学习及使用体验。

Java Web 是使用 Java 技术来解决 Web 领域相关问题的技术栈，主要包括 Web 服务端和 Web 客户端两部分。本书以中小型项目为载体，带领读者切身感受 Java Web 开发的实际过程，可以让读者深刻体会 Java Web 核心技术在项目开发中的具体应用。全书内容不是枯燥的语法和陌生的术语，而是一步一步地引导读者实现一个个热门的项目，从而激发读者学习软件开发的兴趣，变被动学习为主动学习。另外，本书的项目开发过程完整，不但适合在学习软件开发时作为中小型项目开发的参考书，而且可以作为毕业设计的项目参考书。

本书内容

本书提供了 Java Web 三大热门方向的项目，共 8 章，具体内容如下。

第 1 篇：Servlet 应用项目。本篇主要通过"明日科技门户网"和"购好物网络商城"两个功能完善的项目，帮助读者快速掌握 Servlet 技术在实际 Web 项目开发中的应用，并让读者体验使用 Servlet 开发 Web 项目的完整过程。

第 2 篇：SSM 框架应用项目。本篇主要通过"员工信息管理系统""好生活个人账本""嗨乐影评平台"3 个项目，详细讲解使用 SSM 框架开发 Web 项目的完整过程，帮助读者快速掌握使用 SSM 框架开发 Web 项目的核心技术。

第 3 篇：Spring Boot 框架应用项目。本篇主要通过"电瓶车品牌信息管理系统""寻物启事网站""明日之星物业管理系统"3 个项目，详细讲解使用 Spring Boot 框架开发 Web 项目的完整过程，帮助读者快速掌握使用 Spring Boot 框架开发 Web 项目的核心技术。

本书特点

☑ **项目典型**。本书精选 8 个热点项目，涉及 Servlet、SSM 框架和 Spring Boot 框架三大 Java Web 重

点应用方向。所有项目均从实际应用角度出发，可以让读者从项目实现学习中积累丰富的开发经验。
- ☑ **流程清晰**。本书项目从软件工程的角度出发，统一采用"开发背景→系统设计→技术准备→项目实现相关→项目运行→源码下载"的流程进行讲解，可以让读者更加清晰地了解项目的完整开发流程。
- ☑ **技术新颖**。本书所有项目的实现技术均采用目前业内推荐使用的最新稳定版本，与时俱进，实用性极强。同时，项目全部配备"技术准备"环节，对项目中用到的 Java Web 基本技术点、高级应用、第三方模块等进行精要讲解，为初级编程人员参与项目开发扫清了障碍。
- ☑ **栏目精彩**。本书根据项目学习的需要，在每个项目讲解过程的关键位置添加了"注意""说明"等特色栏目，点拨项目的开发要点和精华，以便读者能更快地掌握相关技术的应用技巧。
- ☑ **源码下载**。本书中的每个项目最后都安排了"源码下载"一节，读者能够通过扫描二维码下载对应项目的完整源码，以方便学习。
- ☑ **项目视频**。本书为每个项目提供了项目精讲微视频，使读者能够更加轻松地搭建、运行、使用项目，并能够随时随地查看和学习。

读者对象

- ☑ 初学编程的自学者
- ☑ 参与项目实训的学生
- ☑ 做毕业设计的学生
- ☑ 参加实习的初级程序员
- ☑ 高等院校的教师
- ☑ IT 培训机构的讲师与学员
- ☑ 程序测试及维护人员
- ☑ 编程爱好者

资源与服务

本书提供了大量的辅助学习资源，同时还提供了专业的知识拓展与答疑服务，旨在帮助读者提高学习效率并解决学习过程中遇到的各种疑难问题。读者需要刮开图书封底的防盗码（刮刮卡），扫描并绑定微信，获取学习权限。

- ☑ **开发环境搭建视频**

搭建环境对于项目开发非常重要，它确保了项目开发在一致的环境下进行，减少了因环境差异导致的错误和冲突。通过搭建开发环境，可以方便地管理项目依赖，提高开发效率。本书提供了开发环境搭建讲解视频，可以引导读者快速准确地搭建本书项目的开发环境。扫描右侧二维码即可观看学习。

开发环境搭建视频

- ☑ **项目精讲视频**

本书每个项目均配有对应的项目精讲微视频，主要针对项目的需求背景、应用价值、功能结构、业务流程、实现逻辑以及所用到的核心技术点进行精要讲解，可以帮助读者了解项目概要，把握项目要领，快速进入学习状态。扫描每章首页的对应二维码即可观看学习。

- ☑ **项目源码**

本书每章一个项目，系统全面地讲解了该项目的设计及实现过程。为了方便读者学习，本书提供了完整的项目源码（包含项目中用到的所有素材，如图片、数据表等）。扫描每章最后的二维码即可下载。

- ☑ **AI 辅助开发手册**

在人工智能浪潮的席卷之下，AI 大模型工具呈现百花齐放之态，辅助编程开发的代码助手类工具不断涌现，可为开发人员提供技术点问答、代码查错、辅助开发等非常实用的服务，极大地提高了编程学习和开发效率。为了帮助读者快速熟悉并使用这些工具，本书专门精心配备了电子版的《AI 辅助开发手册》，不仅为读者提供了各个主流大语言模型的使用指南，而且详细讲解了文心快码（Baidu Comate）、通义灵码、腾讯云 AI 代码助手、iFlyCode 等专业的智能代码助手的使用方法。扫描右侧二维码即可阅读学习。

AI 辅助开发手册

☑ **代码查错器**

为了进一步帮助读者提升学习效率，培养良好的编码习惯，本书配备了由明日科技自主开发的代码查错器。读者可以将本书的项目源码保存为对应的 txt 文件，存放到代码查错器的对应文件夹中，然后自己编写相应的实现代码并与项目源码进行比对，快速找出自己编写的代码与源码不一致或者发生错误的地方。代码查错器配有详细的使用说明文档，扫描右侧二维码即可下载。

代码查错器

☑ **Java 开发资源库**

本书配备了强大的线上 Java 开发资源库，包括技术资源库、技巧资源库、实例资源库、项目资源库、源码资源库、视频资源库。扫描右侧二维码，可登录明日科技网站，获取 Java 开发资源库一年的免费使用权限。

Java 开发资源库

☑ **Java 面试资源库**

本书配备了 Java 面试资源库，精心汇编了大量企业面试真题，是求职面试的绝佳指南。扫描本书封底的"文泉云盘"二维码即可获取。

☑ **教学 PPT**

本书配备了精美的教学 PPT，可供高校教师和培训机构讲师备课使用，也可供读者做知识梳理。扫描本书封底的"文泉云盘"二维码即可下载。另外，登录清华大学出版社网站（www.tup.com.cn），可在本书对应页面查阅教学 PPT 的获取方式。

☑ **学习答疑**

在学习过程中，读者难免会遇到各种疑难问题。本书配有完善的新媒体学习矩阵，包括 IT 今日热榜（实时提供最新技术热点）、微信公众号、学习交流群、400 电话等，可为读者提供专业的知识拓展与答疑服务。扫描右侧二维码，根据提示操作，即可享受答疑服务。

学习答疑

致读者

本书由明日科技 Java 开发团队组织编写，主要编写人员有赵宁、王小科、张鑫、王国辉、刘书娟、高春艳、赛奎春、田旭、葛忠月、杨丽、李颖、程瑞红、张颖鹤等。明日科技是一家专业从事软件开发、教育培训以及软件开发教育资源整合的高科技公司，其编写的图书非常注重选取软件开发中的必需、常用内容，同时也很注重内容的易学性、学习的方便性以及相关知识的拓展性，深受读者喜爱。其编写的图书多次荣获"全行业优秀畅销品种""全国高校出版社优秀畅销书"等奖项，多个品种长期位居同类图书销售排行榜的前列。

在编写本书的过程中，我们始终本着科学、严谨的态度，力求精益求精，但疏漏之处在所难免，敬请广大读者批评指正。

感谢您购买本书，希望本书能成为您的良师益友，成为您步入编程高手之路的踏脚石。

宝剑锋从磨砺出，梅花香自苦寒来。祝读书快乐！

编 者
2024 年 11 月

目　录

第1篇　Servlet 应用项目

第1章　明日科技门户网 ... 2
——Servlet + JSP + MySQL

- 1.1　开发背景 ... 2
- 1.2　系统设计 ... 3
  - 1.2.1　开发环境 ... 3
  - 1.2.2　业务流程 ... 3
  - 1.2.3　功能结构 ... 3
- 1.3　技术准备 ... 4
  - 1.3.1　技术概览 ... 4
  - 1.3.2　Servlet 技术 ... 4
  - 1.3.3　JSP 技术 ... 7
- 1.4　数据库设计 ... 11
  - 1.4.1　数据库概述 ... 11
  - 1.4.2　数据表设计 ... 11
- 1.5　首页模块设计 ... 12
  - 1.5.1　上部企业 Logo 和导航栏设计 ... 12
  - 1.5.2　中部轮播图片信息设计 ... 13
  - 1.5.3　下部功能栏设计 ... 14
  - 1.5.4　底部版权信息栏设计 ... 15
- 1.6　新闻模块设计 ... 15
  - 1.6.1　配置文件的编写 ... 16
  - 1.6.2　显示新闻列表 ... 16
  - 1.6.3　查看新闻内容 ... 18
- 1.7　后端管理员登录模块设计 ... 19
- 1.8　后端新闻管理模块设计 ... 21
  - 1.8.1　添加新闻 ... 22
  - 1.8.2　删除新闻 ... 24
  - 1.8.3　修改新闻 ... 25
- 1.9　后端管理员信息模块设计 ... 27

- 1.10　项目运行 ... 29
- 1.11　源码下载 ... 30

第2章　购好物网络商城 ... 31
——Servlet + JSP + MySQL

- 2.1　开发背景 ... 31
- 2.2　系统设计 ... 32
  - 2.2.1　开发环境 ... 32
  - 2.2.2　业务流程 ... 32
  - 2.2.3　功能结构 ... 33
- 2.3　技术准备 ... 33
  - 2.3.1　技术概览 ... 33
  - 2.3.2　调用支付宝完成支付操作 ... 34
- 2.4　数据库设计 ... 34
  - 2.4.1　数据库概述 ... 34
  - 2.4.2　数据表设计 ... 35
- 2.5　数据库公共类的编写 ... 37
- 2.6　会员注册模块设计 ... 38
  - 2.6.1　会员模型类的编写 ... 38
  - 2.6.2　会员数据库操作接口及其实现类的编写 ... 40
  - 2.6.3　会员注册页面的编写 ... 42
- 2.7　会员登录模块设计 ... 42
  - 2.7.1　会员登录页面的编写 ... 43
  - 2.7.2　生成验证码的编写 ... 46
  - 2.7.3　编写会员登录处理页 ... 47
- 2.8　首页模块设计 ... 47
  - 2.8.1　实现显示最新上架商品的功能 ... 49
  - 2.8.2　实现显示打折商品的功能 ... 49
  - 2.8.3　实现显示热门商品的功能 ... 50

2.9 购物车模块设计 50	2.9.5 实现商品订单提交功能 55
2.9.1 购物车商品模型类的编写 51	2.9.6 实现清空购物车功能 57
2.9.2 实现查看商品详细信息的功能 ... 51	2.9.7 实现继续购物功能 58
2.9.3 实现添加购物车的功能 52	2.10 项目运行 59
2.9.4 实现查看购物车的功能 54	2.11 源码下载 60

第 2 篇　SSM 框架应用项目

第 3 章　员工信息管理系统 62
——SSM + JSP + MySQL

3.1 开发背景 62	3.10 修改员工信息 89
3.2 系统设计 63	3.10.1 控制器类设计 90
3.2.1 开发环境 63	3.10.2 服务类设计 90
3.2.2 业务流程 63	3.10.3 DAO 层设计 90
3.2.3 功能结构 64	3.11 删除员工信息 91
3.3 技术准备 64	3.11.1 控制器类设计 91
3.3.1 技术概览 64	3.11.2 服务类设计 91
3.3.2 Spring 64	3.11.3 DAO 层设计 92
3.3.3 Spring MVC 65	3.12 批量删除员工信息 92
3.3.4 MyBatis 67	3.12.1 控制器类设计 93
3.3.5 SSM 框架 68	3.12.2 服务类设计 93
3.4 数据库设计 72	3.12.3 DAO 层设计 93
3.4.1 数据库概述 72	3.13 项目运行 94
3.4.2 数据表设计 72	3.14 源码下载 94
3.5 实体类设计 72	
3.6 工具类设计 75	## 第 4 章　好生活个人账本 95
3.7 Mapper 接口和 Example 类设计 76	——SSM + JSP + MySQL
3.7.1 Mapper 接口设计 76	4.1 开发背景 95
3.7.2 Example 类设计 84	4.2 系统设计 96
3.8 查询员工信息模块设计 86	4.2.1 开发环境 96
3.8.1 控制器类设计 87	4.2.2 业务流程 96
3.8.2 服务类设计 87	4.2.3 功能结构 97
3.8.3 DAO 层设计 87	4.3 技术准备 97
3.9 新增员工信息模块设计 88	4.3.1 技术概览 97
3.9.1 控制器类设计 88	4.3.2 Spring IoC 98
3.9.2 服务类设计 89	4.3.3 Spring AOP 100
3.9.3 DAO 层设计 89	4.4 数据库设计 102
	4.4.1 数据库概述 102
	4.4.2 数据表设计 102
	4.5 SSM 框架的主要配置文件 103

4.5.1 Spring 的配置文件 ………………… 103	5.2.1 开发环境 ………………… 126
4.5.2 Spring MVC 的配置文件 …………… 104	5.2.2 业务流程 ………………… 126
4.5.3 MyBatis 的配置文件 ……………… 105	5.2.3 功能结构 ………………… 127
4.6 登录拦截器设计 ………………………… 105	5.3 技术准备 …………………………………… 128
4.7 实体类设计 ……………………………… 106	5.4 数据库设计 ……………………………… 128
4.7.1 用户类 …………………………… 106	5.4.1 数据库概述 ………………… 128
4.7.2 收支类型类 ……………………… 107	5.4.2 数据表设计 ………………… 129
4.7.3 收支明细类 ……………………… 108	5.5 SSM 框架的主要配置文件 ……………… 130
4.8 Mapper 接口设计 ……………………… 109	5.5.1 Spring 的配置文件 ………… 130
4.8.1 UserMapper 接口 ………………… 109	5.5.2 Spring MVC 的配置文件 …… 131
4.8.2 ShouzhiCategoryMapper 接口 …… 110	5.5.3 MyBatis 的配置文件 ……… 132
4.8.3 ShouzhiRecordMapper 接口 ……… 111	5.6 实体类设计 ……………………………… 133
4.9 用户登录模块设计 ……………………… 112	5.6.1 电影评论类 ………………… 133
4.9.1 用户控制器类设计 ……………… 113	5.6.2 电影信息类 ………………… 134
4.9.2 用户服务类设计 ………………… 113	5.6.3 电影类型类 ………………… 136
4.9.3 用户 DAO 层设计 ………………… 114	5.6.4 用户信息类 ………………… 137
4.10 用户注册模块设计 …………………… 114	5.7 Mapper 接口设计 ……………………… 138
4.10.1 用户控制器类设计 ……………… 114	5.7.1 TCommentDao 接口 ……… 138
4.10.2 用户服务类设计 ………………… 115	5.7.2 TMovieDao 接口 …………… 139
4.10.3 用户 DAO 层设计 ………………… 115	5.7.3 TSortDao 接口 ……………… 144
4.11 收支明细模块设计 …………………… 115	5.7.4 TUserinfoDao 接口 ………… 146
4.11.1 收支明细控制器类设计 ………… 116	5.8 首页模块设计 …………………………… 147
4.11.2 收支明细服务类设计 …………… 117	5.8.1 首页页面设计 ……………… 148
4.11.3 收支明细 DAO 层设计 …………… 118	5.8.2 控制器类设计 ……………… 149
4.12 收入记账模块设计 …………………… 119	5.8.3 服务类设计 ………………… 149
4.12.1 收支明细控制器类设计 ………… 119	5.8.4 DAO 层设计 ………………… 150
4.12.2 收支明细服务类设计 …………… 120	5.9 "更多"模块设计 ……………………… 150
4.12.3 收支明细 DAO 层设计 …………… 120	5.9.1 "更多"页面设计 ………… 151
4.13 支出记账模块设计 …………………… 121	5.9.2 控制器类设计 ……………… 152
4.13.1 收支明细控制器类设计 ………… 121	5.9.3 服务类设计 ………………… 153
4.13.2 其他功能模块设计 ……………… 122	5.9.4 DAO 层设计 ………………… 154
4.14 退出登录模块设计 …………………… 123	5.10 用户登录模块设计 …………………… 155
4.15 项目运行 ……………………………… 123	5.10.1 控制器类设计 …………… 155
4.16 源码下载 ……………………………… 124	5.10.2 服务类设计 ……………… 155
	5.10.3 DAO 层设计 ……………… 155
第 5 章 嗨乐影评平台 …………………… 125	5.11 用户注册模块设计 …………………… 156
——SSM + JSP + MySQL	5.11.1 控制器类设计 …………… 156
5.1 开发背景 ………………………………… 125	5.11.2 服务类设计 ……………… 156
5.2 系统设计 ………………………………… 126	5.11.3 DAO 层设计 ……………… 157

5.12 详情模块设计 157	5.13.3 服务类设计 162
5.12.1 "详情"页面设计 158	5.13.4 DAO 层设计 162
5.12.2 控制器类设计 159	5.14 电影管理模块设计 162
5.12.3 服务类设计 159	5.14.1 后台分页显示电影信息（支持模糊查询）
5.12.4 DAO 层设计 160	设计 164
5.13 写评论模块设计 160	5.14.2 添加电影信息设计 165
5.13.1 评论模态框设计 161	5.15 项目运行 167
5.13.2 控制器类设计 161	5.16 源码下载 167

第3篇　Spring Boot 应用项目

第 6 章　电瓶车品牌信息管理系统 170	6.10 DAO 层设计 189
——Spring Boot + Vue + MySQL	6.11 分页插件模块设计 191
6.1 开发背景 170	6.11.1 分页插件的页面设计 191
6.2 系统设计 171	6.11.2 分页插件配置类设计 192
6.2.1 开发环境 171	6.12 查询电瓶车品牌信息模块设计 192
6.2.2 业务流程 171	6.12.1 查询模块的页面设计 193
6.2.3 功能结构 171	6.12.2 查询模块控制器类设计 194
6.3 技术概览 172	6.12.3 查询模块服务类设计 195
6.4 Spring Boot 技术基础 172	6.13 新增电瓶车品牌信息模块设计 195
6.4.1 IDEA 关联 Maven 173	6.13.1 新增模块的页面设计 196
6.4.2 pom.xml 文件 173	6.13.2 新增模块控制器类设计 197
6.4.3 配置文件的格式 174	6.13.3 新增模块服务类设计 198
6.4.4 注解 175	6.14 删除电瓶车品牌信息模块设计 198
6.4.5 启动类 177	6.14.1 删除模块的页面设计 198
6.4.6 处理 HTTP 请求 177	6.14.2 删除模块控制器类设计 199
6.4.7 Service 层 178	6.14.3 删除模块服务类设计 199
6.5 Vue.js 技术基础 179	6.15 项目运行 199
6.5.1 应用程序实例及选项 179	6.16 源码下载 200
6.5.2 常用指令 181	
6.6 数据库设计 184	第 7 章　寻物启事网站 201
6.7 添加依赖和配置信息 184	——Spring Boot + Vue + MySQL
6.7.1 在 pom.xml 文件中添加依赖 ... 185	7.1 开发背景 201
6.7.2 在 application.yml 文件中添加配置信息 ... 186	7.2 系统设计 202
6.8 工具类设计 186	7.2.1 开发环境 202
6.8.1 全局异常处理类 186	7.2.2 业务流程 202
6.8.2 通用返回类 187	7.2.3 功能结构 203
6.9 实体类设计 188	7.3 技术准备 203

7.4 数据库设计 204	7.15.3 控制器类设计 233
7.4.1 数据库概述 204	7.15.4 服务类设计 234
7.4.2 数据表设计 204	7.15.5 DAO层设计 234
7.5 添加依赖和配置信息 205	7.16 分类管理模块设计 234
7.5.1 在pom.xml文件中添加依赖 .. 205	7.16.1 数据传输对象设计 235
7.5.2 在application.yml文件中添加配置信息 207	7.16.2 分类管理页面设计 236
7.6 实体类设计 207	7.16.3 控制器类设计 238
7.6.1 用户信息类 207	7.16.4 服务类设计 239
7.6.2 失物信息类 208	7.16.5 DAO层设计 239
7.7 登录模块设计 209	7.17 寻物启事审核模块设计 240
7.7.1 展示层对象设计 209	7.17.1 寻物启事审核页面 240
7.7.2 登录页面设计 210	7.17.2 控制器类设计 244
7.7.3 控制器类设计 211	7.17.3 服务类设计 244
7.7.4 服务类设计 212	7.17.4 DAO层设计 244
7.7.5 DAO层设计 212	7.18 项目运行 244
7.8 头部导航链接设计 213	7.19 源码下载 245
7.9 查看失物信息模块设计 214	**第8章 明日之星物业管理系统** 246
7.9.1 数据传输对象设计 215	——Spring Boot + Vue + MySQL
7.9.2 寻物启事页面设计 216	8.1 开发背景 246
7.9.3 控制器类设计 217	8.2 系统设计 247
7.9.4 服务类设计 218	8.2.1 开发环境 247
7.9.5 DAO层设计 218	8.2.2 业务流程 247
7.10 发布寻物启事模块设计 218	8.2.3 功能结构 248
7.10.1 发布寻物启事页面设计 219	8.3 技术准备 249
7.10.2 控制器类设计 221	8.4 数据库设计 249
7.10.3 服务类设计 222	8.4.1 数据库概述 249
7.10.4 DAO层设计 222	8.4.2 数据表设计 249
7.11 联系管理员模块设计 222	8.5 添加依赖和配置信息 251
7.12 修改用户信息模块设计 224	8.5.1 添加依赖 251
7.12.1 个人中心页面设计 224	8.5.2 添加配置信息 253
7.12.2 控制器类设计 226	8.6 实体类设计 254
7.12.3 服务类设计 226	8.6.1 用户信息类 254
7.12.4 DAO层设计 227	8.6.2 报修信息类 255
7.13 退出登录模块设计 227	8.6.3 投诉信息类 256
7.14 左侧导航链接设计 228	8.6.4 收费信息类 257
7.15 用户管理模块设计 229	8.7 登录模块设计 257
7.15.1 展示层对象设计 230	8.7.1 登录页面设计 258
7.15.2 用户管理页面设计 230	8.7.2 控制器类设计 260

- 8.7.3 登录类对象设计 261
- 8.7.4 服务类设计 261
- 8.7.5 DAO 层设计 262
- 8.8 侧边栏（面向住户）设计 262
- 8.9 缴纳费用模块设计 264
  - 8.9.1 缴纳费用页面设计 264
  - 8.9.2 控制器类设计 266
  - 8.9.3 服务类设计 266
  - 8.9.4 DAO 层设计 266
- 8.10 申请报修模块设计 267
  - 8.10.1 申请报修页面设计 267
  - 8.10.2 控制器类设计 269
  - 8.10.3 服务类设计 269
  - 8.10.4 DAO 层设计 269
- 8.11 报修管理（面向住户）模块设计 270
  - 8.11.1 报修管理（面向住户）页面设计 270
  - 8.11.2 控制器类设计 272
  - 8.11.3 服务类设计 272
  - 8.11.4 DAO 层设计 272
- 8.12 发起投诉模块设计 272
  - 8.12.1 发起投诉页面设计 273
  - 8.12.2 控制器类设计 274
  - 8.12.3 服务类设计 275
  - 8.12.4 DAO 层设计 275
- 8.13 投诉管理（面向住户）模块设计 275
  - 8.13.1 投诉管理（面向住户）页面设计 276
  - 8.13.2 控制器类设计 277
  - 8.13.3 服务类设计 278
  - 8.13.4 DAO 层设计 278
- 8.14 侧边栏（面向管理员）设计 278
- 8.15 报修管理（面向管理员）模块设计 280
  - 8.15.1 报修管理（面向管理员）页面设计 281
  - 8.15.2 控制器类设计 281
  - 8.15.3 服务类设计 281
  - 8.15.4 DAO 层设计 282
- 8.16 投诉管理（面向管理员）模块设计 282
  - 8.16.1 投诉管理（面向管理员）页面设计 283
  - 8.16.2 控制器类设计 283
  - 8.16.3 服务类设计 284
  - 8.16.4 DAO 层设计 285
- 8.17 收费项目管理模块设计 285
  - 8.17.1 收费项目管理页面设计 286
  - 8.17.2 控制器类设计 289
  - 8.17.3 服务类设计 289
  - 8.17.4 DAO 层设计 290
- 8.18 物业人员管理模块设计 290
  - 8.18.1 物业人员管理页面设计 291
  - 8.18.2 控制器类设计 294
  - 8.18.3 服务类设计 295
  - 8.18.4 DAO 层设计 295
- 8.19 住户信息管理模块设计 296
  - 8.19.1 住户信息管理页面设计 297
  - 8.19.2 控制器类设计 301
  - 8.19.3 服务类设计 301
  - 8.19.4 DAO 层设计 302
- 8.20 退出登录模块设计 302
- 8.21 项目运行 304
- 8.22 源码下载 304

第1篇

Servlet 应用项目

Java Servlet 是使用 Java 编写的、运行在 Web 服务器上的程序，主要负责接收客户端请求并生成动态响应。具体而言，Servlet 主要用于读取客户端发送的数据（如 HTML 表单数据），处理这些数据（可能需要访问数据库或进行计算），并将结果发送给客户端。虽然 Servlet 是一门传统的技术，并且当下的使用率很低，但学习 Servlet 是很有必要的，这是因为在 SSM 框架应用项目或者 Spring Boot 应用项目中有很多部分都用到了 Servlet 的思想。换言之，读者学习 Servlet，相当于在夯实基础，会更容易学习、理解本书后面的关于 SSM 框架和 Spring Boot 的内容。

本篇将使用 Servlet、JSP 和 MySQL 等技术开发两个 Java Web 应用项目，具体如下：
- ☑ 明日科技门户网
- ☑ 购好物网络商城

第1章 明日科技门户网

——Servlet + JSP + MySQL

在广泛利用信息资源、加速实现数字化的潮流中,越来越多的企业开设了自己的门户网站。所谓门户网站,是指一个集成某类综合性的数据资源或者互联网资源的信息管理平台。企业的门户网站通常把统一的页面提供给用户,并建立企业对客户、企业对内部员工和企业对企业的信息通道,使企业能够高效地向内或者向外发布各种信息。本章将使用 Java Web 开发中的 MySQL 数据库、Servlet、JSP 等关键技术开发一个企业级门户网站项目——明日科技门户网。

项目微视频

本项目的核心功能及实现技术如下:

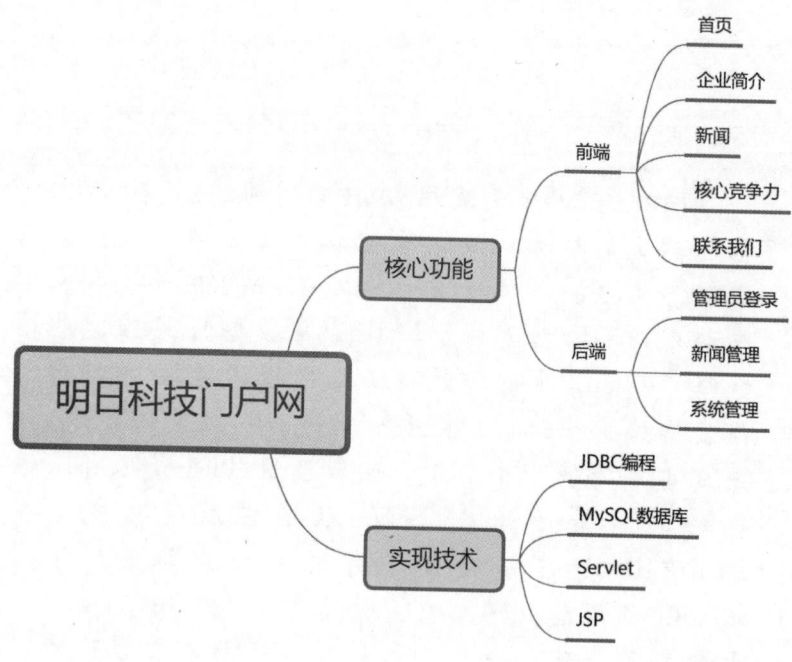

1.1 开发背景

明日科技门户网是一个连接企业内部和外部的信息管理平台,能够为企业发布的信息资源提供访问入口。无论是企业的员工、客户、供应商或者合作伙伴,都可以通过访问该门户网站各取所需。它既可以帮助用户快速了解企业的基本信息,又可以为企业的数据资源提供便利的展示平台,以达到"企业商务社区"的服务效果。明日科技门户网极大程度地改进了企业的业务形式和管理模式,实现了企业内部、外部的便捷沟通,提高了源整合效率,推进了企业的数字化进程。

明日科技门户网将实现以下目标:

☑ 各页面简洁、美观、层次分明;

- ☑ 首页导航栏里的功能明确，让用户一目了然，快速享受指定的信息服务；
- ☑ 首页通过轮播的图片，凸显企业价值观；
- ☑ 用户能够通过浏览新闻页面上发布的新闻了解公司的动态；
- ☑ 管理员可以对网站进行后端管理，包括新闻管理和查看管理员信息。

1.2 系统设计

1.2.1 开发环境

本项目的开发及运行环境如下：
- ☑ 操作系统：推荐 Windows 10、11 及以上，兼容 Windows 7（SP1）。
- ☑ 开发工具：IntelliJ IDEA。
- ☑ 开发语言：Java EE。
- ☑ 数据库：MySQL 8.0。
- ☑ Web 服务器：Tomcat 9.0 及以上版本。

1.2.2 业务流程

启动项目后，明日科技门户网首页会被自动打开。在明日科技门户网首页上，用户通过单击导航栏中各个功能的超链接，能够快速获取"企业简介""新闻""核心竞争力""联系我们"等信息；管理员通过登录明日科技门户网的后端，既能够对新闻进行管理，又能够查看管理员信息。明日科技门户网的业务流程如图1.1所示。

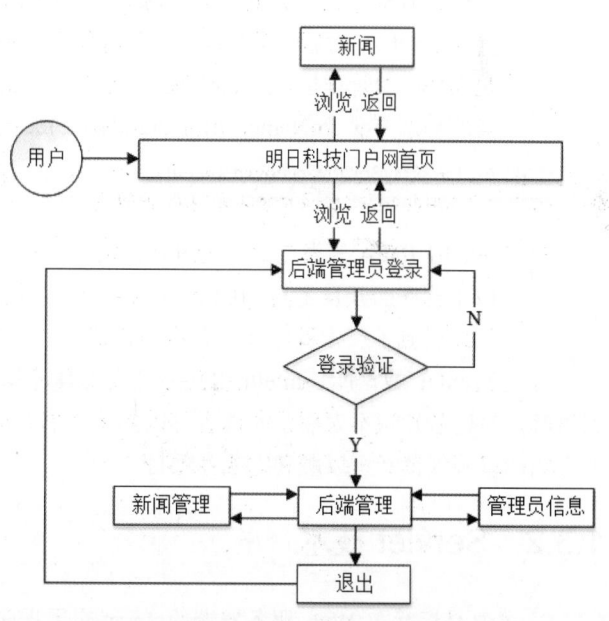

图1.1 明日科技门户网的业务流程图

说明

明日科技门户网具有一个导航栏，其中包括首页、企业简介、新闻、核心竞争力和联系我们5个模块，这5个模块的设计过程都大同小异。限于本书篇幅，这里仅讲解首页模块和新闻模块的设计过程。读者可以参考这两个模块的设计过程，自行根据提供的源码学习其他模块的设计过程。

1.2.3 功能结构

本项目的功能结构已经在章首页中给出。作为一个门户网站，本项目实现的具体功能如下：
- ☑ 首页：设置导航栏、轮播图片、功能栏和版权信息栏。
- ☑ 新闻：展示公司动态，展示指定的公司新闻。
- ☑ 后端管理员登录：管理员输入正确的用户名（mr）和密码（mrsoft）后即可成功登录明日科技门户网后端。
- ☑ 后端新闻管理：管理员在明日科技门户网后端能够对新闻执行添加、修改、删除等操作。

- 后端查看管理员信息：管理员在明日科技门户网后端能够查看管理员的用户名、密码、创建时间等信息。

1.3 技术准备

1.3.1 技术概览

- MySQL 数据库：MySQL 数据库是一个中小型关系型数据库管理系统，由于其体积小、速度快、总体拥有成本低，尤其是开放源码这一特点，许多大中小型网站都是为了降低网站运营成本而选择 MySQL 作为后端数据库。本章将使用 JDBC 技术操作 MySQL 数据库，例如，DriverManager 类是 JDBC 的管理层，用于管理数据库中的驱动程序。在操作指定数据库之前，需要使用 Java 中 Class 类的静态方法 forName(String className)加载指定数据库的驱动程序。代码如下：

```
Class.forName(MySqlDriver).newInstance();
conn = DriverManager.getConnection(MySqlURL);
```

- IntelliJ IDEA 开发工具：IntelliJ IDEA，简称 IDEA，是由 JetBrains 公司推出的一款用于设计 Java 应用程序的开发工具。IDEA 在代码补全、代码提示、代码重构、代码审查等方面表现得尤为突出，因此受到了广大程序开发人员的青睐。

有关 MySQL 数据库、IntelliJ IDEA 开发工具等基础知识在《Java 从入门到精通（第 7 版）》中有详细的讲解，对这些知识不太熟悉的读者可以参考该书对应的内容。下面将对 Servlet 和 JSP 这两个技术进行必要介绍，以确保读者可以顺利完成本项目。

1.3.2 Servlet 技术

Servlet 是运行在 Web 服务器端的 Java 应用程序，由 Servlet 容器对其进行管理，当用户对容器发送 HTTP 请求时，容器将通知相应的 Servlet 对象进行处理，完成用户与程序之间的交互。在 Servlet 编程中，Servlet API 提供了标准的接口和类，这些接口和类为 HTTP 请求与程序响应提供了丰富的方法。

- Servlet 接口：Servlet 的运行需要 Servlet 容器的支持，Servlet 容器通过调用 Servlet 对象提供了标准的 API 接口，对请求进行处理。在 Servlet 开发中，任何一个 Servlet 对象都要直接或间接地实现 javax.servlet.Servlet 接口。在该接口中包含 5 个方法，其功能及作用如表 1.1 所示。

表 1.1 Servlet 接口中的方法及说明

方　　法	说　　明
public void init(ServletConfig config)	Servlet 实例化后，Servlet 容器调用该方法来完成初始化工作
public void service(ServletRequest request, ServletResponse response)	用于处理客户端的请求
public void destroy()	当 Servlet 对象从 Servlet 容器中被移除时，容器调用该方法，以便释放资源
public ServletConfig getServletConfig()	用于获取 Servlet 对象的配置信息，返回 ServletConfig 对象
public String getServletInfo()	返回有关 Servlet 的信息，它是纯文本格式的字符串，如作者、版本等

- ServletConfig 接口：ServletConfig 接口位于 javax.servlet 包中，它封装了 Servlet 的配置信息，在 Servlet 初始化期间被传递。每一个 Servlet 都有且只有一个 ServletConfig 对象。该对象定义了 4 个方法，如表 1.2 所示。

表 1.2　ServletConfig 接口中的方法及说明

方　　法	说　　明
public String getInitParameter(String name)	返回 String 类型名称为 name 的初始化参数值
public Enumeration getInitParameterNames()	获取所有初始化参数名的枚举集合
public ServletContext getServletContext()	获取 Servlet 上下文对象
public String getServletName()	返回 Servlet 对象的实例名

- HttpServletRequest 接口：HttpServletRequest 接口位于 javax.servlet.http 包中，继承了 javax.servlet.ServletRequest 接口，是 Servlet 中的重要对象，在开发过程中较为常用，其常用方法及说明如表 1.3 所示。

表 1.3　HttpServletRequest 接口的常用方法及说明

方　　法	说　　明
public String getContextPath()	返回请求的上下文路径，此路径以 "/" 开关
public Cookie[] getCookies()	返回请求中发送的所有 cookie 对象，返回值为 cookie 数组
public String getMethod()	返回请求所使用的 HTTP 类型，如 get、post 等
public String getQueryString()	返回请求中参数的字符串形式，如请求 MyServlet?username=mr，则返回 username=mr
public String getRequestURI()	返回主机名到请求参数之间的字符串形式
public StringBuffer getRequestURL()	返回请求的 URL，此 URL 中不包含请求的参数。注意此方法返回的数据类型为 StringBuffer
public String getServletPath()	返回请求 URI 中的 Servlet 路径的字符串，不包含请求中的参数信息
public HttpSession getSession()	返回与请求关联的 HttpSession 对象

- HttpServletResponse 接口：HttpServletResponse 接口位于 javax.servlet.http 包中，它继承了 javax.servlet.ServletResponse 接口，同样是一个非常重要的对象，其常用方法及说明如表 1.4 所示。

表 1.4　HttpServletResponse 接口的常用方法及说明

方　　法	说　　明
public void addCookie(Cookie cookie)	向客户端写入 cookie 信息
public void sendError(int sc)	发送一个错误状态码为 sc 的错误响应到客户端
public void sendError(int sc, String msg)	发送一个包含错误状态码及错误信息的响应到客户端，参数 sc 为错误状态码，参数 msg 为错误信息
public void sendRedirect(String location)	使用客户端重定向到新的 URL，参数 location 为新的地址

- GenericServlet 类：在编写一个 Servlet 对象时，必须实现 javax.servlet.Servlet 接口，在 Servlet 接口中包含 5 个方法，也就是说创建一个 Servlet 对象要实现这 5 个方法，这样操作非常不方便。javax.servlet.GenericServlet 类简化了此操作，实现了 Servlet 接口。代码如下：

```
public abstract class GenericServlet
        extends Object
        implements Servlet, ServletConfig, Serializable
```

说明

　　GenericServlet 类是一个抽象类,分别实现了 Servlet 接口与 ServletConfig 接口。该类实现了除 service() 之外的其他方法。在创建 Servlet 对象时,可以继承 GenericServlet 类来简化程序中的代码,但需要实现 service() 方法。

- ☑ HttpServlet 类：HttpServlet 类仍然是一个抽象类,实现了 service() 方法,并针对 HTTP 1.1 中定义的 7 种请求类型提供了相应的方法——doGet() 方法、doPost() 方法、doPut() 方法、doDelete() 方法、doHead() 方法、doTrace() 方法和 doOptions() 方法。在这 7 个方法中,除了对 doTrace() 方法与 doOptions() 方法进行简单实现,HttpServlet 类并没有对其他方法进行实现,需要开发人员在使用过程中根据实际需要对其进行重写。HttpServlet 类继承了 GenericServlet 类,通过其对 GenericServlet 类的扩展,可以很方便地对 HTTP 请求进行处理及响应。代码如下：

```
public abstract class HttpServlet
    extends GenericServlet implements Serializable
```

说明

　　GenericServlet 类实现了 javax.servlet.Servlet 接口,为程序的开发提供了方便。但在实际开发过程中,大多数的应用都是使用 Servlet 处理 HTTP 协议的请求,并对请求做出响应,所以通过继承 GenericServlet 类仍然不是很方便。javax.servlet.http.HttpServlet 类对 GenericServlet 类进行了扩展,为 HTTP 请求的处理提供了灵活的方法。

　　为了保证 Servlet 对象正常运行,需要对其进行适当的配置,以告知 Web 容器哪一个请求应调用哪一个 Servlet 对象处理,对 Servlet 起到一个注册的作用。Servlet 的配置包含在 web.xml 文件中,主要通过以下两步进行设置：

　　(1) 声明 Servlet 对象。在 web.xml 文件中,通过<servlet>标签声明一个 Servlet 对象,在此标签下包含两个主要子元素,分别为<servlet-name>与<servlet-class>。其中,<servlet-name>元素用于指定 Servlet 的名称,该名称可以为自定义的名称；<servlet-class>元素用于指定 Servlet 对象的完整位置,包含 Servlet 对象的包名与类名。其声明语句如下：

```
<servlet>
    <servlet-name>SimpleServlet</servlet-name>
    <servlet-class>com.lyq.SimpleServlet</servlet-class>
</servlet>
```

　　(2) 映射 Servlet。在 web.xml 文件中声明了 Servlet 对象后,需要映射访问 Servlet 的 URL。该操作使用<servlet- mapping>标签进行配置。<servlet-mapping>标签包含两个子元素,分别为<servlet-name>与<url-pattern>。其中,<servlet-name>元素与<servlet>标签中的<servlet-name>元素相对应,不可以随意命名。<url-pattern>元素用于映射访问 URL。其配置方法如下：

```
<servlet-mapping>
    <servlet-name>SimpleServlet</servlet-name>
    <url-pattern>/SimpleServlet</url-pattern>
</servlet-mapping>
```

　　下面通过一个示例演示 Servlet 的创建及配置。

　　(1) 创建名为 MyServlet 的 Servlet 对象,它继承了 HttpServlet 类。在该类中重写 doGet() 方法,用于处理 HTTP 的 get 请求,通过 PrintWriter 对象进行简单输出。关键代码如下：

```
public class MyServlet extends HttpServlet {
    public void doGet(HttpServletRequest request, HttpServletResponse response)
            throws ServletException, IOException {
        response.setContentType("text/html");
        response.setCharacterEncoding("GBK");
        PrintWriter out = response.getWriter();
        out.println("<HTML>");
        out.println("<HEAD><TITLE>Servlet 实例</TITLE></HEAD>");
        out.println("<BODY>");
        out.print("Servlet 实例：  ");
        out.print(this.getClass());
        out.println("</BODY>");
        out.println("</HTML>");
        out.flush();
        out.close();
    }
}
```

（2）在 web.xml 文件中对 MyServlet 进行配置，其中访问 URL 的相对路径为/servlet/MyServlet。关键代码如下：

```
<servlet>
    <servlet-name>MyServlet</servlet-name>
    <servlet-class>com.lyq.MyServlet</servlet-class>
</servlet>
<servlet-mapping>
    <servlet-name>MyServlet</servlet-name>
    <url-pattern>/servlet/MyServlet</url-pattern>
</servlet-mapping>
```

1.3.3 JSP 技术

在进行 Java Web 应用开发时，JSP 页面是必不可少的。JSP 页面是指扩展名为.jsp 的文件。在一个 JSP 页面中，可以包括指令标识、HTML 代码、JavaScript 代码、嵌入的 Java 代码、注释和 JSP 动作标识等内容。

指令标识主要用于设定整个 JSP 页面范围内都有效的相关信息，它是被服务器解释并执行的，不会产生任何内容输出到网页中。也就是说，指令标识对于客户端浏览器是不可见的。

page 是 JSP 页面最常用的指令，用于定义整个 JSP 页面的相关属性，这些属性在 JSP 被服务器解析成 Servlet 时会转换为相应的 Java 程序代码。page 指令的语法格式如下：

```
<%@ page attr1="value1" attr2="value2" ……%>
```

include 是 JSP 的另一条指令标识。通过该指令可以在一个 JSP 页面中包含另一个 JSP 页面。不过该指令是静态包含，也就是说被包含文件中所有内容会被原样包含到该 JSP 页面中，即使被包含文件中有 JSP 代码，在包含时也不会被编译执行。使用 include 指令，最终将生成一个文件，所以在被包含和包含的文件中，不能有相同名称的变量。include 指令的语法格式如下：

```
<%@ include file="path"%>
```

例如，应用 include 指令包含网站 Banner 和版权信息栏，具体步骤如下：

（1）编写一个名称为 top.jsp 的文件，用于放置网站的 Banner 信息和导航条。这里将 Banner 信息和导航栏设计为一张图片。这样完成 top.jsp 文件，只需要在该页面通过<img>标记引入图片即可。top.jsp 文件的代码如下：

```
<%@ page pageEncoding="GB18030"%>
<img src="images/banner.JPG">
```

（2）编写一个名称为 copyright.jsp 的文件，用于放置网站的版权信息。copyright.jsp 文件的代码如下：

```jsp
<%@ page pageEncoding="GB18030"%>
<%
String copyright=" All Copyright &copy; 2024 吉林省明日科技有限公司";
%>
<table width="778" height="61" border="0" cellpadding="0" cellspacing="0" background="images/copyright.JPG">
  <tr>
    <td> <%= copyright %></td>
  </tr>
</table>
```

（3）创建一个名称为 index.jsp 的文件，在该页面中包括 top.jsp 和 copyright.jsp 文件，从而实现一个完整的页面。index.jsp 文件的代码如下：

```jsp
<%@ page language="java" contentType="text/html; charset=GB18030" pageEncoding="GB18030"%>
<html>
  <head>
    <meta http-equiv="Content-Type" content="text/html; charset=GB18030">
    <title>使用文件包含 include 指令</title>
  </head>
  <body style="margin:0px;">
    <%@ include file="top.jsp"%>
    <table width="781" height="279" border="0" cellpadding="0" cellspacing="0" background="images/center.JPG">
      <tr>
        <td> </td>
      </tr>
    </table>
    <%@ include file="copyright.jsp"%>
  </body>
</html>
```

taglib 指令标识声明该页面中所使用的标签库，同时引用标签库，并指定标签的前缀。在页面中引用标签库后，就可以通过前缀来引用标签库中的标签。taglib 指令的语法格式如下：

```jsp
<%@ taglib prefix="tagPrefix" uri="tagURI" %>
```

JSP 中的脚本标识包括 3 部分，即 JSP 表达式、声明标识和代码片段。

JSP 表达式用于向页面中输出信息，其语法格式如下：

```jsp
<%= 表达式%>
```

例如，使用 JSP 表达式在页面中输出信息。代码如下：

```jsp
<%String manager="mr"; %>                              //定义保存管理员名的变量
管理员：<%=manager %>                                   //输出结果为：管理员：mr
<%="管理员："+manager %>                                //输出结果为：管理员：mr
<%= 5+6 %>                                             //输出结果为：11
<%String url="126875.jpg"; %>                          //定义保存文件名称的变量
<img src="images/<%=url %>">                           //输出结果为：<img src="images/126875.jpg">
```

声明标识用于在 JSP 页面中定义全局的变量或方法。通过声明标识定义的变量和方法可以被整个 JSP 页面访问，所以通常使用该标识定义整个 JSP 页面需要引用的变量或方法。声明标识的语法格式如下：

```jsp
<%! 声明变量或方法的代码 %>
```

例如，通过声明标识声明一个全局变量和全局方法。代码如下：

```jsp
<%!
    int number = 0;                                     //声明全局变量
```

```
    int count() {                              //声明全局方法
        number++;                              //累加 number
        return number;                         //返回 number 的值
    }
%>
```

代码片段就是在 JSP 页面中嵌入的 Java 代码或是脚本代码。代码片段将在页面请求的处理期间被执行,通过 Java 代码可以定义变量或是流程控制语句等;而通过脚本代码可以应用 JSP 的内置对象在页面输出内容、处理请求和响应、访问 session 会话等。代码片段的语法格式如下:

```
<% Java 代码或是脚本代码 %>
```

例如,编写一个名称为 index.jsp 的文件,在该页面中,先通过代码片段将输出九九乘法表的文本连接成一个字符串,然后通过 JSP 表达式输出该字符串。index.jsp 文件的代码如下:

```
<body>
<%
    String str = "";                           //声明保存九九乘法表的字符串变量
    //连接生成九九乘法表的字符串
    for(int i = 1; i <= 9; i++) {              //外循环
        for(int j = 1; j <= i; j++) {          //内循环
            str += j + "*" + i + "=" + j * i;
            str += " ";                   //加入空格符
        }
        str += "<br>";                         //加入换行符
    }
%>
<table width="440" height="85" border="1" cellpadding="0" cellspacing="0" style="font:9pt;"
    bordercolordark="#666666" bordercolorlight="#FFFFFF" bordercolor="#FFFFFF">
  <tr>
    <td height="30" align="center">九九乘法表</td>
  </tr>
  <tr>
    <td style="padding:3pt">
        <%=str%>                               //输出九九乘法表
    </td>
  </tr>
</table>
</body>
```

JSP 的动作标识<jsp:include>用于向当前页面中包含其他的文件。被包含的文件可以是动态文件,也可以是静态文件。<jsp:include>动作标识有两种语法格式。第一种语法格式如下:

```
<jsp:include page="url" flush="false|true" />
```

第二种语法格式如下:

```
<jsp:include page="url" flush="false|true" >
    子动作标识<jsp:param>
</jsp:include>
```

例如,应用<jsp:include>标识包含网站 Banner 和版权信息栏,具体步骤如下:

(1)编写一个名称为 top.jsp 的文件,用于放置网站的 Banner 信息和导航条。这里将 Banner 信息和导航栏设计为一张图片。这样完成 top.jsp 文件,只需要在该页面中通过<img>标记引入图片即可。top.jsp 文件的代码如下:

```
<%@ page pageEncoding="GB18030"%>
<img src="images/banner.JPG">
```

(2)编写一个名称为 copyright.jsp 的文件,用于放置网站的版权信息。copyright.jsp 文件的代码如下:

```
<%@ page pageEncoding="GB18030"%>
<%
String copyright=" All Copyright &copy; 2024 吉林省明日科技有限公司";
%>
<table width="778" height="61" border="0" cellpadding="0" cellspacing="0" background="images/copyright.JPG">
  <tr>
    <td> <%= copyright %></td>
  </tr>
</table>
```

（3）创建一个名称为 index.jsp 的文件，在该页面中包括 top.jsp 和 copyright.jsp 文件，从而实现一个完整的页面。index.jsp 文件的代码如下：

```
<%@ page language="java" contentType="text/html; charset=GB18030" pageEncoding="GB18030"%>
<html>
  <head>
    <meta http-equiv="Content-Type" content="text/html; charset=GB18030">
    <title>使用&lt;jsp:include&gt;动作标识包含文件</title>
  </head>
  <body style="margin:0px;">
    <jsp:include page="top.jsp"/>
      <table width="781" height="279" border="0" cellpadding="0" cellspacing="0" background="images/center.JPG">
        <tr>
          <td> </td>
        </tr>
      </table>
    <jsp:include page="copyright.jsp"/>
  </body>
</html>
```

通过<jsp:forward>动作标识可以将请求转发到其他的 Web 资源，如另一个 JSP 页面、HTML 页面、Servlet 等。执行请求转发后，当前页面将不再被执行，而是去执行该标识指定的目标页面。<jsp:forward>动作标识有两种语法格式。第一种语法格式如下：

```
<jsp:forward page="url"/>
```

第二种语法格式如下：

```
<jsp:forward page="url">
        子动作标识<jsp:param>
</jsp:forward>
```

例如，应用<jsp:forward>标识将页面转发到用户登录页面，具体步骤如下：

（1）创建一个名称为 index.jsp 的文件，该文件为中转页，用于通过<jsp:forward>动作标识将页面转发到用户登录页面（login.jsp）。index.jsp 文件的代码如下：

```
<%@ page language="java" contentType="text/html; charset=GB18030" pageEncoding="GB18030"%>
<html>
  <head>
    <meta http-equiv="Content-Type" content="text/html; charset=GB18030">
    <title>中转页</title>
  </head>
  <body>
    <jsp:forward page="login.jsp"/>
  </body>
</html>
```

（2）编写 login.jsp 文件，在该文件中添加用于收集用户登录信息的表单及表单元素。代码如下：

```
<%@ page language="java" contentType="text/html; charset=GB18030" pageEncoding="GB18030"%>
<html>
```

```
<head>
  <meta http-equiv="Content-Type" content="text/html; charset=GB18030">
  <title>用户登录</title>
</head>
<body>
  <form name="form1" method="post" action="">
    用户名： <input name="name" type="text" id="name" style="width: 120px"><br>
    密  码： <input name="pwd" type="password" id="pwd" style="width: 120px"> <br>
    <br>
    <input type="submit" name="Submit" value="提交">
  </form>
</body>
</html>
```

JSP 的动作标识 `<jsp:param>` 可以作为其他标识的子标识，用于为其他标识传递参数。语法格式如下：

`<jsp:param name="参数名" value="参数值" />`

例如，通过 `<jsp:param>` 标识为 `<jsp:forward>` 标识指定参数，代码如下：

```
<jsp:forward page="modify.jsp">
    <jsp:param name="userId" value="7"/>
</jsp:forward>
```

在上面的代码中，实现了在请求转发到 modify.jsp 页面的同时，传递了参数 userId，其参数值为 7。

1.4 数据库设计

1.4.1 数据库概述

本项目采用的数据库名称为 webdb，数据库包含两张数据表，如表 1.5 所示。

表 1.5 明日科技门户网的数据库结构

数据库名	表 名	表 说 明
webdb	admin	管理员表
	news	新闻表

1.4.2 数据表设计

admin 和 news 两张表都采用主键自增原则，下面将详细介绍这两张表的结构设计。

☑ admin（管理员表）：主要用于存储管理员的信息。该数据表的结构如表 1.6 所示。

表 1.6 admin 表结构

字 段 名 称	数 据 类 型	长 度	是否主键	说 明
AdminID	INT	11	主键	自动编号 ID
AdminName	VARCHAR	32		管理员名称
AdminPwd	VARCHAR	64		管理员密码
AdminType	SMALLINT	6		管理员类别
LastLoginTime	VARCHAR	50		上次登录时间

☑ news（新闻表）：主要用于存储公司新闻的信息。该数据表的结构如表 1.7 所示。

表 1.7 news 表结构

字 段 名 称	数 据 类 型	长　度	是 否 主 键	说　　明
NewsID	INT	11	主键	自动编号 ID
NewsTitle	VARCHAR	60		新闻标题
NewsContent	LONGTEXT			新闻详情
NewsTime	VARCHAR	50		发布时间
AdminName	VARCHAR	32		发布人名称

1.5 首页模块设计

明日科技门户网首页是由上部的企业 Logo 和导航栏、中部的轮播图片信息、下部的功能栏、底部的版权信息栏这 4 个部分构成的，其效果如图 1.2 所示。

图 1.2 明日科技门户网首页的效果图

1.5.1 上部企业 Logo 和导航栏设计

各网页上部均是由企业 Logo 和导航信息构成的。其中，导航信息包括"首页""企业简介""新闻""核心竞争力""联系我们"等。上部企业 Logo 和导航栏的效果如图 1.3 所示。

图 1.3 上部企业 Logo 和导航栏的效果图

为了统一明日科技门户网每个页面的上部企业 Logo 和导航栏，将其设置为一个公共使用的页面，即 common_header.jsp。因为 common_header.jsp 不仅是公共页面，而且会被其他页面引用，所以在编码时不能包含 JSP 自带的各种形式的标签。也就是说，在创建 common_header.jsp 完毕后，需将内部自动生成的代码全部清空，重新编码。common_header.jsp 的代码如下：

```jsp
<%@ page language="java" import="java.util.*" pageEncoding="UTF-8" %>
<!--头部-->
    <div class="header_bg">
        <div class="wrap">
            <div class="header">
                <div class="logo">
                    <a href="index.jsp"><img src="img/logo.png" alt=""> </a>
                </div>
                <div class="pull-icon">
                    <a id="pull"></a>
                </div>
                <div class="cssmenu">
                    <ul>
                        <li>
                            <a href="index.jsp">首页</a>
                        </li>
                        <li>
                            <a href="about.jsp">企业简介</a>
                        </li>
                        <li>
                            <a href="newsFrontList.jsp">新闻</a>
                        </li>
                        <li>
                            <a href="content.jsp">核心竞争力</a>
                        </li>
                        <li class="last">
                            <a href="contact.jsp">联系我们</a>
                        </li>
                    </ul>
                </div>
                <!--清除浮动-->
                <div class="clear"></div>
            </div>
        </div>
    </div>
<!--头部-->
```

1.5.2　中部轮播图片信息设计

在本项目的 WebContent/front/css 文件夹内，存储的是全部页面的 CSS 样式，其中的 main.css 就是首页的 CSS 样式。为了实现首页轮播图片的效果，需要向 WebContent/front 文件夹内的 index.jsp 引入 main.css。代码如下：

```html
<link href="css/main.css" rel="stylesheet" type="text/css" media="all" >
```

在编码时，以第一个图片 img1.jpg 为例，首先编写外围的<div>和</div>标签，然后在标签内编写 img1.jpg 的引入方法，以此类推。这样既节省时间，又可以防止错误发生。代码如下：

```html
<!--轮播 -->
<div id="fwslider" style="height: 554px;">
```

```html
<div class="slider_container">
    <div class="slide" style=" opacity: 1; z-index: 0; display: none;">
        <!--第一张图片-->
        <img id="img1" src="img/img1.jpg">
    </div>
    <div class="slide" style="opacity: 1; z-index: 1; display: block;">
        <!--第二张图片-->
        <img id="img2" src="img/img2.jpg">
    </div>
    <div class="slide" style="opacity: 1; z-index: 1; display: block;">
        <!--第三张图片-->
        <img id="img3" src="img/img3.jpg">
    </div>
    <div class="slide" style=" opacity: 1;z-index: 0; display: none;">
        <!--第四张图片-->
        <img id="img4" src="img/img4.jpg">
    </div>
    <div class="slide" style=" opacity: 1;z-index: 0; display: none;">
        <!--第五张图片-->
        <img id="img5" src="img/img5.jpg">
    </div>
</div>
</div>
<!--轮播 -->
```

1.5.3 下部功能栏设计

明日科技门户网首页下部的功能栏为图片信息，如图 1.4 所示。

图 1.4 功能栏展示

在 index.jsp 中，在用于实现上部的导航栏和中部的轮播图片的代码编写完毕后，开始编写用于实现下部的功能栏的代码。在使用 HTML 标签库中的<div>标签进行分层显示时，<div>标签使用 class 属性方法来进行页面效果定义，根据属性名称进行 CSS 样式的展现，起到美化页面、突出页面布局等作用。<img>为图片引入标签，用于展示图片，其中的 src 属性用于标识图片所处的系统位置。代码如下：

```html
<div class="main_bg">
    <div class="business">
        业务领域 BUSINESS
    </div>
    <div class="wrap w_72">
        <div class="grids_1_of_3">
            <div class="grid_1_of_3   images_1_of_3">
                <img src="img/pic1.png" >
            </div>
            <div class="grid_1_of_3   images_1_of_3">
                <img src="img/pic2.png" >
            </div>
            <div class="grid_1_of_3   images_1_of_3">
                <img src="img/pic3.png" >
            </div>
```

```
            <div class="grid_1_of_3  images_1_of_3">
                <img src="img/pic4.png" >
            </div>
            <div class="grid_1_of_3  images_1_of_3"  style="background: none">
                <img src="img/pic5.png" >
            </div>
            <div class="clear"></div>
        </div>
    </div>
</div>
```

1.5.4 底部版权信息栏设计

明日科技门户网首页的底部是版权信息栏，版权信息是对该网站的版权归属的明确说明，用于提醒用户当前网站是受版权保护的。版权信息页面设计如图 1.5 所示。

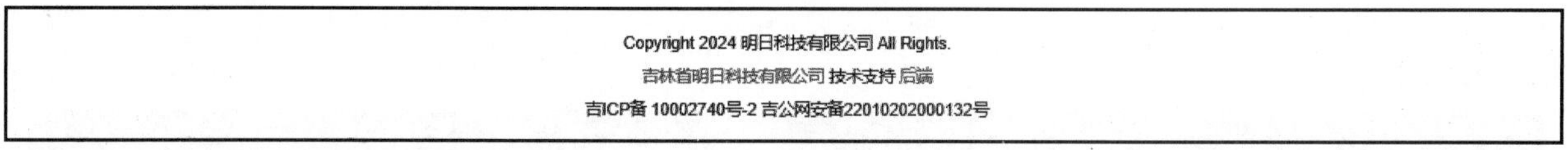

```
                      Copyright 2024 明日科技有限公司 All Rights.
                            吉林省明日科技有限公司 技术支持 后端
                      吉ICP备 10002740号-2 吉公网安备22010202000132号
```

图 1.5 版权信息

在 index.jsp 的头部，定义页面变量 path 和 basePath，通过 request 对象获取所需地址信息。代码如下：

```jsp
<%
    String path = request.getContextPath();
    String basePath = request.getScheme() + "://"
            + request.getServerName() + ":" + request.getServerPort()
            + path;
%>
```

在 index.jsp 内加入<div>标签，将页面进行层次化管理。使用<a>标签，将页面某部分说明文字变成可以点击，并访问所设置的地址信息。代码如下：

```html
<div class="address">
    Copyright 2024 明日科技有限公司 All Rights.
    <br>
    <a href="http://www.mingrisoft.com">吉林省明日科技有限公司</a> 技术支持
    <a href="<%=basePath%>/admin/login.jsp">后端</a>
    <br>
    吉 ICP 备 10002740 号-2  吉公网安备 22010202000132 号
</div>
```

1.6 新闻模块设计

如图 1.3 所示，在明日科技门户网首页的导航栏中，包括首页、企业简介、新闻、核心竞争力和联系我们这 5 个模块。在第 1.5 节中，已经介绍了首页模块的设计过程。对于其余 4 个模块，它们各自的设计过程都大同小异。下面将以"新闻"模块为例，讲解其他模块的设计过程。

新闻模块是明日科技门户网的交流模块，如图 1.6 所示。在新闻模块中，管理员能够通过后端进行新闻的发布和修改，用户能够在前端进行新闻的访问和查询。

Java Web 项目开发全程实录（第 2 版）

图 1.6　新闻模块的效果图

1.6.1　配置文件的编写

　　数据库和数据表建立完毕后，需进行配置文件的编写，允许本项目连接数据库后对数据表进行相关操作。本项目的配置文件是 src/com 文件夹内的 DBConfig.property，其中包含了数据库类型、数据库驱动程序、连接 MySQL 8.0 数据库的字符串等。代码如下：

```
#数据库类型:1 为 MYSQL,默认为 1。
DBType=1

#MySQL 数据库连接信息
#MySQL 数据库驱动程序
MySQLDriver=org.gjt.mm.mysql.Driver

# MySQL 数据库连接字符串。
# 127.0.0.1:3306 为数据库地址和端口
# webdb 为数据库名，user 为登录用户名，password 为登录密码。请自行更改。
MySQLURL=jdbc:mysql://127.0.0.1:3306/webdb?user=root&password=root&useUnicode=true&characterEncoding=UTF-8&useSSL=false&serverTimezone=Asia/Shanghai&zeroDateTimeBehavior=CONVERT_TO_NULL&allowPublicKeyRetrieval=true
```

1.6.2　显示新闻列表

　　在本项目的 src/com/mingrisoft 文件夹下，创建了 News.java 文件，在该文件中编写的是用于显示新闻列表的代码。在 News.java 文件中，定义 ListNewsFront()方法，用于查询新闻列表。该列表的题头分别为"新闻标题""发布人""发布时间""详情"。如果 ListNewsFront()方法中的结果集不为空，就把结果集中的数据按照上述 4 个题头予以分别显示，如图 1.7 所示。

新闻标题	发布人	发布时间	详情
启蒙思维	主管	2024-03-22	详情
编程助手	管理员	2024-02-01	详情

图 1.7　新闻列表的效果图

ListNewsFront()方法的代码如下：

```java
//新闻列表查询方法
public String ListNewsFront(String sPage, String strPage) {
    try {                                                         //用于获取系统运行异常信息

        Connection Conn = DBConn.getConn();                       //建立数据库连接
        Statement stmt = Conn.createStatement();                  //创建数据查询
        ResultSet rs = null;                                      //定义结果集

        //定义本方法返回字符串数据
        StringBuffer sb = new StringBuffer();

        int i;                                                    //定义数字型变量
        int intPage = 1;                                          //定义数字型变量并赋值 1
        int intPageSize = 5;                                      //定义数字型变量并赋值 5

        //创建 sql 语句查询 News 表全部信息
        String sSql = "select * from News order by NewsID desc";
        //通过执行 sql 语句得到查询结果
        rs = stmt.executeQuery(sSql);

        if (!rs.next()) {                                         //判定当查询结果为空
            //返回属性添加字符串数据用于页面显示
            sb.append("<tr height=\"25\" bgcolor=\"#d6dff7\"    class=\"info1\"><td colspan=\"5\">\r\n");
            //返回属性添加字符串数据用于页面显示
            sb.append("<div align=\"center\"><b>没有记录!</b></div></td></tr>\r\n");
        } else {                                                  //判定当查询结果为不为空
            //将传入参数 strPage 进行数据格式转换
            intPage = Fun.StrToInt(strPage);
            //将传入参数 sPage 进行数据处理
            sPage = Fun.CheckReplace(sPage);
            if (intPage == 0) {                                   //判定 intPage 为 0
                intPage = 1;                                      //参数 intPage 赋值为 1
            }
            //计算当前页面显示新闻条数
            rs.absolute((intPage - 1) * intPageSize + 1);
            i = 0;                                                //参数 i 赋值为 0
            //i 属性小于页面显示条数并且查询结果集不为空，进行循环方法
            while (i < intPageSize && !rs.isAfterLast()) {
                //定义数字型变量并赋值 News 表里的 NewsID 属性
                int NewsID = rs.getInt("NewsID");
                //定义数字型变量并赋值 News 表里的 NewsTitle 属性
                String NewsTitle = rs.getString("NewsTitle");
                //定义数字型变量并赋值 News 表里的 NewsTime 属性
                String NewsTime = rs.getString("NewsTime");
                //定义数字型变量并赋值 News 表里的 AdminName 属性
                String AdminName = rs.getString("AdminName");
                //返回属性添加字符串数据用于页面显示<tr>表示换行
                sb.append("<tr>");
                //返回属性添加字符串数据用于页面显示新闻标题
                sb.append("<td>" + NewsTitle + "</td>");
                //返回属性添加字符串数据用于页面显示用户名
                sb.append("<td >" + AdminName + "</td>");
```

```
                //返回属性添加字符串数据用于页面显示新闻时间
                sb.append("<td >" + NewsTime + "</td>");
                //返回属性添加字符串数据用于页面显示详情按钮
                sb.append("<td ><a style=\"color:#3F862E\" target=\"_blank\" href=\"newsFrontDetail.jsp?newsId="
                        + NewsID + "\">详情</a></td></tr>");
                rs.next();                                              //判定是否存在下一条信息
                i++;                                                    //i 属性数值自增 1
            }
            //拼写字符串数据用于列表最下方的分页方法
            sb.append(Fun.PageFront(sPage, rs, intPage, intPageSize));
        }
        rs.close();                                                     //关闭结果集
        stmt.close();                                                   //关闭查询
        Conn.close();                                                   //关闭数据连接
        return sb.toString();                                           //返回字符串数据
    } catch (Exception e) {                                             //得到系统运行异常
        return "No";                                                    //如果系统异常方法返回字符"No"
    }
}
```

1.6.3 查看新闻内容

News.java 文件包含另一个重要方法：FrontNewsDetail()方法，该方法用于查询新闻内容。在新闻列表中，每一条新闻在题头"详情"下对应的都是"详情"超链接。用户单击与某一条新闻对应的"详情"超链接后，页面就会发生跳转，并在新的页面中显示此条新闻的具体内容，如图 1.8 所示。

图 1.8 显示新闻内容的效果图

FrontNewsDetail()方法的代码如下：

```
public String FrontNewsDetail(String s0) {
    try {                                                               //捕获系统异常
        Connection Conn = DBConn.getConn();                             //建立数据库连接
```

```java
        Statement stmt = Conn.createStatement();          //创建数据查询
        ResultSet rs = null;                              //定义结果集
        int NewsID = Fun.StrToInt(s0);                    //将参数 s0 进行转换
        if (NewsID == 0) {                                //s0 判定等于 0
            return "No";                                  //返回字符"No"
        } else {                                          //s0 判定不等于 0
            try {
                //创建 sql 语句查询 News 表全部信息
                String sql = "select * from News where NewsID=" + NewsID;
                rs = stmt.executeQuery(sql);              //得到执行 sql 结果
                //定义本方法返回字符串数据
                StringBuffer sb = new StringBuffer();

                int i = 0;                                //定义起始数据
                //i 属性小于页面显示条数并且查询结果集不为空，进行循环方法
                while (i < 1 && !rs.isAfterLast()) {
                    rs.next();                            //判定是否存在
                    //定义数字型变量并赋值 News 表里的 NewsTitle 属性
                    String NewsTitle = rs.getString("NewsTitle");
                    //定义数字型变量并赋值 News 表里的 NewsContent 属性
                    String NewsContent = rs.getString("NewsContent");

                    String[] content = NewsContent.split("#");  //以#号进行分割
                    //返回属性添加字符串数据用于页面显示新闻标题
                    sb.append("<br><h2 style=\"font-size:28px;margin-left:30%\">" + NewsTitle + "</h2>");

                    for (int j = 0; j < content.length; j++) {  //循环
                        //返回属性添加字符串数据用于页面显示内容信息
                        sb.append("<p>" + content[j] + "</p>");
                    }
                    rs.next();                            //判定是否存在
                    i++;                                  //自增
                }

                rs.close();                               //关闭结果集
                stmt.close();                             //关闭查询
                Conn.close();                             //关闭数据连接
                return sb.toString();                     //返回字符串数据
            } catch (Exception e) {                       //得到系统运行异常
                Conn.rollback();                          //JDBC 回滚机制
                Conn.close();                             //关闭数据库连接
                return "No";                              //返回字符串"No"
            }
        }
    } catch (Exception e) {                               //得到系统运行异常

        return "No";                                      //系统异常后返回字符"No"
    }

}
```

1.7 后端管理员登录模块设计

如图 1.2 所示，在明日科技门户网首页的底部版权信息栏中，有一个"后端"超链接。明日科技门户网的管理员可以通过单击这个超链接，进入如图 1.9 所示的明日科技门户网后端的登录页面。管理员在输入正确的用户名（mr）和密码（mrsoft）后，即可登录明日科技门户网的后端。

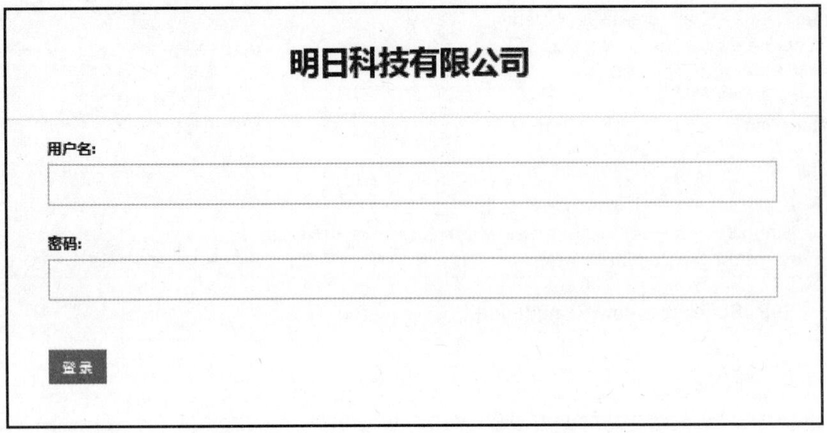

图 1.9　明日科技门户网后端的登录页面

本项目把位于 WebContent/admin 下的 login.jsp 作为明日科技门户网后端的登录页面。在 login.jsp 中，需要对管理员输入的用户名和密码进行验证，并根据验证结果弹出相应的提示信息。代码如下：

```jsp
<%
response.setHeader("Pragma", "No-cache");
response.setHeader("Cache-Control", "no-cache");
response.setDateHeader("Expires", 0);
request.setCharacterEncoding("utf-8");                              //设置编码方式为 utf-8
//处理表单，并进行异常处理
String Action = request.getParameter("Action");

if (Action != null && Action.equals("Login")) {
    String Page1 = (String) request.getHeader("Referer");           //得到页面地址来源
    String Page2 = request.getRequestURL().toString();              //得到当前页面地址
    String methon = request.getMethod();

    if (methon.equals("POST") && Page2.equals(Page1)) {
        try {
            Function Fun = new Function();
            Login login = new Login();
            String IP = request.getRemoteAddr();                    //得到客户端 Ip 地址
            String User = request.getParameter("User");             //得到登录用户名
            String Pwd = request.getParameter("Pwd");               //得到登录密码
            if (login.LoginCheck(User, Pwd)) {
                session.setAttribute("Login", "Yes");
                //session.setAttribute("AdminID",Integer.toString(login.AdminID));
                //session.setAttribute("AdminType",Integer.toString(login.AdminType));
                session.setAttribute("AdminName", User);
                out.println
                ("<SCRIPT LANGUAGE='JavaScript'>alert('登录成功！');location.href='news.jsp';</SCRIPT>");
                return;
            } else {
                out.println
                ("<SCRIPT LANGUAGE='JavaScript'>alert('用户名或密码不正确！');location.href='login.jsp';</SCRIPT>");
                return;
            }
        } catch (Exception e) {
            out.println("<SCRIPT LANGUAGE='JavaScript'>alert('服务器异常！');location.href='login.jsp';</SCRIPT>");
            return;
        }
    } else {
        response.sendError(403, "禁止访问");
        return;
```

```
        }
    }
%>
```

在上述代码中，引用了 com.mingrisoft 下的 Login.java 文件中的 LoginCheck()方法。该方法用于获取对管理员输入的用户名和密码进行验证的结果。在该方法中，包含了两个字符串类型的参数，其中 s1 表示管理员的用户名，s2 表示管理员的密码。LoginCheck()方法的代码如下：

```
public boolean LoginCheck(String s1, String s2) {
    try {
        Connection Conn = DBConn.getConn();
        boolean OK = true;
        OK = Fun.CheckLogin(Conn, s1, s2);
        return OK;
    } catch (SQLException e) {
        return false;
    }
}
```

在 LoginCheck()方法中，引用了 com.mingrisoft 下的 Function.java 文件中的 CheckLogin()方法。该方法用于对管理员输入的用户名和密码进行验证。在该方法中，判断管理员在明日科技门户网后端的登录页面中输入的用户名和密码与数据库存储的管理员的用户名和密码是否一致。代码如下：

```
public boolean CheckLogin(Connection conn, String s1, String s2) throws SQLException {
    Statement stmt = conn.createStatement();
    ResultSet rs = null;
    boolean OK = true;
    int AdminID = 0;
    int AdminType = 0;
    String AdminPwd = "";
    String User = CheckReplace(s1);
    String Pwd = CheckReplace(s2);
    String Sql = "select * from Admin where AdminName='" + User + "'";
    rs = stmt.executeQuery(Sql);
    if (!rs.next()) {
        OK = false;
    } else {
        AdminPwd = rs.getString("AdminPwd");
        if (Pwd.equals(AdminPwd)) {

            AdminID = rs.getInt("AdminID");
            AdminType = rs.getInt("AdminType");
            OK = true;
        } else {
            OK = false;
        }
    }
    return OK;
}
```

1.8 后端新闻管理模块设计

管理员输入正确的用户名和密码后，即可登录明日科技门户网的后端。在明日科技门户网的后端，管理员能够对新闻进行管理，如添加未发布新闻、修改或者删除已发布的新闻等，如图 1.10 所示。

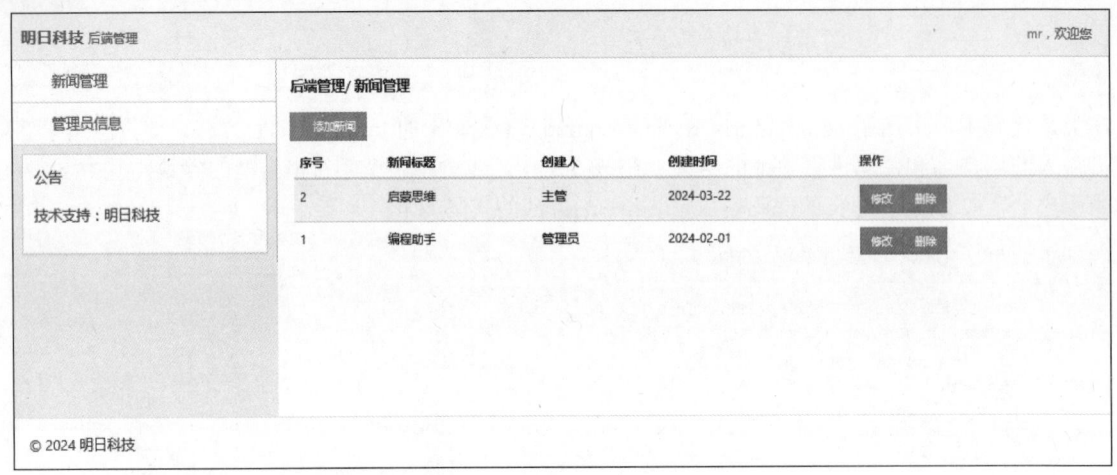

图 1.10　新闻管理的效果图

1.8.1　添加新闻

本项目 WebContent/admin 文件夹下的 newsAdd.jsp 用于显示添加新闻的窗口。当管理员在新闻管理页面中单击"添加新闻"按钮时，会弹出用于添加新闻的窗口，如图 1.11 所示。

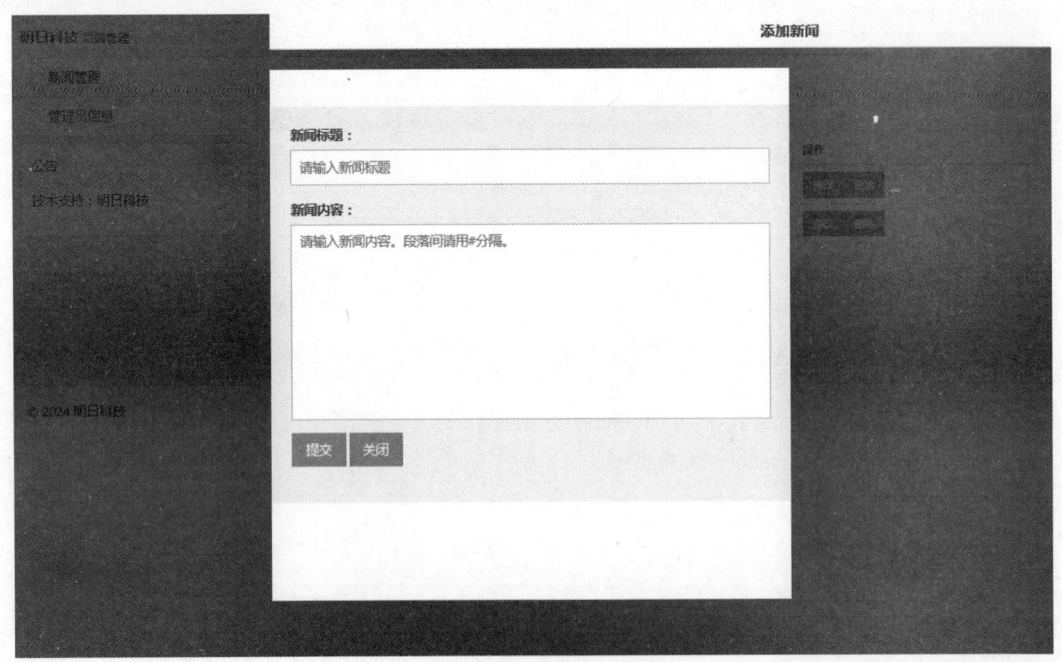

图 1.11　用于添加新闻的窗口

管理员在把新闻标题和新闻内容编写完毕后，单击"提交"按钮，即可成功发布此条新闻。代码如下：

```
<%
    request.setCharacterEncoding("UTF-8");                              //设置编码方式为 utf-8
    News News1 = new News();                                            //引入新闻方法类

    String AdminName = (String)session.getAttribute("AdminName");       //从页面缓存中提取当前用户
    String Action = request.getParameter("Action");                     //从页面请求中获取命令
```

```
        if (Action!=null && Action.equals("Add"))                    //判断用户是否存在
        {
            String IP = request.getRemoteAddr();                     //得到客户端 Ip 地址
            String [] s = new String[2];                             //定义数组
            s[0] = request.getParameter("NewsTitle");                //填写数组第一位
            s[1] = request.getParameter("NewsContent");              //填写数组第二位
            String sNews = News1.AddNews(s,AdminName,IP);            //获取方法返回值
            if (sNews.equals("Yes"))                                 //判断返回值
            {
                out.print("<script>alert('添加新闻成功!');location.href='news.jsp';</script>");   //页面输出
                return;                                              //结束
            }
            else                                                     //或者
            {
                out.print("<script>alert('添加新闻失败!');location.href='news.jsp';</script>");   //页面输出
                return;                                              //结束
            }
        }
%>
```

在上述代码中，引用了 com.mingrisoft 下的 News.java 文件中的 AddNews()方法。该方法用于把图 1.12 中的新闻标题和新闻内容添加到数据库中，并返回执行添加新闻操作后的结果。AddNews()方法的代码如下：

```
public String AddNews(String[] s, String s1, String s2) {
    try {
        Connection Conn = DBConn.getConn();
        Statement stmt = Conn.createStatement();
        ResultSet rs = null;
        String sSql = "select * from News order by NewsID desc";
        rs = stmt.executeQuery(sSql);
        int z = 0;
        int newNum = 0;
        if (!rs.next()) {
            newNum = 1;
        } else {
            while (z < 1 && !rs.isAfterLast()) {
                int NewsID = rs.getInt("NewsID");
                newNum = NewsID + 1;
                break;
            }
        }
        for (int i = 0; i < s.length; i++) {
            if (i != 1)
                s[i] = Fun.getStrCN(Fun.CheckReplace(s[i]));
            else
                s[i] = Fun.getStrCN(s[i]);
        }
        SimpleDateFormat format1 = new SimpleDateFormat("yyyy-MM-dd HH:mm");
        String newsTime = format1.format(new Date());
        StringBuffer sql = new StringBuffer();
        sql.append("insert into News (NewsID,NewsTitle,NewsContent,NewsTime,AdminName) "
            + "values (" + "'" + newNum + "'," + "'" + s[0] + "'," + "'"
            + s[1] + "'," + "'" + newsTime + "'," + "'" + s1 + "')");
        System.out.println(sql);
        try {
            Conn.setAutoCommit(false);
            stmt.execute(sql.toString());
            Conn.commit();
            Conn.setAutoCommit(true);
```

```
                stmt.close();
                Conn.close();
                return "Yes";
            } catch (Exception e) {
                Conn.rollback();
                e.printStackTrace();
                Conn.close();
                return "添加成功!";
            }
        } catch (Exception e) {
            e.printStackTrace();
            return "添加失败";
        }
}
```

1.8.2 删除新闻

本项目 WebContent/admin 文件夹下的 newsDel.jsp 用于显示删除新闻的窗口。当管理员在新闻管理页面中单击某条新闻后的"删除"按钮时，会弹出用于确定删除新闻的对话框，如图 1.12 所示。

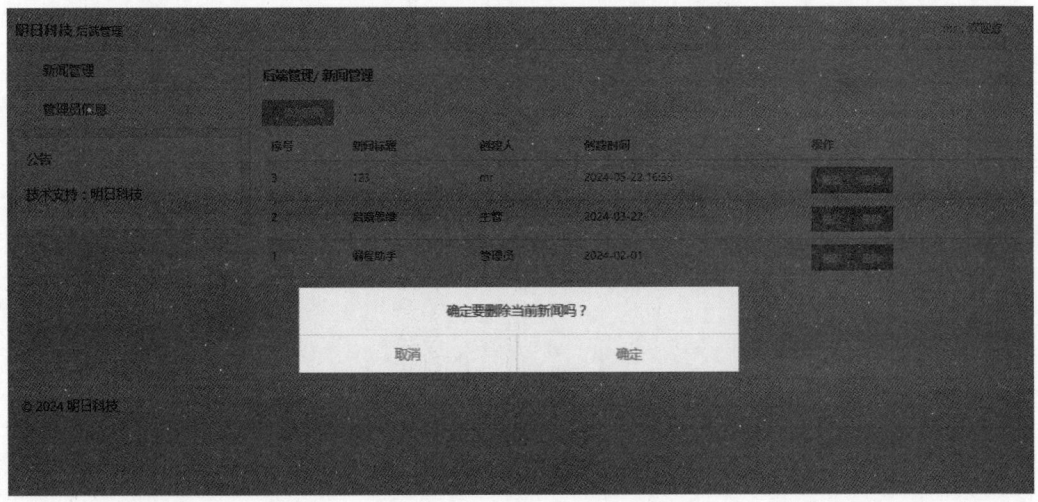

图 1.12 用于确定删除新闻的对话框

如果管理员单击"确定"按钮，就会删除当前新闻；如果管理员单击"取消"按钮，就不会删除当前新闻。newsDel.jsp 代码如下：

```
<%
    News News1 = new News();                                              //引入新闻方法类
    String IP = request.getRemoteAddr();                                  //得到客户端 Ip 地址
    String AdminName = (String) session.getAttribute("AdminName");        //从页面缓存中提取当前用户
    String NewsID = request.getParameter("NewsID");                       //从页面请求中获取编号
    if (News1.DelNews(NewsID, AdminName, IP))                             //判断返回值
        out.print("<script>alert('删除新闻成功!');location.href='news.jsp';</script>");  //页面输出
    else {                                                                //否则
        out.print("<script>alert('删除新闻失败!');location.href='news.jsp';</script>");  //页面输出
    }
%>
```

在上述代码中，引用了 com.mingrisoft 下的 News.java 文件中的 DelNews() 方法。该方法用于从数据库中删除某条新闻，并返回执行删除某条新闻操作后的结果。DelNews() 方法的代码如下：

```
public boolean DelNews(String s0, String s1, String s2) {
    try {
        Connection Conn = DBConn.getConn();
        Statement stmt = Conn.createStatement();
        int NewsID = Fun.StrToInt(s0);
        if (NewsID == 0)
            return false;
        else {
            try {
                String sql = "delete from News where NewsID=" + NewsID;
                Conn.setAutoCommit(false);
                stmt.executeUpdate(sql);
                Conn.commit();
                Conn.setAutoCommit(true);
                stmt.close();
                Conn.close();
                return true;
            } catch (Exception e) {
                Conn.rollback();
                Conn.close();
                return false;
            }
        }
    } catch (Exception e) {
        e.printStackTrace();
        return false;
    }
}
```

1.8.3 修改新闻

本项目 WebContent/admin 文件夹下的 newsEdit.jsp 用于显示修改新闻的窗口。当管理员在新闻管理页面中单击某条新闻后的"修改"按钮时，会弹出用于修改新闻的窗口，如图 1.13 所示。

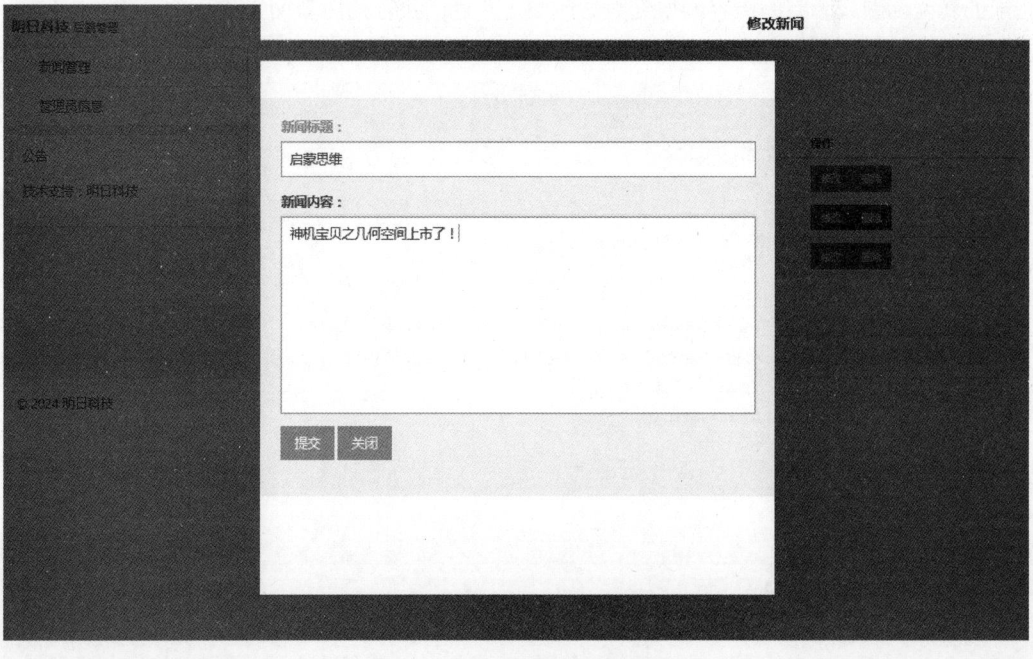

图 1.13　用于修改新闻的窗口

管理员在修改新闻标题或者新闻内容完毕后，单击"提交"按钮，即可成功发布此条新闻。newsEdit.jsp 的代码如下：

```jsp
<%
    request.setCharacterEncoding("utf-8");                              //设置编码方式为 utf-8
    News News1 = new News();                                            //引入新闻方法类
    String [] sa = null;                                                //定义数组
    String NewsID = request.getParameter("newsId");                     //从页面请求中获取编号
    String Action = request.getParameter("Action");                     //从页面请求中获取命令
    if (Action!=null && Action.equals("Edit"))                          //判断是否存在
    {
        String IP = request.getRemoteAddr();                            //得到客户端 Ip 地址
        String AdminName = (String)session.getAttribute("AdminName");   //从页面缓存中提取当前用户
        String [] s = new String[2];                                    //定义数组
        s[0] = request.getParameter("upd_NewsTitle");                   //填写数组第一位
        s[1] = request.getParameter("upd_NewsContent");                 //填写数组第二位

        String sOK = News1.EditNews(s,NewsID,AdminName,IP);             //获取方法返回值
        if (sOK.equals("Yes"))                                          //判断返回值
        {
            out.println("<script>alert('修改新闻成功!');location.href='news.jsp';</script>");  //页面输出
            return;                                                     //结束
        }
        else                                                            //否则
        {
            out.println("<script>alert('修改新闻失败!');location.href='news.jsp';</script>");  //页面输出
            return;                                                     //结束
        }
    }
%>
```

在上述代码中，引用了 com.mingrisoft 下的 News.java 文件中的 EditNews() 方法。该方法用于把图 1.13 中修改完毕的新闻标题或者新闻内容添加到数据库中，并返回执行修改新闻操作后的结果。EditNews() 方法的代码如下：

```java
public String EditNews(String[] s, String s0, String s1, String s2) {
    try {
        Connection Conn = DBConn.getConn();
        Statement stmt = Conn.createStatement();
        for (int i = 0; i < s.length; i++) {
            s[i] = Fun.getStrCN(Fun.CheckReplace(s[i]));
        }
        int NewsID = Fun.StrToInt(s0);
        StringBuffer sql = new StringBuffer();
        sql.append("update News set NewsTitle='" + s[0] + "'" + " ,NewsContent='" + s[1]
                + "'" + " where NewsID='" + NewsID + "'");
        stmt.executeUpdate(sql.toString());
        stmt.close();
        Conn.close();
        return "Yes";
    } catch (Exception e) {
        e.printStackTrace();
        System.out.print(e.getMessage());
        return "编辑错误!";
    }
}
```

1.9 后端管理员信息模块设计

在明日科技门户网的后端，管理员除了能够对新闻进行管理，还能够查看管理员信息。管理员信息页面的效果如图 1.14 所示。

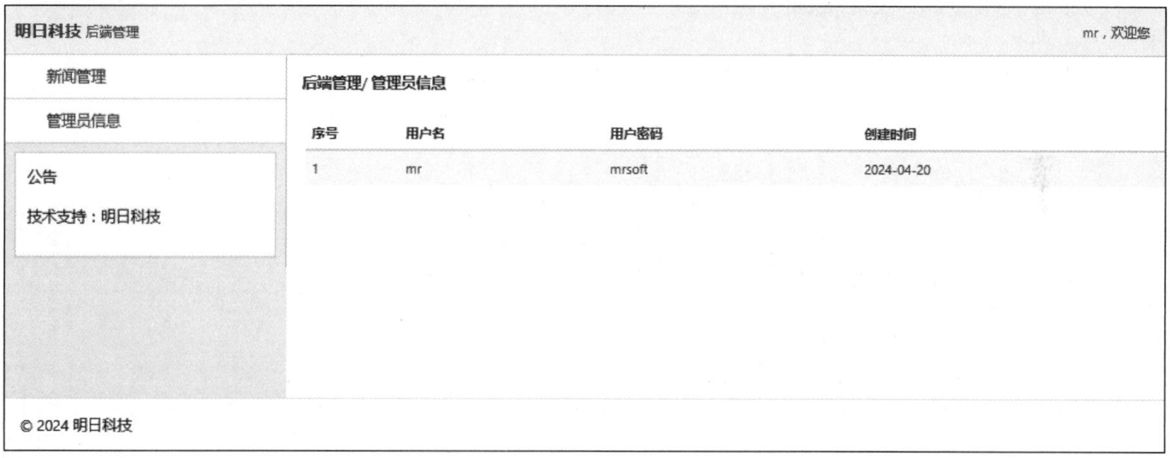

图 1.14 管理员信息页面的效果图

> **说明**
> 后端管理员信息模块没有实现添加管理员信息、删除管理员信息、修改管理员信息等功能，只实现了查看管理员信息的功能。如果读者想要实现添加管理员信息、删除管理员信息、修改管理员信息等功能，可以参考后端新闻管理模块中的添加新闻、删除新闻、修改新闻等功能的实现过程来自行实现。这两个模块的实现过程是如出一辙的。

本项目 WebContent/admin 文件夹下的 adminUser.jsp 用于显示管理员信息页面。当管理员在明日科技门户网的后端单击"管理员信息"时，就会在管理员信息页面上显示管理员信息。在 adminUser.jsp 中，包含一个脚本程序，其作用是接收从服务器返回的数据，代码如下：

```jsp
<%
request.setCharacterEncoding("UTF-8");
User user = new User();
String strPage = request.getParameter("intPage");
String sPage = request.getContextPath() + request.getServletPath()+ "?";
String sOK = user.ListUser(sPage, strPage);
if (sOK.equals("No")) {
    out.println("数据服务器出现错误！");
} else {
    out.println(sOK);
}
%>
```

在上述代码中，引用了 com.mingrisoft 下的 User.java 文件中的 ListUser() 方法。该方法首先把用于显示管理员信息（用户名、用户密码和创建时间）的代码存储在一个 StringBuffer 类的对象中，然后把该对象表示为字符串，并返回给 adminUser.jsp 中的脚本程序。ListUser() 方法的代码如下：

```java
public String ListUser(String sPage, String strPage) {
    try {
        Connection Conn = DBConn.getConn();
        Statement stmt = Conn.createStatement();
        ResultSet rs = null;
        StringBuffer sb = new StringBuffer();
        int i;
        int intPage = 1;
        int intPageSize = 5;
        String sSql = "select * from Admin order by AdminID desc";
        rs = stmt.executeQuery(sSql);
        if (!rs.next()) {
            sb.append
            ("<tr height=\"25\" bgcolor=\"#d6dff7\" class=\"info1\"><td colspan=\"4\">\r\n");
            sb.append("<div align=\"center\"><b>没有管理员信息！</b></div></td></tr>\r\n");
        } else {
            intPage = Fun.StrToInt(strPage);
            sPage = Fun.CheckReplace(sPage);
            if (intPage == 0)
                intPage = 1;
            rs.absolute((intPage - 1) * intPageSize + 1);
            i = 0;
            while (i < intPageSize && !rs.isAfterLast()) {
                int AdminID = rs.getInt("AdminID");
                String AdminName = rs.getString("AdminName");
                String AdminPwd = rs.getString("AdminPwd");
                String LastLoginTime = rs.getString("LastLoginTime");
                sb.append("<tr>");
                sb.append("<td class=\"table-id\">" + AdminID + "</td>");
                sb.append("<td>" + AdminName + "</td>");
                sb.append("<td class=\"table-title\">" + AdminPwd + "</td>");
                sb.append("<td class=\"table-title\">" + LastLoginTime + "</td>");
                sb.append("</tr>");
                rs.next();
                i++;
            }
            sb.append(Fun.Page(sPage, rs, intPage, intPageSize));
        }
        rs.close();
        stmt.close();
        Conn.close();
        return sb.toString();
    } catch (Exception e) {
        return "No";
    }
}
```

在 ListUser()方法中，引用了 com.mingrisoft 下的 Function.java 文件中的 Page()方法。该方法用于拼接分页显示管理员信息（用户名、用户密码和创建时间）的代码。Page()方法的代码如下：

```java
public String Page(String sPage, ResultSet rs, int intPage, int intPageSize) {
    String s = null;
    int i = 0;
    try {
        rs.last();
        int intRowCount = rs.getRow();
        int intPageCount;
        if (intRowCount % intPageSize == 0)
            intPageCount = intRowCount / intPageSize;
        else
            intPageCount = (int) Math.floor(intRowCount / intPageSize) + 1;
        if (intPageCount == 0)
```

```
                    intPageCount = 1;
                if (intPage < 1)
                    intPage = 1;
                if (intPage > intPageCount)
                    intPage = intPageCount;
                if (intRowCount > intPageSize) {
                    s = "<table class=\"am-table am-table-striped\" width=\"90%\" "
                        + "border=\"0\" align=\"center\" cellpadding=\"2\" cellspacing=\"0\"><tr>";
                    s = s + "<td width=\"80%\" height=\"30\" class=\"chinese\"><span class=\"chinese\">";
                    s = s + "当前第" + intPage + "页/共" + intPageCount + "页,    共"
                        + intRowCount + "条记录,    " + intPageSize + "条/页";
                    int showye = intPageCount;
                    if (showye > 10)
                        showye = 10;
                    for (i = 1; i <= showye; i++)
                        ;
                    s = s + "</span></td>";
                    s = s + "<td width=\"20%\">";
                    s = s + "<table width=\"100%\" border=\"0\">";
                    s = s + "<tr><td><div align=\"right\"><span class=\"chinese\">";
                    s = s + "<select id=\"ipage\" name=\"ipage\" class=\"chinese\" "
                        + "onChange=\"jumpMenu('self',this,0)\">";
                    s = s + "<option value=\"\" selected>请选择</option>";
                    for (i = 1; i <= intPageCount; i++) {
                        String sSelect = i == intPage ? "SELECTED" : "";
                        s = s + "<option value=\"" + sPage + "intPage=" + i + "\"" + sSelect
                            + ">第" + i + "页</option>";
                    }
                    s = s + "</select></span></div>";
                    s = s + "</td></tr></table>";
                    return s + "</td></tr></table>";
                }
                return "";
            } catch (Exception e) {
            }
            return "分页出错!";
        }
```

1.10 项目运行

通过前述步骤，设计并完成了"明日科技门户网"项目的开发。下面运行本项目，以检验我们的开发成果。如图 1.15 所示，在 IntelliJ IDEA 中，单击 ▶ 快捷图标，即可运行本项目。

图 1.15　IntelliJ IDEA 的快捷图标

成功运行该项目，会自动打开如图 1.16 所示的"明日科技门户网"首页。在首页上，用户通过单击导航栏中的"企业简介""新闻""核心竞争力""联系我们"等超链接，能够快速获取与指定超链接相对应的信息；管理员登录了明日科技门户网的后端以后，即可对管理员和新闻进行管理。这样，我们就成功地检验了该项目的运行。

本项目应用了许多比较常用的 Java Web 技术。例如，采用 Servlet、JSP、CSS 等技术完成了各网页模块的设计；采用 JDBC 编程操作 MySQL 数据库、Servlet 页面跳转等技术完成了新闻模块和后端新闻维护模块等的设计。

图 1.16　成功运行项目后进入首页

1.11　源码下载

虽然本章详细地讲解了如何编码实现"明日科技门户网"项目的各个功能,但给出的都是代码片段,而非源码。为了方便读者学习,本书提供了完整的项目源码,扫描右侧二维码即可下载。

第 2 章
购好物网络商城

——Servlet + JSP + MySQL

随着互联网技术的高速发展，电子商务已经成为现代生活不可或缺的重要组成部分，各类购物网站不断涌现。购物网站是买卖双方用于交易商品的电子商务平台。卖家可以在此平台上开设网店，发布出售商品的信息，低成本、高效率地为消费者提供商品及服务；买家可以在此平台上选择并购买喜欢的商品，足不出户地享受购物的快乐。本章将使用 Java Web 开发中的 MySQL 数据库、Servlet、JSP 等关键技术开发一个购物网站项目——购好物网络商城。

项目微视频

本项目的核心功能及实现技术如下：

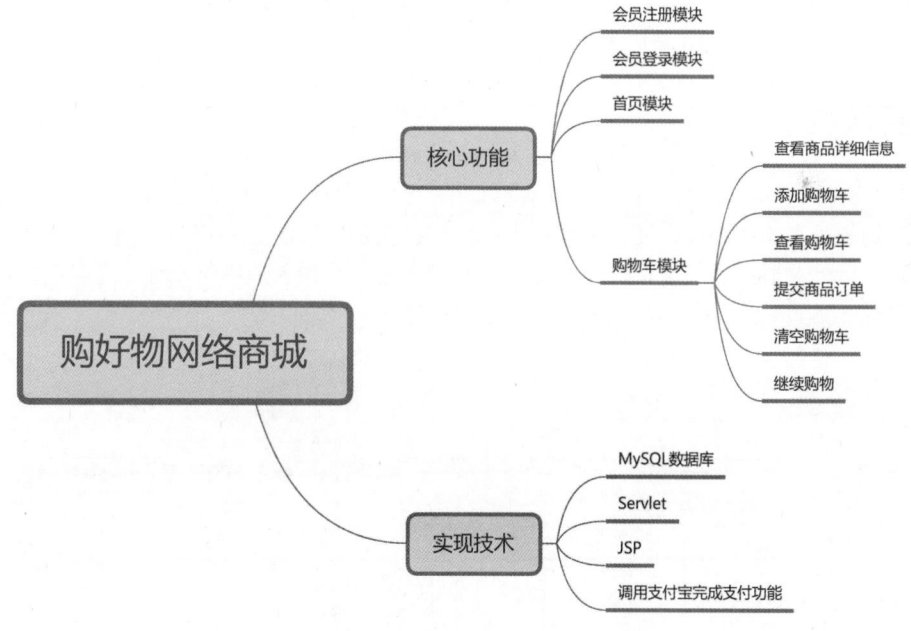

2.1 开发背景

与传统的线下购物相比，网上购物的优点十分突出，它不受时间和地点的限制，消费者可以随时随地浏览和购买商品，这种便捷性使得它越来越受到消费者的青睐。加之配套的物流服务与网上支付技术越来越完善，网上购物不仅能让消费者十分便捷地购买到心仪的商品，并且能让消费者享受到送货上门的优质服务，极大程度地提升了消费者的购物体验。如今，网上购物已经成为一种主流的购物形式，各大购物网站都在积极优化网站的功能，力求为用户提供更好的购物服务。为此，本章使用 Java Web 技术开发一个名为"购好物网络商城"的购物网站。

购好物网络商城将实现以下目标：

☑ 首页展示商品类别和主要商品；

- ☑ 商品详情页展示某件商品的详细信息；
- ☑ 支持会员注册和会员登录的功能；
- ☑ 支持购物车功能，通过购物车可以完成支付和提交订单操作。

2.2 系统设计

2.2.1 开发环境

本项目的开发及运行环境如下：
- ☑ 操作系统：推荐 Windows 10、11 及以上，兼容 Windows 7（SP1）。
- ☑ 开发工具：IntelliJ IDEA。
- ☑ 开发语言：Java EE。
- ☑ 数据库：MySQL 8.0。
- ☑ Web 服务器：Tomcat 9.0 及以上版本。

2.2.2 业务流程

启动项目后，购好物网络商城的首页会被自动打开。在购好物网络商城的首页上，用户既可以浏览热门商品、最新上架的商品和折扣商品，又可以通过导航栏切换页面，进而去浏览图书类、家电类、服装类和电子类的对应商品。用户注册或登录会员后，不仅可以把喜欢的商品加入购物车，而且能够在购物车中完成支付和提交订单操作。购好物网络商城的业务流程如图 2.1 所示。

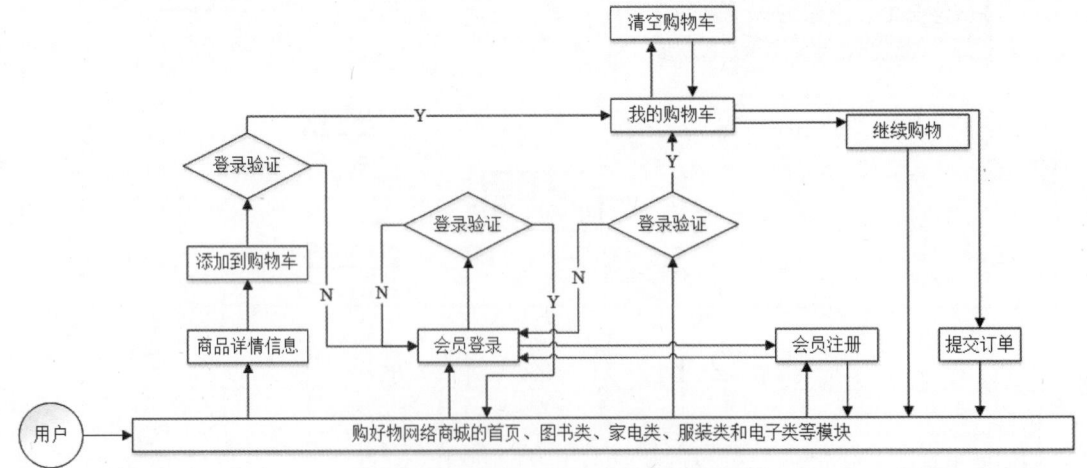

图 2.1 购好物网络商城的业务流程图

> **说明**
> 购好物网络商城具有一个导航栏，其中包括首页、图书类、家电类、服装类和电子类 5 个模块，这 5 个模块的设计过程都大同小异。限于本书篇幅，这里仅讲解首页模块的设计过程。读者可以参考首页模块的设计过程，自行根据提供的源码学习其他模块的设计过程。

2.2.3 功能结构

本项目的功能结构已经在章首页中给出。作为一个网上购物网站，本项目实现的具体功能如下：
- 首页模块：展示热门商品、最新上架的商品和折扣商品，以及通过导航栏链接商品分类页面。
- 会员注册模块：允许新用户注册，成为购好物网络商城的会员。
- 会员登录模块：验证用户是否是购好物网络商城的会员，并允许会员登录。
- 购物车模块：用户通过该模块可以实现查看某件商品的详细信息、把喜欢的商品添加到购物车、查看购物车、完成支付并提交订单、清空购物车、继续购物等操作。

2.3 技术准备

2.3.1 技术概览

- MySQL 数据库：数据库（Database）就是一个存储数据的仓库。通过数据库管理系统，可以有效地存储、组织和管理数据。MySQL 就是这样的一个关系型数据库管理系统（RDBMS），它可以称得上是目前运行速度最快的 SQL 语言数据库管理系统。本章将使用 JDBC 技术操作 MySQL 数据库，如使用 Class 类的静态方法 forName(String className)加载指定数据库的驱动程序。代码如下：

```java
public Connection conn = null;                                              //数据库连接对象
public Statement stmt = null;                                               //Statement 对象，用于执行 SQL 语句
public ResultSet rs = null;                                                 //结果集对象
private static String dbClassName = "com.mysql.cj.jdbc.Driver";             //驱动类的类名
private static String dbUrl = "jdbc:mysql://127.0.0.1:3306/db_shop?user=root&password=root"
        + "&useUnicode=true&characterEncoding=UTF-8&useSSL=false&serverTimezone=Asia/Shanghai"
        + "&zeroDateTimeBehavior=CONVERT_TO_NULL&allowPublicKeyRetrieval=true";  //访问 MySQL 数据库的路径
public static Connection getConnection() {
    Connection conn = null;                                                 //声明数据库连接对象
    try {                                                                   //捕捉异常
        Class.forName(dbClassName).newInstance();                           //装载数据库驱动
        conn = DriverManager.getConnection(dbUrl);                          //获取数据库连接对象
    } catch (Exception ee) {                                                //处理异常
        ee.printStackTrace();                                               //输出异常信息
    }
    return conn;                                                            //返回数据库连接对象
}
```

- Servlet 技术：Servlet 实质上就是按 Servlet 规范编写的 Java 类，它采用 Java 语言编写，具有 Java 语言的优点。与 Java 语言的区别是，Servlet 对象对 Web 应用进行了封装，提供了 Servlet 对 Web 应用的编程接口，以处理相应的 HTTP 请求，如处理提交数据、会话跟踪、读取和设置 HTTP 头信息等。例如，本章使用 Servlet 技术完成了会员登录模块的设计。
- JSP 技术：JSP 是一种动态网页技术标准，JSP 页面通常是由 Java 代码片段和 JSP 指令标识构成的。在接收到用户请求时，服务器首先会处理 Java 代码片段，然后生成处理结果返回客户端，客户端将呈现相应的页面效果。本章的会员注册、会员登录、首页、购物车等模块所呈现的页面效果均是采用 JSP 技术设计的。

有关 MySQL 数据库等基础知识以及开发工具 IntelliJ IDEA 在《Java 从入门到精通（第 7 版）》中有详细的讲解，对这些知识不太熟悉的读者可以参考该书对应的内容。Servlet 和 JSP 这两个技术在本书第 1 章

的"技术准备"中有详细的介绍,对这两个技术不太熟悉的读者可以参考本书第1章对应的内容。下面将对"调用支付宝完成支付操作"技术进行必要介绍,以确保读者可以顺利完成本项目。

2.3.2 调用支付宝完成支付操作

在购物车页面中,单击"结账"按钮,会弹出支付对话框,在该对话框中,扫描二维码将调用支付宝完成支付操作。实现此功能的基本步骤如下:

- ☑ 注册支付宝企业账户:进入支付宝开发平台(蚂蚁金服开放平台)。注册成为"商家",选择"企业账户",按照向导进行操作即可。
- ☑ 完成支付宝实名认证:注册支付宝企业账户后,会要求用户进行实名认证。准备以下资料后,单击"企业实名信息填写"按钮,按照向导操作即可完成实名认证。
  - ➢ 营业执照影印件;
  - ➢ 对公银行账户,可以是基本户或一般户;
  - ➢ 法定代表人的身份证影印件。

说明

如果是代理人,除以上资料外,还需要准备代理人的身份证影印件和企业委托书,必须盖有公司公章或者财务专用章。

- ☑ 申请支付套餐:支付宝提供了多种支付套餐。一般情况下,选择"即时到账"套餐,该套餐可以让用户在线向购物平台的支付宝账号支付资金,并且交易资金即时到账。申请"即时到账"套餐后,会进入审核阶段。通常情况下,2~5天会有申请结果。
- ☑ 生成与配置密钥:在调用支付宝完成支付操作时,需要提供商户的私钥和支付宝的公钥,使用支付宝提供的一键生成工具即可获取。
- ☑ 下载Demo:前面的工作准备就绪后,就可以测试支付功能了。这时,可以下载支付宝开发平台提供的即时到账交易接口的Demo,根据Demo中的说明进行测试即可。

2.4 数据库设计

2.4.1 数据库概述

本项目采用的数据库名称为db_shop,数据库包含7张数据表,如表2.1所示。

表2.1 购好物网络商城的数据库结构

数据库名	表 名	表 说 明
db_shop	tb_goods	商品信息表
	tb_manager	管理员信息表
	tb_member	会员信息表
	tb_order	订单信息主表
	tb_order_detail	订单信息明细表
	tb_subtype	商品小分类信息表
	tb_supertype	商品大分类信息表

2.4.2 数据表设计

在本项目中发挥主要作用的数据表有 tb_supertype、tb_subtype、tb_goods、tb_order、tb_order_detail、tb_manager 和 tb_member，下面将详细介绍这 7 张数据表的结构设计。

- ☑ tb_supertype（商品大分类信息表）：主要用于保存商品大分类信息，也就是父分类，其结构如表 2.2 所示。

表 2.2 tb_supertype 表结构

字 段 名 称	数 据 类 型	长　　度	是 否 主 键	说　　明
ID	INT		主键	ID 号
TypeName	VARCHAR	50		分类名称

- ☑ tb_subtype（商品小分类信息表）：主要用于保存商品小分类信息，也就是子分类，其结构如表 2.3 所示。

表 2.3 tb_subtype 表结构

字 段 名 称	数 据 类 型	长　　度	是 否 主 键	说　　明
ID	INT		主键	ID 号
superType	INT			父类 ID 号
TypeName	VARCHAR	50		分类名称

- ☑ tb_goods（商品信息表）：主要用于保存商品信息，其结构如表 2.4 所示。

表 2.4 tb_goods 表结构

字 段 名 称	数 据 类 型	长　　度	是 否 主 键	说　　明
ID	BIGINT		主键	商品 ID
typeID	INT			类别 ID
goodsName	VARCHAR	200		商品名称
introduce	TEXT			商品简介
price	DECIMAL	19(4)		定价
nowPrice	DECIMAL	19(4)		现价
picture	VARCHAR	100		图片文件
INTime	DATETIME			录入时间
newGoods	INT			是否新品，1 为是，默认 0
sale	INT			是否特价，1 为是，默认 0
hit	INT			浏览次数

- ☑ tb_order（订单信息主表）：主要用于保存订单的概要信息，其结构如表 2.5 所示。

表 2.5 tb_order 表结构

字 段 名 称	数 据 类 型	长　　度	是 否 主 键	说　　明
OrderID	BIGINT		主键	订单编号
bnumber	SMALLINT			品种数
username	VARCHAR	15		用户名

续表

字段名称	数据类型	长度	是否主键	说明
recevieName	VARCHAR	15		收货人
address	VARCHAR	100		收货地址
tel	VARCHAR	20		联系电话
OrderDate	DATETIME			订单日期
bz	VARCHAR	200		备注

☑ tb_order_detail（订单信息明细表）：主要用于保存订单的详细信息，其结构如表 2.6 所示。

表 2.6 tb_order_detail 表结构

字段名称	数据类型	长度	是否主键	说明
ID	BIGINT		主键	ID 号
orderID	BIGINT			与 tb_Order 表的 OrderID 字段关联
goodsID	BIGINT			商品 ID
price	DECIMAL	19(4)		价格
number	INT			数量

☑ tb_manager（管理员信息表）：主要用于保存管理员信息，其结构如表 2.7 所示。

表 2.7 tb_manager 表结构

字段名	数据类型	长度	是否主键	说明
ID	INT		主键	ID 号
manager	VARCHAR	30		管理员名称
PWD	VARCHAR	30		密码

☑ tb_member（会员信息表）：主要用于保存会员信息，其结构如表 2.8 所示。

表 2.8 tb_member 表结构

字段名称	数据类型	长度	是否主键	说明
ID	INT		主键	ID 号
userName	VARCHAR	20		账户名称
trueName	VARCHAR	20		真实姓名
passWord	VARCHAR	20		密码
city	VARCHAR	20		所在城市
address	VARCHAR	100		地址
postcode	VARCHAR	6		邮编
cardNO	VARCHAR	24		证件号码
cardType	VARCHAR	20		证件类型
grade	INT			等级
Amount	DECIMAL	19(4)		预存金额
tel	VARCHAR	20		联系电话
email	VARCHAR	100		邮箱
freeze	INT			是否被冻结，1 为是，默认 0

2.5 数据库公共类的编写

数据库公共类 ConnDB 被保存在 com.tools 包中,其中包含了 5 个成员变量,代码如下:

```
public class ConnDB {
    public Connection conn = null;                                      //数据库连接对象
    public Statement stmt = null;                                       //Statement 对象,用于执行 SQL 语句
    public ResultSet rs = null;                                         //结果集对象
    private static String dbClassName = "com.mysql.cj.jdbc.Driver";     //驱动类的类名
    private static String dbUrl = "jdbc:mysql://127.0.0.1:3306/db_shop?user=root&password=root"
            + "&useUnicode=true&characterEncoding=UTF-8&useSSL=false&serverTimezone=Asia/Shanghai"
            + "&zeroDateTimeBehavior=CONVERT_TO_NULL&allowPublicKeyRetrieval=true";  //访问 MySQL 数据库的路径
}
```

在 ConnDB 类中,编写用于连接数据库的 getConnection()方法,用于根据指定的数据库驱动获取数据库连接对象。如果连接失败,则输出异常信息。该方法返回一个数据库连接对象。getConnection()方法的代码如下:

```
public static Connection getConnection() {
    Connection conn = null;                                 //声明数据库连接对象
    try {                                                   //捕捉异常
        Class.forName(dbClassName).newInstance();           //装载数据库驱动
        conn = DriverManager.getConnection(dbUrl);          //获取数据库连接对象
    } catch (Exception ee) {                                //处理异常
        ee.printStackTrace();                               //输出异常信息
    }
    return conn;                                            //返回数据库连接对象
}
```

在 ConnDB 类中,编写用于更新数据的 executeUpdate()方法,该方法的返回值为 int 型的整数,表示更新数据的行数。executeUpdate()方法的代码如下:

```
public int executeUpdate(String sql) {
    int result = 0;                                         //更新数据的记录条数
    try {                                                   //捕捉异常
        conn = getConnection();                             //获取数据库连接
        //创建用于执行 SQL 语句的 Statement 对象
        stmt = conn.createStatement(ResultSet.TYPE_SCROLL_INSENSITIVE, ResultSet.CONCUR_READ_ONLY);
        result = stmt.executeUpdate(sql);                   //执行 SQL 语句
    } catch (SQLException ex) {                             //处理异常
        result = 0;                                         //指定更新数据的记录条数为 0,表示没有更新数据
        ex.printStackTrace();                               //输出异常信息
    }
    try {                                                   //捕捉异常
        stmt.close();                                       //关闭用于执行 SQL 语句的 Statement 对象
    } catch (SQLException ex1) {                            //处理异常
        ex1.printStackTrace();                              //输出异常信息
    }
    return result;                                          //返回更新数据的记录条数
}
```

在 ConnDB 类中,编写用于查询数据的 executeQuery()方法。在该方法中,首先调用 getConnection()方法获取数据库连接对象,然后通过该对象的 createStatement()方法创建一个 Statement 对象,并且调用该对象的 executeQuery()方法执行指定的 SQL 语句,进而实现查询数据的功能。executeQuery()方法的代码如下:

```
public ResultSet executeQuery(String sql) {
    try {                                                   //捕捉异常
```

```
            conn = getConnection();                                    //获取数据库连接
            //创建用于执行 SQL 语句的 Statement 对象
            stmt = conn.createStatement(ResultSet.TYPE_SCROLL_INSENSITIVE, ResultSet.CONCUR_READ_ONLY);
            rs = stmt.executeQuery(sql);                               //执行 SQL 语句
        } catch (SQLException ex) {                                    //处理异常
            ex.printStackTrace();                                      //输出异常信息
        }
        return rs;                                                     //返回查询结果
    }
```

在 ConnDB 类中，编写用于关闭数据库连接的 close()方法。在该方法中，首先关闭结果集对象，然后关闭 Statement 对象，最后再关闭数据库连接对象。close()方法的代码如下：

```
public void close() {
    try {                                                              //捕捉异常
        if (rs != null) {
            rs.close();                                                //关闭结果集对象
        }
        if (stmt != null) {
            stmt.close();                                              //关闭 Statement 对象
        }
        if (conn != null) {
            conn.close();                                              //关闭数据库连接对象
        }
    } catch (Exception e) {                                            //处理异常
        e.printStackTrace(System.err);                                 //输出异常信息
    }
}
```

2.6 会员注册模块设计

会员注册模块主要用于实现新用户成为购好物网络商城会员的功能。在会员注册页面中，用户需要先根据实际情况填写当前用户的信息，再单击"同意协议并注册"按钮。程序会自动验证用户输入的会员信息是否唯一。如果唯一，就把填写完成的会员信息保存到数据库中，完成会员注册的操作；否则给出提示。会员注册页面的效果如图 2.2 所示。

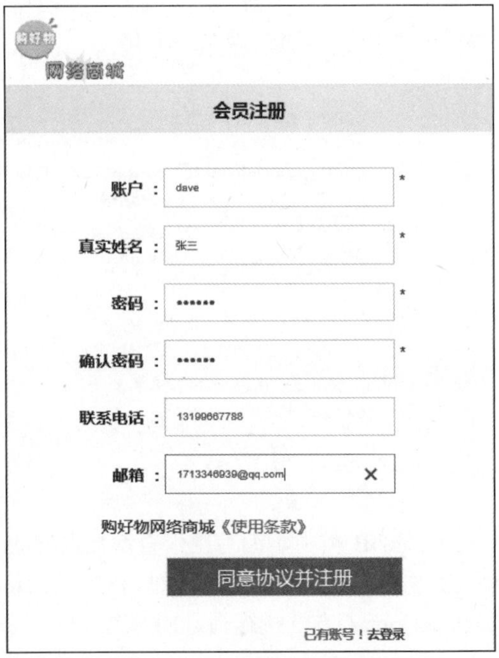

2.6.1 会员模型类的编写

会员模型类 Member 被保存在 com.model 包中，其中包含会员 ID、账户、真实姓名、密码、所在城市、地址、邮编、证件号码、证件类型、联系电话、邮箱、新密码等属性。此外，还要为这些属性添加 Getters 和 Setters 方法。这样，不仅可以控制对上述属性的访问和修改，还可以使代码更加安全，并且可以维护。创建会员模型类 Member 的代码如下：

图 2.2 会员注册页面

```java
public class Member {
    private Integer ID = Integer.valueOf("-1");              //会员 ID 属性
    private String username = "";                            //账户属性
    private String truename = "";                            //真实姓名属性
    private String pwd = "";                                 //密码属性
    private String city = "";                                //所在城市属性
    private String address = "";                             //地址属性
    private String postcode = "";                            //邮编属性
    private String cardno = "";                              //证件号码属性
    private String cardtype = "";                            //证件类型属性
    private String tel = "";                                 //联系电话属性
    private String email = "";                               //邮箱属性
    private String newPwd = "";                              //新密码

    public String getNewPwd() {
        return newPwd;
    }
    public void setNewPwd(String newPwd) {
        this.newPwd = newPwd;
    }
    public Integer getID() {
        return ID;
    }
    public void setID(Integer iD) {
        ID = iD;
    }
    public String getUsername() {
        return username;
    }
    public void setUsername(String username) {
        this.username = username;
    }
    public String getTruename() {
        return truename;
    }
    public void setTruename(String truename) {
        this.truename = truename;
    }
    public String getPwd() {
        return pwd;
    }
    public void setPwd(String pwd) {
        this.pwd = pwd;
    }
    public String getCity() {
        return city;
    }
    public void setCity(String city) {
        this.city = city;
    }
    public String getAddress() {
        return address;
    }
    public void setAddress(String address) {
        this.address = address;
    }
    public String getPostcode() {
        return postcode;
    }
    public void setPostcode(String postcode) {
```

```java
        this.postcode = postcode;
    }
    public String getCardno() {
        return cardno;
    }
    public void setCardno(String cardno) {
        this.cardno = cardno;
    }
    public String getCardtype() {
        return cardtype;
    }
    public void setCardtype(String cardtype) {
        this.cardtype = cardtype;
    }
    public String getTel() {
        return tel;
    }
    public void setTel(String tel) {
        this.tel = tel;
    }
    public String getEmail() {
        return email;
    }
    public void setEmail(String email) {
        this.email = email;
    }
}
```

2.6.2 会员数据库操作接口及其实现类的编写

会员数据库操作接口 MemberDao 被保存在 com.dao 包中，其中定义一个 insert()方法（用于保存会员信息）、一个 select()方法（用于查询会员信息）、一个 update()方法（用于修改会员信息）和一个 delete()方法（用于删除会员信息）。需要注意的是，这里只定义了方法，没有实现方法。会员数据库操作接口 MemberDao 的代码如下：

```java
public interface MemberDao {
    public int insert(Member m);                //保存会员信息
    public List select();                       //查询会员信息
    public int update(Member m);                //修改会员信息
    public int delete(Member m);                //删除会员信息
}
```

创建接口后，必须实现该接口。在 com.dao 包中创建一个 MemberDao 接口的实现类，类名为 MemberDaoImpl。在实现 MemberDao 接口中的各个方法前，需要创建数据库连接类的对象和字符串操作类的对象。代码如下：

```java
private ConnDB conn = new ConnDB();             //创建数据库连接类的对象
private ChStr chStr = new ChStr();              //创建字符串操作类的对象
```

在 MemberDaoImpl 类中，使用 Insert into 语句实现用于保存会员信息的 insert()方法；使用 select 语句实现用于查询会员信息 select()方法；使用 update 语句实现用于修改会员信息 update()方法；使用 delete 语句实现用于删除会员信息 delete()方法。MemberDaoImpl 类的代码如下：

```java
@Override
public int insert(Member m) {
    int ret = -1;                               //用于记录更新记录的条数
    try {                                       //捕捉异常
```

```java
            if(m.getUsername()!=null){
                String sql = "Insert into tb_Member (UserName,TrueName,PassWord,City,address,postcode,"
                    + "CardNO,CardType,Tel,Email) values('"
                    + m.getUsername() + "','" + chStr.chStr(m.getTruename()) + "','"
                    + chStr.chStr(m.getPwd()) + "','" + chStr.chStr(m.getCity()) + "','" + chStr.chStr(m.getAddress())
                    + "','" + chStr.chStr(m.getPostcode()) + "','" + m.getCardno() + "','"
                    + chStr.chStr(m.getCardtype()) + "','" + chStr.chStr(m.getTel()) + "','"+ m.getEmail()
                    + "')";                                            //用于实现保存会员信息的 SQL 语句
                ret = conn.executeUpdate(sql);                         //执行 SQL 语句实现保存会员信息到数据库
            }else{
                ret = 0;                                               //表示注册失败
            }
        } catch (Exception e) {                                        //处理异常
            e.printStackTrace();                                       //输出异常信息
            ret = 0;                                                   //设置变量的值为 0,表示保存会员信息失败
        }
        conn.close();                                                  //关闭数据库的连接
        return ret;                                                    //返回更新记录的条数
}

@Override
public List select() {
    Member form = null;                                                //声明会员对象
    List list = new ArrayList();                                       //创建一个 List 集合对象,用于保存会员信息
    String sql = "select * from tb_member";                            //查询全部会员信息的 SQL 语句
    ResultSet rs = conn.executeQuery(sql);                             //执行查询操作
    try {                                                              //捕捉异常
        while (rs.next()) {
            form = new Member();                                       //实例化一个会员对象
            form.setID(Integer.valueOf(rs.getString(1)));              //获取会员 ID
            list.add(form);                                            //把会员信息添加到 List 集合对象中
        }
    } catch (SQLException ex) {                                        //处理异常
    }
    conn.close();                                                      //关闭数据库的连接
    return list;
}

//执行删除操作
public int delete(Member m) {
    String sql = "delect from tb_member where ID=" + m.getID();
    int ret = conn.executeUpdate(sql);
    conn.close();
    return 0;
}

//执行修改操作
public int update(Member m) {
    int ret = -1;
    try {
        String sql = "update tb_member set TrueName='" + chStr.chStr(m.getTruename()) + "',UserName='"
            + chStr.chStr(m.getUsername()) + "',PassWord='" + chStr.chStr(m.getNewPwd()) + "',City='"
            + chStr.chStr(m.getCity()) + "',address='" + chStr.chStr(m.getAddress()) + "',postcode='"
            + chStr.chStr(m.getPostcode()) + "',CardNO='" + chStr.chStr(m.getCardno()) + "',CardType='"
            + chStr.chStr(m.getCardtype()) + "',Tel='" + chStr.chStr(m.getTel()) + "',Email='"
            + chStr.chStr(m.getEmail()) + "' where ID=" + m.getID();
        ret = conn.executeUpdate(sql);
        System.out.println(sql);
    } catch (Exception e) {
        e.printStackTrace();
        ret = 0;
    }
```

```
        conn.close();                                          //关闭数据库连接
        return ret;
}
```

2.6.3 会员注册页面的编写

本项目采用一个 JSP 文件作为会员注册页面，这里指定的 JSP 文件是位于 WebContent/front 下的 register_deal.jsp。在 register_deal.jsp 文件中添加代码，用于创建 ConnDB、MemberDaoImpl 和 Member 类的对象，并且通过 "<jsp:setProperty name="member" property="*"/>" 对 Member 类的所有属性进行赋值，用于获取用户填写的注册信息。代码如下：

```jsp
<%-- 创建 ConnDB 类的对象 --%>
<jsp:useBean id="conn" scope="page" class="com.tools.ConnDB" />
<%-- 创建 MemberDaoImpl 类的对象 --%>
<jsp:useBean id="ins_member" scope="page" class="com.dao.MemberDaoImpl" />
<%-- 创建 Member 类的对象，并对 Member 类的所有属性进行赋值 --%>
<jsp:useBean id="member" scope="request" class="com.model.Member">
    <jsp:setProperty name="member" property="*" />
</jsp:useBean>
```

判断输入的账号是否存在，如果存在则给予提示，否则调用 MemberDaoImpl 类的 insert()方法，将填写的会员信息保存到数据库中。代码如下：

```jsp
<%
    request.setCharacterEncoding("UTF-8");                     //设置请求的编码为 UTF-8
    String username = member.getUsername();                    //获取会员账号
    ResultSet rs = conn.executeQuery("select * from tb_Member where username='"
        + username + "'");
    if (rs.next()) {                                           //如果结果集中有数据
        out.println("<script language='javascript'>alert('该账号已经存在，请重新注册！');"
                + "window.location.href='register.jsp';</script>");
    } else {
        int ret = 0;                                           //记录更新记录条数的变量
        ret = ins_member.insert(member);                       //将填写的会员信息保存到数据库
        if (ret != 0) {
            session.setAttribute("username", username);        //将会员账号保存到 Session 中
            out.println("<script language='javascript'>alert('会员注册成功！');"
                + "window.location.href='index.jsp';</script>");
        } else {
            out.println("<script language='javascript'>alert('会员注册失败！');"
                + "window.location.href='register.jsp';</script>");
        }
    }
%>
```

2.7 会员登录模块设计

会员登录模块主要用于验证用户是否为购好物网络商城的会员，并允许会员登录商城，进行后续相关操作。在会员登录页面中，填写会员账户、密码和验证码（如果验证码看不清楚，可以单击验证码图片刷新验证码），单击"登录"按钮，即可完成会员登录的操作，如图 2.3 所示。如果没有输入账户、密码或者验证码，程序都将给予提示。此外，账户、密码或者验证码输入错误，程序也将给予提示。

购好物网络商城 第 2 章

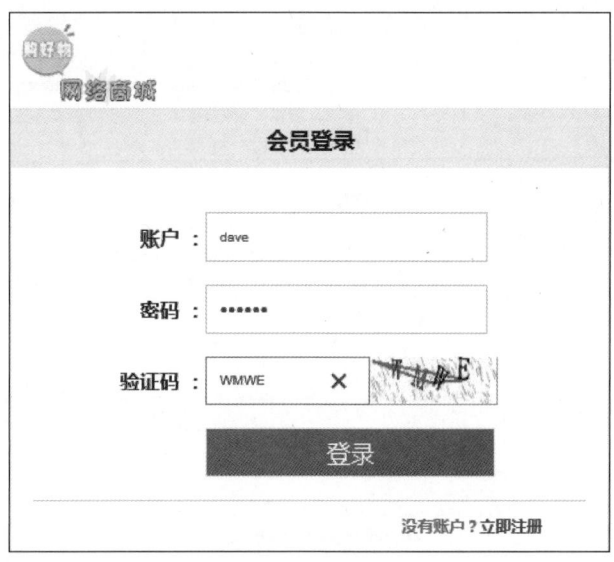

图 2.3　会员登录页面

2.7.1　会员登录页面的编写

本项目采用一个 JSP 文件作为会员登录页面,这里指定的 JSP 文件是位于 WebContent/front 下的 login.jsp。在 login.jsp 中,没有业务逻辑,主要使用<div>标签规划页面区域。login.jsp 的代码如下:

```
<%@ page contentType="text/html; charset=UTF-8"%>
<!DOCTYPE html>
<html xmlns="http://www.w3.org/1999/xhtml">
<head>
<meta http-equiv="content-type" content="text/html; charset=UTF-8">
<title>登录-购好物网络商城</title>
<link rel="stylesheet" href="css/mr-01.css" type="text/css">
</head>

<body>
    <div id="mr-mainbody" class="container mr-mainbody">
        <div class="row">
            <!-- 主体内容 -->
            <div id="mr-content" class="mr-content col-xs-12">
                <div class="login-wrap" style="margin-bottom: 60px; margin-top: 50px">
                    <div style="max-width: 540px; margin: 0 auto;">
                        <a href="index.jsp" title="点击返回首页"><img src="images/51logo.png"></a>
                    </div>
                    <div class="login">
                        <div class="page-header" style="pause: 0px;"> <h1 class="login_h1">
                        会员登录</h1> </div>
                        <!-- 会员登录表单 -->
                        <form action="login_check.jsp" method="post" class="form-horizontal">
                            <fieldset>
                                <div class="form-group">
                                    <div class="col-sm-4 control-label">
                                        <label id="username-lbl" for="username" class="required">
                                        账户：</label>
                                    </div>
                                    <div class="col-sm-8">
```

43

```html
                                        <!-- 账户文本框 -->
                                        <input type="text" name="username" id="username" value=""
                                            size="38" class="validate-username required"
                                            required="required" autofocus="">
                                    </div>
                                </div>
                                <div class="form-group">
                                    <div class="col-sm-4 control-label">
                                        <label id="password-lbl" for="password" class="required">
                                            密码：</label>
                                    </div>
                                    <div class="col-sm-8">
                                        <!-- 密码文本框 -->
                                        <input type="password" name=PWD id="password" value=""
                                            class="validate-password required" size="38" maxlength="99"
                                            required="required" aria-required="true">
                                    </div>
                                </div>
                                <div class="form-group">
                                    <div class="col-sm-4 control-label">
                                        <label id="password-lbl" for="password" class="required">
                                            验证码：</label>
                                    </div>
                                    <div class="col-sm-8" style="clear: none;">
                                        <!-- 验证码文本框 -->
                                        <input type="text" name="checkCode" id="checkCode" value=""
                                            class="validate-password required" style="float: left;"
                                            title="验证码区分大小写" size="18" maxlength="4"
                                            required="required" aria-required="true">
                                            <!-- 显示验证码 -->
                                            <img src="../CheckCode" name="checkCode"
                                                onClick="myReload()" width="116" height="43"
                                                class="img_checkcode" id="img_checkCode" />
                                    </div>
                                </div>
                                <div class="form-group">
                                    <div class="col-sm-offset-4 col-sm-8">
                                        <button type="submit" class="btn btn-primary login">登录</button>
                                    </div>
                                </div>
                                <div class="form-group"
                                    style="border-top: 1px solid #D9D9D9; margin: 20px;">
                                    <label
                                        style="float: right; color: #858585; margin-right: 40px;
                                        margin-top: 10px; font-size: 14px;">没有账户？<a
                                        href="register.jsp">立即注册</a></label>
                                </div>
                            </fieldset>
                        </form>
                    </div>
                </div>
            </div>
            <!-- //主体内容 -->
        </div>
    </div>
    <script language="javascript">
        //刷新验证码
        function myReload() {
```

```
                    document.getElementById("img_checkCode").src = document.getElementById("img_checkCode").src
                    + "?nocache=" + new Date().getTime();
            }
    </script>
</body>
</html>
```

在向会员登录页面添加验证码时,引用了 com.tools 下的用于生成验证码的 CheckCode.java 文件。在该文件中,包含一个 getRandColor()方法,该方法通过随机生成 RGB 颜色中的 r 值、g 值和 b 值获取随机颜色。getRandColor()方法代码如下:

```java
public Color getRandColor(int s, int e) {
    Random random = new Random();
    if (s > 255)
        s = 255;
    if (e > 255)
        e = 255;
    int r = s + random.nextInt(e - s);                  //随机生成 RGB 颜色中的 r 值
    int g = s + random.nextInt(e - s);                  //随机生成 RGB 颜色中的 g 值
    int b = s + random.nextInt(e - s);                  //随机生成 RGB 颜色中的 b 值
    return new Color(r, g, b);
}
```

在 CheckCode.java 文件中,还包含一个用于生成验证码的 Servlet 的 service()方法,该方法用于处理来自客户端的请求(生成验证码),并将处理结果(验证码的图片)返回客户端。service()方法的代码如下:

```java
public void service(HttpServletRequest request, HttpServletResponse response) throws ServletException, IOException {
    response.setHeader("Pragma", "No-cache");
    response.setHeader("Cache-Control", "No-cache");
    response.setDateHeader("Expires", 0);
    //指定生成的响应是图片
    response.setContentType("image/jpeg");
    int width = 116;                                    //指定验证码的宽度
    int height = 33;                                    //指定验证码的高度
    
    BufferedImage image = new BufferedImage(width, height, BufferedImage.TYPE_INT_RGB);
    Graphics g = image.getGraphics();                   //获取 Graphics 类的对象
    Random random = new Random();                       //实例化一个 Random 对象
    Font mFont = new Font("宋体", Font.BOLD, 22);       //通过 Font 构造字体
    g.fillRect(0, 0, width, height);                    //绘制验证码背景
    g.setFont(mFont);                                   //设置字体
    g.setColor(getRandColor(180, 200));                 //设置颜色
    //画随机的线条
    for (int i = 0; i < 100; i++) {
        int x = random.nextInt(width - 1);
        int y = random.nextInt(height - 1);
        int x1 = random.nextInt(3) + 1;
        int y1 = random.nextInt(6) + 1;
        g.drawLine(x, y, x + x1, y + y1);               //绘制直线
    }
    /*************************** 画一条折线 ***********************************/
    //创建一个供画笔选择线条粗细的对象
    BasicStroke bs = new BasicStroke(2f, BasicStroke.CAP_BUTT, BasicStroke.JOIN_BEVEL);
    Graphics2D g2d = (Graphics2D) g;                    //通过 Graphics 类的对象创建一个 Graphics2D 类的对象
    g2d.setStroke(bs);                                  //改变线条的粗细
    g.setColor(Color.GRAY);                             //设置当前颜色为预定义颜色中的灰色
    int lineNumber = 4;                                 //指定端点的个数
    int[] xPoints = new int[lineNumber];                //定义保存 x 轴坐标的数组
```

```
        int[] yPoints = new int[lineNumber];              //定义保存 x 轴坐标的数组
        //通过循环为 x 轴坐标和 y 轴坐标的数组赋值
        for (int j = 0; j < lineNumber; j++) {
            xPoints[j] = random.nextInt(width - 1);
            yPoints[j] = random.nextInt(height - 1);
        }
        g.drawPolyline(xPoints, yPoints, lineNumber);     //绘制折线
        /****************************************************************/
        String sRand = "";
        //输出随机的验证文字
        for (int i = 0; i < 4; i++) {
            char ctmp = (char) (random.nextInt(26) + 65); //生成 A~Z 的字母
            sRand += ctmp;
            Color color = new Color(20 + random.nextInt(110), 20 + random.nextInt(110), 20 + random.nextInt(110));
            g.setColor(color);                            //设置颜色
            /** **随机缩放文字并将文字旋转指定角度* */
            //将文字旋转指定角度
            Graphics2D g2d_word = (Graphics2D) g;
            AffineTransform trans = new AffineTransform();
            trans.rotate(random.nextInt(45) * 3.14 / 180, 22 * i + 8, 7);
            //缩放文字
            float scaleSize = random.nextFloat() + 0.8f;
            if (scaleSize > 1f)
                scaleSize = 1f;
            trans.scale(scaleSize, scaleSize);            //进行缩放
            g2d_word.setTransform(trans);
            /** ********************* */
            g.drawString(String.valueOf(ctmp), width / 6 * i + 23, height / 2);

        }
        //将生成的验证码保存到 Session 中
        HttpSession session = request.getSession(true);
        session.setAttribute("randCheckCode", sRand);
        g.dispose();
        ImageIO.write(image, "JPEG", response.getOutputStream());
}
```

2.7.2 生成验证码的编写

在配置一个用于生成验证码的 Servlet 时，主要是通过<servlet>标记先配置 Servlet 文件，再通过<servlet-mapping>标记配置一个映射路径，用于使用该 Servlet。因此，在本项目的 WebContent/WEB-INF/web.xml 文件中添加如下代码：

```
<servlet>
    <servlet-name>CheckCode</servlet-name>
    <servlet-class>com.tools.CheckCode</servlet-class>
</servlet>
<servlet-mapping>
    <servlet-name>CheckCode</servlet-name>
    <url-pattern>/CheckCode</url-pattern>
</servlet-mapping>
```

打开会员登录页面 login.jsp，使用<img>标记显示验证码，并且调用 myReload()方法，实现单击该验证码图片时重新获取一个验证码图片功能。代码如下：

```
<img src="../CheckCode" name="img_checkCode" onClick="myReload()" width="116"
    height="43" class="img_checkcode" id="img_checkCode" />
```

2.7.3 编写会员登录处理页

在实现会员登录的功能时，除了会员登录页面，还需要给表单设置一个处理页 login_check.jsp（位于 WebContent/front 下），该处理页用来将会员输入的账户和密码与数据库中的数据进行匹配，并给出提示。

在 login_check.jsp 文件中，导入 java.sql 包中的 ResultSet 类，并且创建 ConnDB 类的对象。代码如下：

```jsp
<%-- 导入 java.sql.ResultSet 类 --%>
<%@ page import="java.sql.ResultSet"%>
<%-- 创建 ConnDB 类的对象 --%>
<jsp:useBean id="conn" scope="page" class="com.tools.ConnDB" />
```

获取当前用户输入的账号和密码，并将其与数据库中保存的账户和密码进行匹配，根据匹配结果给予相应的提示，并跳转到指定页面。代码如下：

```jsp
<%
String username = request.getParameter("username");           //获取账户
String checkCode = request.getParameter("checkCode");         //获取验证码
if (checkCode.equals(session.getAttribute("randCheckCode").toString())) {
        try {                                                 //捕获异常
                ResultSet rs = conn.executeQuery("select * from tb_Member where username='" + username + "'");
                if (rs.next()) {                              //如果找到相应的账号
                        String PWD = request.getParameter("PWD");  //获取密码
                        if (PWD.equals(rs.getString("password"))) {  //如果输入的密码和获取的密码一致
                                //把当前的账户保存到 session 中，实现登录
                                session.setAttribute("username", username);
                                response.sendRedirect("index.jsp");  //跳转到前台首页
                        } else {
                                out.println(
"<script language='javascript'>alert('您输入的用户名或密码错误，请与管理员联系!');"
                                        +"window.location.href='login.jsp';</script>");
                        }
                } else {
                        out.println(
                                "<script language='javascript'>alert('您输入的用户名或密码错误，或您的账户"+
                                "已经被冻结，请与管理员联系!');window.location.href='login.jsp';</script>");
                }
        } catch (Exception e) {                               //处理异常
                out.println(
                        "<script language='javascript'>alert('您的操作有误!');"
                        +"window.location.href='login.jsp';</script>");
        }
        conn.close();                                         //关闭数据库连接
} else {
        out.println("<script language='javascript'>alert('您输入的验证码错误!');history.back();</script>");
}
%>
```

2.8 首页模块设计

当用户访问购好物网络商城时，首先进入的就是首页。在首页中，用户既可以浏览热门商品、最新上架的商品和折扣商品，又可以浏览图书类、家电类、服装类和电子类的商品。购好物网络商城首页的效果如图 2.4 所示。

Java Web 项目开发全程实录（第2版）

图 2.4 购好物网络商城的首页

2.8.1 实现显示最新上架商品的功能

本项目采用一个 JSP 文件用于显示最新上架的商品，这里指定的 JSP 文件是位于 WebContent/front 下的 index.jsp。在 index.jsp 文件中，因为在实现查询最新上架商品时，需要访问数据库，所以需要导入 java.sql.ResultSet 类并创建 com.tools.ConnDB 类的对象。代码如下：

```jsp
<%@ page import="java.sql.ResultSet"%>              <%-- 导入 java.sql.ResultSet 类 --%>
<jsp:useBean id="conn" scope="page" class="com.tools.ConnDB" />  <%-- 创建 com.tools.ConnDB 类的对象 --%>
```

在 index.jsp 文件中，调用 ConnDB 类的 executeQuery()方法执行 SQL 语句，用于从数据表中查询最新上架商品，这里需要编写一个连接查询的 SQL 语句。另外，还需要定义保存商品信息的变量。代码如下：

```jsp
<%
    /* 最新上架商品信息 */
    ResultSet rs_new = conn.executeQuery(
            "select t1.ID, t1.GoodsName,t1.price,t1.picture,t2.TypeName "
            +"from tb_goods t1,tb_subType t2 where t1.typeID=t2.ID and "
            +"t1.newGoods=1 order by t1.INTime desc limit 12");     //查询最新上架商品信息
    int new_ID = 0;                                                 //保存最新上架商品 ID 的变量
    String new_goodsname = "";                                      //保存最新上架商品名称的变量
    float new_nowprice = 0;                                         //保存最新上架商品价格的变量
    String new_picture = "";                                        //保存最新上架商品图片的变量
    String typeName = "";                                           //保存商品分类的变量
%>
```

在 index.jsp 文件中，通过循环显示从数据库获取的 12 条最新上架商品信息，将获取的商品信息显示到页面的"最新上架商品展示区"。代码如下：

```jsp
<%
while (rs_new.next()) {                                 //设置一个循环
    new_ID = rs_new.getInt(1);                          //获取最新上架商品的 ID
    new_goodsname = rs_new.getString(2);                //获取最新上架商品的商品名称
    new_nowprice = rs_new.getFloat(3);                  //获取最新上架商品的价格
    new_picture = rs_new.getString(4);                  //获取最新上架商品的图片
    typeName = rs_new.getString(5);                     //获取最新上架商品的类别
%>
```

2.8.2 实现显示打折商品的功能

用于显示打折商品的 JSP 文件也是位于 WebContent/front 下的 index.jsp。在 index.jsp 文件中，调用 ConnDB 类的 executeQuery()方法执行 SQL 语句，用于从数据表中查询打折商品，这里也需要编写一个连接查询的 SQL 语句。另外，还需要定义保存商品信息的变量。代码如下：

```jsp
/* 打折商品信息 */
ResultSet rs_sale = conn.executeQuery(
        "select t1.ID, t1.GoodsName,t1.price,t1.nowPrice,t1.picture,t2.TypeName "
        +"from tb_goods t1,tb_subType t2 where t1.typeID=t2.ID and t1.sale=1 "
        +"order by t1.INTime desc limit 12");           //查询打折商品信息
int sale_ID = 0;                                        //保存打折商品 ID 的变量
String s_goodsname = "";                                //保存打折商品名称的变量
float s_price = 0;                                      //保存打折商品的原价格的变量
float s_nowprice = 0;                                   //保存打折商品的打折后价格的变量
String s_introduce = "";                                //保存打折商品简介的变量
String s_picture = "";                                  //保存打折商品图片的变量
```

在 index.jsp 文件中，通过循环显示从数据库获取的 12 条打折商品信息，将获取的商品信息显示到页面

的"打折商品展示区"。代码如下:

```
<%
    while (rs_sale.next()) {                              //设置一个循环
        sale_ID = rs_sale.getInt(1);                      //获取打折商品的ID
        s_goodsname = rs_sale.getString(2);               //获取打折商品的商品名称
        s_price = rs_sale.getFloat(3);                    //获取打折商品的原价
        s_nowprice = rs_sale.getFloat(4);                 //获取打折商品的现价
        s_picture = rs_sale.getString(5);                 //获取打折商品的图片
        typeName = rs_sale.getString(6);                  //获取打折商品的类别
%>
```

2.8.3 实现显示热门商品的功能

热门商品是指商城中点击率最高的商品,这里将获取并显示两件商品。用于显示热门商品的JSP文件依然是位于WebContent/front下的index.jsp。在index.jsp文件中,调用ConnDB类的executeQuery()方法执行SQL语句,用于从数据表中查询点击率最高的两件商品,这里需要编写一个倒序排列的SQL语句。另外,还需要定义保存商品信息的变量。代码如下:

```
/* 热门商品信息 */
ResultSet rs_hot = conn
        .executeQuery("select ID,GoodsName,nowprice,picture "
        +"from tb_goods order by hit desc limit 2");      //查询热门商品信息
int hot_ID = 0;                                           //保存热门商品ID的变量
String hot_goodsName = "";                                //保存热门商品名称的变量
float hot_nowprice = 0;                                   //保存热门商品价格的变量
String hot_picture = "";                                  //保存热门商品图片的变量
```

在index.jsp文件中,通过循环显示从数据库获取的两条热门商品信息,将获取的商品信息显示到页面的"热门商品展示区"。代码如下:

```
<%
    while (rs_hot.next()) {                               //设置一个循环
        hot_ID = rs_hot.getInt(1);                        //获取商品ID
        hot_goodsName = rs_hot.getString(2);              //获取商品名称
        hot_nowprice = rs_hot.getFloat(3);                //获取商品价格
        hot_picture = rs_hot.getString(4);                //获取商品图片
%>
```

2.9 购物车模块设计

会员登录购好物网络商城后,使页面跳转至"我的购物车"页面的方式有两种:一种是单击首页右上角的"我的购物车"按钮;另一种是在某件商品的详细信息页面上单击"添加到购物车"按钮。在"我的购物车"页面上,会员能够看到已经被添加到购物车的商品信息。

会员如果想要为购物车里的商品提交订单,那么需要首先填写物流信息,然后单击"我的购物车"页面上的"结账"按钮,接着用支付宝扫描对话框中的二维码进行支付,再单击对话框上的"支付"按钮完成支付操作,最后程序会提交订单并在对话框中显示订单号。

会员如果想要继续购物,就单击"我的购物车"页面上的"继续购物"按钮,页面会跳转至购好物网络商城的首页。

会员如果想要清空购物车,就单击"我的购物车"页面上的"清空购物车"按钮,这时在"我的购物车"页面上显示的商品详细信息会被删除。在清空购物车后,会员可以单击"购物车为空"页面上的"继续购物"

按钮,将页面跳转至购好物网络商城的首页。

2.9.1 购物车商品模型类的编写

在 com.model 包中,创建一个购物车商品模型类 Goodselement。在该类中,添加 3 个公有类型的属性,分别表示商品 ID、当前价格和数量。代码如下:

```
package com.model;

public class Goodselement {
    public int ID;                                      //商品 ID
    public float nowprice;                              //当前价格
    public int number;                                  //数量
}
```

2.9.2 实现查看商品详细信息的功能

会员登录购好物网络商城后,单击首页上的任何商品名称或者商品图片,都会显示该商品的详细信息页面,如图 2.5 所示。在显示某件商品详细信息的页面上,会员可以浏览该商品的名称、当前价格、商品描述等信息。下面将介绍查看某件商品详细信息的页面的实现过程。

图 2.5 商品的详细信息页面

WebContent/front 下的 goodsDetail.jsp 文件用于显示某件商品的详细信息。在 goodsDetail.jsp 文件中，导入 java.sql 包中的 ResultSet 类，并且创建 ConnDB 类的对象。代码如下：

```jsp
<%@ page import="java.sql.ResultSet"%>                              <%-- 导入 java.sql.ResultSet 类 --%>
<jsp:useBean id="conn" scope="page" class="com.tools.ConnDB" />     <%-- 创建 com.tools.ConnDB 类的对象 --%>
```

在 goodsDetail.jsp 文件中，根据获取的商品 ID 查询商品信息。具体方法：首先获取商品 ID，然后根据该商品 ID 从数据表中获取所需的商品信息，如果找到对应的商品，则将商品信息保存到相应的变量中，最后关闭数据库连接。代码如下：

```jsp
<%
    int typeSystem = 0;                                              //保存商品类型 ID 的变量
    int ID = Integer.parseInt(request.getParameter("ID"));           //获取商品 ID
    if (ID > 0) {
        ResultSet rs = conn.executeQuery("select ID,GoodsName,Introduce,nowprice,picture, "
            + " price,typeID from tb_goods where ID=" + ID);         //根据 ID 查询商品信息
        String goodsName = "";                                       //保存商品名称的变量
        float nowprice = (float) 0.0;                                //保存商品现价的变量
        float price = (float) 0.0;                                   //保存商品原价的变量
        String picture = "";                                         //保存商品图片的变量
        String introduce = "";                                       //保存商品描述的变量
        if (rs.next()) {                                             //如果找到对应的商品信息
            goodsName = rs.getString(2);                             //获取商品名称
            introduce = rs.getString(3);                             //获取商品描述
            nowprice = rs.getFloat(4);                               //获取商品现价
            picture = rs.getString(5);                               //获取商品图片
            price = rs.getFloat(6);                                  //获取商品原价
            typeSystem = rs.getInt(7);                               //获取商品类别 ID
        }
    conn.close();                                                    //关闭数据库连接
%>
```

2.9.3 实现添加购物车的功能

如图 2.5 所示，在某件商品的详细信息页面上单击"添加到购物车"按钮后，该商品将被添加至购物车，页面会跳转至"我的购物车"页面，如图 2.6 所示。下面将介绍添加购物车的实现过程。

在 goodsDetail.jsp 文件中，通过"添加到购物车"按钮的 onclick 属性，调用自定义的 JavaScript 函数 addCart()，用于验证商品数量是否合法。如果不合法，则给出提示，并且返回某件商品的详细信息页面；如果合法，就将该商品添加到购物车。代码如下：

```html
<script src="js/jquery.1.3.2.js" type="text/javascript"></script>
<script type="text/javascript">
    function addCart() {
        var num = $('#shuliang').val();                              //获取输入的商品数量
        //验证输入的数量是否合法
        if (num < 1) {                                               //如果输入的数量不合法
            alert('数量不能小于1! ');
            return;
        }
        //调用添加购物车页面，实现将该商品添加到购物车
        window.location.href="cart_add.jsp?goodsID=<%=ID%>&num="+num;
    }
</script>
```

> **说明**
> 在上面的代码段中，cart_add.jsp 是用于将商品添加到购物车的处理页面，后面的问号"?"用于标识它之后是要传递的参数，多个参数间用"&"分隔。

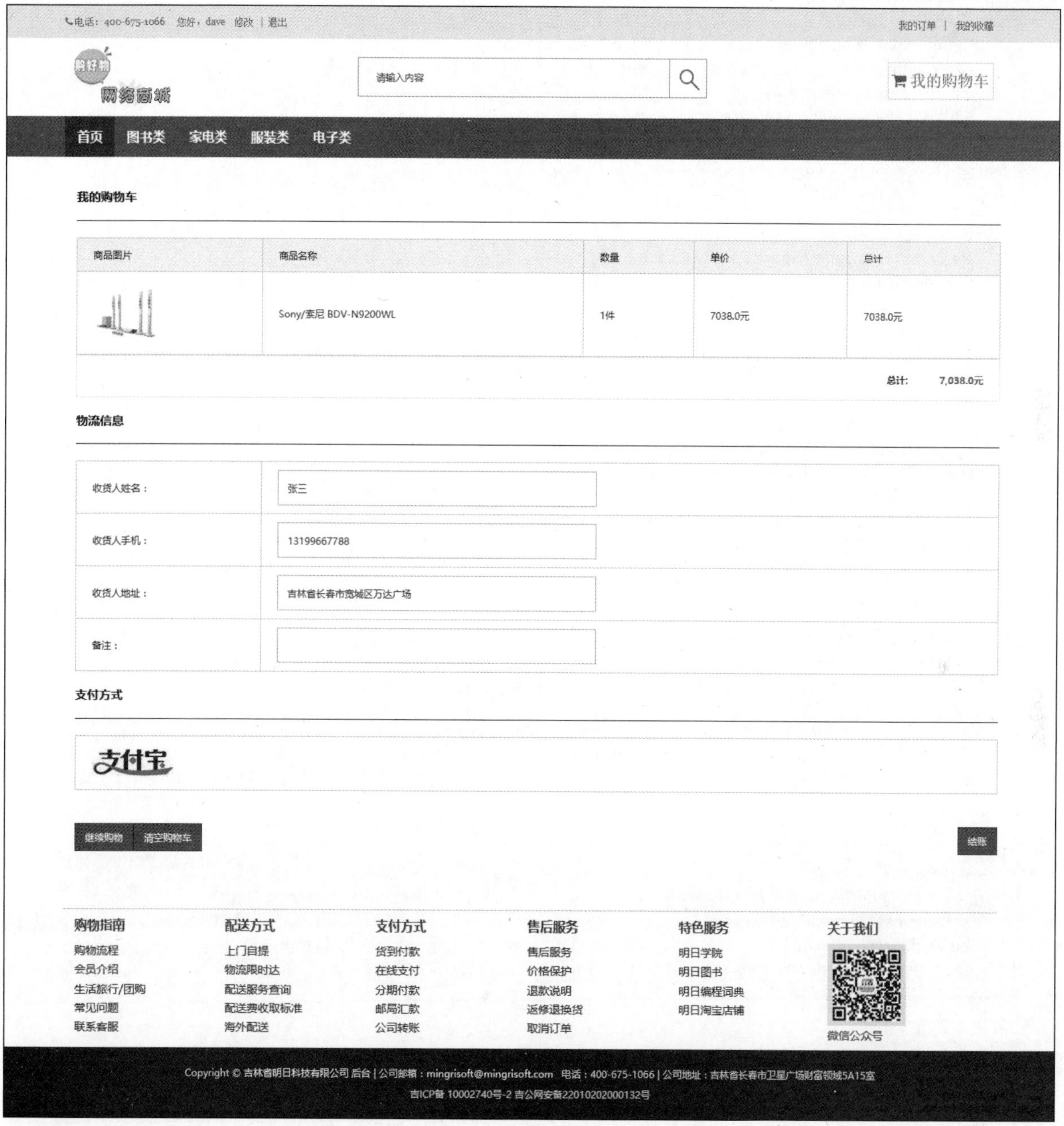

图 2.6 "我的购物车"页面

WebContent/front 下的 cart_add.jsp 文件用于显示"我的购物车"页面。在 goodsDetail.jsp 文件中，导入 java.sql 包中的 ResultSet 类、向量类，以及购物车商品模型类，并且创建 ConnDB 类的对象。

```
<%@ page import="java.sql.ResultSet"%>                          <%-- 导入 java.sql.ResultSet 类 --%>
<%@ page import="java.util.Vector"%>                            <%-- 导入 Java 的向量类 --%>
<%@ page import="com.model.Goodselement"%>                      <%-- 导入购物车商品模型类 --%>
<jsp:useBean id="conn" scope="page" class="com.tools.ConnDB"/>  <%-- 创建 ConnDB 类的对象 --%>
```

在 cart_add.jsp 文件中，为了实现添加到购物车功能，首先获取会员账号和商品量，并判断会员是否登录。如果没有登录，则重定向到会员登录页面进行登录。然后将商品基本信息保存到购物车商品模型类的对象 mygoodselement 中。接着把该商品添加到购物车中。最后使页面跳转到"我的购物车"页面。代码如下：

53

```jsp
<%
    String username=(String)session.getAttribute("username");          //获取会员账号
    String num = (String) request.getParameter("num");                 //获取商品数量
    //如果没有登录，将跳转到登录页面
    if (username == null || username == "") {
        response.sendRedirect("login.jsp");                            //重定向页面到会员登录页面
        return;                                                        //返回
    }
    int ID = Integer.parseInt(request.getParameter("goodsID"));        //获取商品ID
    String sql = "select * from tb_goods where ID=" + ID;              //定义根据商品ID查询商品信息的SQL语句
    ResultSet rs = conn.executeQuery(sql);                             //根据商品ID查询商品
    float nowprice = 0;                                                //定义保存商品价格的变量
    if (rs.next()) {                                                   //如果查询到指定商品
        nowprice = rs.getFloat("nowprice");                            //获取该商品的价格
    }
    //创建保存购物车内商品信息的模型类的对象mygoodselement
    Goodselement mygoodselement = new Goodselement();
    mygoodselement.ID = ID;                                            //将商品ID保存到mygoodselement对象中
    mygoodselement.nowprice = nowprice;                                //将商品价格保存到mygoodselement对象中
    mygoodselement.number = Integer.parseInt(num);                     //将购买数量保存到mygoodselement对象中
    boolean Flag = true;                                               //记录购物车内是否已经存在所要添加的商品
    Vector cart = (Vector) session.getAttribute("cart");               //获取购物车对象
    if (cart == null) {                                                //如果购物车对象为空
        cart = new Vector();                                           //创建一个购物车对象
    } else {
        //判断购物车内是否已经存在所购买的商品
        for (int i = 0; i < cart.size(); i++) {
            Goodselement goodsitem = (Goodselement) cart.elementAt(i); //获取购物车内的一个商品
            if (goodsitem.ID == mygoodselement.ID) {                   //如果当前要添加的商品已经在购物车中
                //直接改变购物数量
                goodsitem.number = goodsitem.number + mygoodselement.number;
                cart.setElementAt(goodsitem, i);                       //重新保存到购物车中
                Flag = false;                                          //设置标记变量Flag为false，代表购物车中存在该商品
            }
        }
    }
    if (Flag)                                                          //如果购物车内不存在该商品
        cart.addElement(mygoodselement);                               //将要购买的商品保存到购物车中
    session.setAttribute("cart", cart);                                //将购物车对象添加到session中
    conn.close();                                                      //关闭数据库的连接
    response.sendRedirect("cart_see.jsp");                             //重定向页面到查看购物车页面
%>
```

2.9.4　实现查看购物车的功能

如图 2.6 所示，已经被添加到购物车的商品显示在"我的购物车"页面的上部，会员可以查看购物车中商品的图片、名称、数量、单价、总计（价格）等信息。下面将介绍查看购物车功能的实现过程。

WebContent/front 下的 cart_see.jsp 文件用于查看购物车。在 cart_see.jsp 文件中，导入查看购物车所能用到的类（主要包括保存结果集的 ResultSet 类、向量类、格式化数字的类，以及购物车商品模型类），并且创建 ConnDB 类的对象。代码如下：

```jsp
<%@ page import="java.sql.ResultSet"%>                    <%-- 导入java.sql.ResultSet类 --%>
<%@ page import="java.util.Vector"%>                      <%-- 导入Java的向量类 --%>
<%@ page import="java.text.DecimalFormat"%>               <%-- 导入格式化数字的类 --%>
<%@ page import="com.model.Goodselement"%>                <%-- 导入购物车商品的模型类 --%>
<jsp:useBean id="conn" scope="page" class="com.tools.ConnDB" />   <%-- 创建com.tools.ConnDB类的对象 --%>
```

在 cart_see.jsp 文件中，需要判断当前用户是否登录。如果没有登录，则进入登录页面进行登录；否则，

获取购物车对象。代码如下：

```jsp
<%
    String username = (String) session.getAttribute("username");        //获取会员账号
    //如果没有登录，将跳转到登录页面
    if (username == "" || username == null) {
        response.sendRedirect("login.jsp");                              //重定向页面到"会员登录"页面
        return;                                                          //返回
    } else {
        Vector cart = (Vector) session.getAttribute("cart");             //获取购物车对象
        if (cart == null || cart.size() == 0) {                          //如果购物车为空
            response.sendRedirect("cart_null.jsp");                      //重定向页面到"购物车为空"页面
        } else {
%>
```

在 cart_see.jsp 文件中，使用循环遍历购物车对象，获取需要显示的商品信息。代码如下：

```jsp
<%
    float sum = 0;
    DecimalFormat fnum = new DecimalFormat("#,##0.0");                   //定义显示金额的格式
    int ID = -1;                                                         //保存商品 ID 的变量
    String goodsname = "";                                               //保存商品名称的变量
    String picture = "";                                                 //保存商品图片的变量
    //遍历购物车中的商品
    for (int i = 0; i < cart.size(); i++) {
        Goodselement goodsitem = (Goodselement) cart.elementAt(i);       //获取一个商品
        sum = sum + goodsitem.number * goodsitem.nowprice;               //计算总计金额
        ID = goodsitem.ID;                                               //获取商品 ID
        if (ID > 0) {
            ResultSet rs_goods = conn.executeQuery("select * from tb_goods where ID=" + ID);
            if (rs_goods.next()) {
                goodsname = rs_goods.getString("goodsname");             //获取商品名称
                picture = rs_goods.getString("picture");                 //获取商品图片
            }
            conn.close();                                                //关闭数据库的连接
        }
    }
%>
```

2.9.5 实现商品订单提交功能

如图 2.6 所示，单击"我的购物车"页面右下角的"结账"按钮，将弹出"支付"对话框，如图 2.7 所示。

图 2.7 "支付"对话框

首先使用支付宝扫描对话框中的二维码进行支付，然后单击对话框右下角的"支付"按钮完成支付操作，程序会提交订单并显示订单号，如图 2.8 所示。下面将介绍商品订单提交的实现过程。

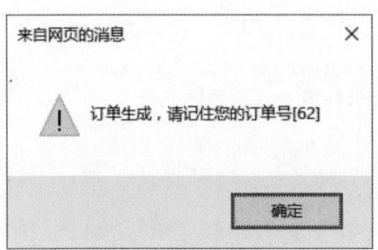

图 2.8　订单生成

用于提交订单的 JSP 文件是 WebContent/front 下的 cart_order.jsp。在 cart_order.jsp 文件中，导入 java.sql 包中的 ResultSet 类、向量类，以及购物车商品模型类，并且创建 ChStr 类和 ConnDB 类的对象。代码如下：

```jsp
<%@ page import="java.sql.ResultSet"%>                              <%-- 导入 java.sql.ResultSet 类 --%>
<%@ page import="java.util.Vector"%>                                <%-- 导入 Java 的向量类 --%>
<%@ page import="com.model.Goodselement"%>                          <%-- 导入购物车商品模型类 --%>
<jsp:useBean id="chStr" scope="page" class="com.tools.ChStr" />     <%-- 创建 ChStr 类的对象 --%>
<jsp:useBean id="conn" scope="page" class="com.tools.ConnDB" />     <%-- 创建 ConnDB 类的对象 --%>
```

在 cart_order.jsp 文件中，为了实现提交订单的功能，首先判断购物车是否为空，不为空时，判断会员账户是否合法，只有会员账户合法，才能够提交订单。在提交订单信息时，需要分别向数据库的订单主表和订单明细表插入数据。代码如下：

```jsp
<%
    if (session.getAttribute("cart") == "") {                       //判断购物车对象是否为空
        out.println(
            "<script language='javascript'>alert('您还没有购物!');"
            +"window.location.href='index.jsp';</script>");
    }
    String Username = (String) session.getAttribute("username");    //获取输入的账户名称
    if (Username != "") {
        try {                                                       //捕捉异常
            ResultSet rs_user = conn.executeQuery("select * from tb_Member where username='"
                + Username + "'");
            if (!rs_user.next()) {                                  //如果获取的账户名称在会员信息表中不存在（表示非法会员）
                session.invalidate();                               //销毁 session
                out.println(
                    "<script language='javascript'>alert('请先登录后，再进行购物!'); "
                    +"window.location.href='index.jsp';</script>");
                return;                                             //返回
            } else {                                                //如果会员合法，则提交订单
                //获取输入的收货人姓名
                String recevieName = chStr.chStr(request.getParameter("recevieName"));
                //获取输入的收货人地址
                String address = chStr.chStr(request.getParameter("address"));
                String tel = request.getParameter("tel");           //获取输入的电话号码
                String bz = chStr.chStr(request.getParameter("bz"));//获取输入的备注
                int orderID = 0;                                    //定义提交订单 ID 的变量
                Vector cart = (Vector) session.getAttribute("cart");//获取购物车对象
                int number = 0;                                     //定义提交商品数量的变量
                float nowprice = (float) 0.0;                       //定义提交商品价格的变量
                float sum = (float) 0;                              //定义商品金额的变量
                float Totalsum = (float) 0;                         //定义商品件数的变量
                boolean flag = true;                                //标记订单是否有效，为 true 表示有效
                int temp = 0;                                       //保存返回自动生成的订单号的变量
                int ID = -1;
```

```java
                        //插入订单主表数据
                        float bnumber = cart.size();
                        String sql = "insert into tb_Order(bnumber,username,recevieName,address, "
                            +"tel,bz) values("+ bnumber + ",'" + Username + "','" + recevieName
                            + "','" + address + "','" + tel+ "','" + bz + "')";
                        temp = conn.executeUpdate_id(sql);              //保存订单主表数据
                        if (temp == 0) {                                //如果返回的订单号为0，表示不合法
                            flag = false;
                        } else {
                            orderID = temp;                             //把生成的订单号赋值给订单ID变量
                        }
                        String str = "";                                //保存插入订单详细信息的SQL语句
                        //插入订单明细表数据
                        for (int i = 0; i < cart.size(); i++) {
                            //获取购物车中的一个商品
                            Goodselement mygoodselement = (Goodselement) cart.elementAt(i);
                            ID = mygoodselement.ID;                     //获取商品ID
                            nowprice = mygoodselement.nowprice;         //获取商品价格
                            number = mygoodselement.number;             //获取商品数量
                            sum = nowprice * number;                    //计算商品金额
                            str = "insert into tb_order_Detail (orderID,goodsID,price,number)"
                                +" values(" + orderID + ","+ ID + "," + nowprice + ","
                                + number + ")";                         //插入订单明细的SQL语句
                            temp = conn.executeUpdate(str);             //保存订单明细
                            Totalsum = Totalsum + sum;                  //累加合计金额
                            if (temp == 0) {                            //如果返回值为0，表示不合法
                                flag = false;
                            }
                        }
                        if (!flag) {                                    //如果订单无效
                            out.println("<script language='javascript'>alert('订单无效');"
                                +"history.back();</script>");
                        } else {
                            session.removeAttribute("cart");            //清空购物车
                            out.println("<script language='javascript'>alert('订单生成，请记住您"
                                +"的订单号[" + orderID
                                + "]');window.location.href='index.jsp';</script>");    //显示生成的订单号
                        }
                        conn.close();                                   //关闭数据库连接
                    }
                } catch (Exception e) {                                 //处理异常
                    out.println(e.toString());                          //输出异常信息
                }
            } else {
                session.invalidate();                                   //销毁session
                out.println(
                    "<script language='javascript'>alert('请先登录后，再进行购物!');"
                    +"window.location.href='index.jsp';</script>");
            }
%>
```

2.9.6 实现清空购物车功能

会员在登录购好物网络商城并把喜欢的商品添加到购物车后，就可以在"我的购物车"页面上查看购物车中的商品信息了。会员如果想要清空购物车，就单击"我的购物车"页面左下角的"清空购物车"按钮，这时在"我的购物车"页面上显示的商品详细信息就会被删除，如图 2.9 所示。下面将介绍清空购物车的实现过程。

图 2.9 购物车为空页面

WebContent/front 下的 cart_clear.jsp 文件用于清空购物车。在 cart_clear.jsp 文件中，只包含了一个脚本程序。该脚本程序的作用有两个：一个是移除 Session；另一个是将页面跳转到"购物车为空"页面。代码如下：

```jsp
<%
    session.removeAttribute("cart");                              //移除 session
    response.sendRedirect("cart_null.jsp");                       //转到购物车为空页面
%>
```

用于显示"购物车为空"页面的 JSP 文件是位于 WebContent/front 下的 cart_null.jsp。如图 2.9 所示，"购物车为空"页面中的内容相对简单：一个是显示"您的购物车为空！"的信息；另一个是显示"继续购物"按钮。代码如下：

```html
<!-- MAIN CONTENT -->
<div id="mr-content" class="mr-content col-xs-12">
    <div id="mrshop" class="mrshop common-home">
        <div class="container_oc">
            <div class="container_oc">
                <div class="breadcrumb"></div>
            </div>
            <div class="row">
                <div id="content_oc" class="col-sm-12" style="min-height:300px;">
                    <h1>我的购物车</h1>
                    <div class="table-responsive cart-info" style="margin-bottom:50px;">您的购物车为空！</div>
                    <div class="buttons">
                        <div class="pull-left">
                            <a href="index.jsp" class="btn btn-primary btn-default">继续购物</a>
                        </div>
                    </div>
                </div>
            </div>
        </div>
    </div>
</div>
```

2.9.7　实现继续购物功能

如图 2.6 和 2.9 所示，在"我的购物车"页面和"购物车为空"页面上，都有"继续购物"按钮。不论是在"我的购物车"页面上，还是在"购物车为空"页面上，会员单击"继续购物"按钮后，页面都会跳转至购好物网络商城的首页。因此，不仅这两个"继续购物"按钮的作用是完全相同的（即跳转页面），而且这两个"继续购物"按钮的实现代码也是完全相同的。代码如下：

```
<div class="pull-left">
    <a href="index.jsp" class="btn btn-primary btn-default">继续购物</a>
</div>
```

2.10 项目运行

通过前述步骤，设计并完成了"购好物网络商城"项目的开发。下面运行本项目，以检验我们的开发成果。如图 2.10 所示，在 IntelliJ IDEA 中，单击 ▶ 快捷图标，即可运行本项目。

图 2.10 IntelliJ IDEA 的快捷图标

成功运行该项目，会自动打开如图 2.11 所示的"购好物网络商城"首页。在首页上，用户注册或登录会员后，单击首页上的任何商品名称或者商品图片，都会显示该商品的详细信息页面。在某件商品的详细信息页面上单击"添加到购物车"按钮后，该商品将被添加至购物车。用户可以在购物车中查看商品的图片、名称、数量、单价、总计（价格）等信息。用户单击"我的购物车"页面上的"结账"按钮，完成支付操作，同时程序会提交订单并生成订单号。用户单击"我的购物车"页面左下角的"继续购物"按钮，页面会跳转至首页。用户单击"我的购物车"页面左下角的"清空购物车"按钮，购物车中的商品就会被删除。这样，我们就成功地检验了该项目的运行。

图 2.11 成功运行项目后进入首页

Servlet 和 JSP 是 Java Web 开发中的两个重要概念，二者之间的交互主要通过表单、超链接、重定向（redirect）等方式进行。JSP 页面可以通过 HTML 表单收集用户输入的数据，这些数据通过 HTTP 请求发送到 Servlet。Servlet 通过 request.getParameter()方法获取这些参数值，并进行相应的处理。这种方式允许 JSP 页面与用户进行交互，并将数据传递给 Servlet 进行处理。JSP 页面中的超链接可以指向其他 JSP 页面或 Servlet。当用户点击超链接时，会根据链接的目标进行页面跳转。如果链接指向 Servlet，则 Servlet 处理请求并返回响应给客户端；如果链接指向 JSP 页面，则直接加载该 JSP 页面。重定向是通过 response.sendRedirect()方法实现的，它会结束当前请求并发送一个新的请求到指定的 URL。这种方式下，原始请求的属性不会保留在新请求中，适用于需要改变网页地址或引导用户到另一个完全不同的页面的情况。通过上述方式，JSP 和 Servlet 可以有效地进行交互，实现动态 Web 页面的生成和数据处理。

2.11　源码下载

源码下载

　　虽然本章详细地讲解了如何编码实现"购好物网络商城"项目的各个功能，但给出的都是代码片段，而非源码。为了方便读者学习，本书提供了完整的项目源码，扫描右侧二维码即可下载。

第 2 篇

SSM 框架应用项目

SSM 框架由 Spring、Spring MVC 和 MyBatis 三个框架整合而成，它通常被分为四层，分别是 DAO 层（Mapper）、Service 层、Controller 层和 View 层。其中，Spring 负责业务对象管理；Spring MVC 负责请求的转发和视图管理；MyBatis 作为数据对象的持久化引擎。正是因为这种分层架构，使得 SSM 框架实现了代码之间的低耦合和高内聚，提高了代码的可维护性和可扩展性。虽然当下 Spring Boot 很流行，并且是开发企业级应用项目的首推框架，但 SSM 框架依然经典，而且并未过时。Spring Boot 的强大之处在于整合了 Spring、Spring MVC、MyBatis 等框架，但是在 Spring Boot 中真正发挥作用的依然是 SSM 等框架，这也正好体现了 SSM 框架的重要性。换言之，具备了扎实的 SSM 框架基础，再来学习 Spring Boot，就能够真正地掌握 Spring Boot 的底层实现机制和原理。

本篇主要使用 SSM 框架、JSP 和 MySQL 等技术开发了 3 个 Java Web 应用项目，具体如下：
- ☑ 员工信息管理系统
- ☑ 好生活个人账本
- ☑ 嗨乐影评平台

第 3 章
员工信息管理系统

——SSM + JSP + MySQL

在互联网高度普及的今天，如果某个企业对员工信息的管理仍处于手工作业阶段，那么不仅容易出现管理纰漏，而且会导致管理效率低下。因此，一个为企业量身定制的、帮助企业实现信息化管理的员工信息管理系统，既能大幅度节约管理成本，又能有效提升管理效率。本章将使用 Java Web 开发中的 SSM 框架、JSP 和 MySQL 数据库等关键技术开发一个简单的员工信息管理系统。

项目微视频

本项目的核心功能及实现技术如下：

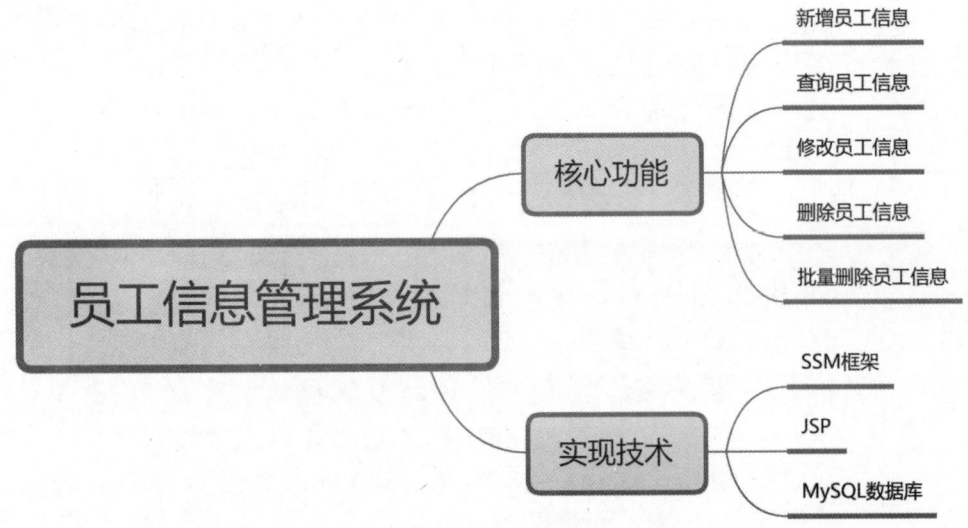

3.1 开发背景

员工信息管理系统是一个主要对企业员工的信息进行集中管理的平台，相当于为企业建立一个可视化的员工信息数据库，其中包含了企业所有员工的序号、姓名、性别、邮箱、所属部门等信息。通过该系统，企业的人事工作人员可以查询当前企业所有在职员工的信息，可以添加新进员工的信息，可以根据具体情况修改在职员工的信息，还可以（批量）删除离职员工的信息。这样，就能够帮助企业在员工信息的管理方面实现办公自动化，进而提高企业的管理效率。

员工信息管理系统将实现以下目标：

- ☑ 页面简洁、功能清晰、操作方便；
- ☑ 企业的人事工作人员可以对企业员工的信息执行增、删、改、查等操作；
- ☑ 企业的人事工作人员能够批量删除多名员工的信息。

3.2 系统设计

3.2.1 开发环境

本项目的开发及运行环境如下：
- ☑ 操作系统：推荐 Windows 10、11 及以上，兼容 Windows 7（SP1）。
- ☑ 开发工具：IntelliJ IDEA。
- ☑ 开发语言：Java EE。
- ☑ 数据库：MySQL 8.0。
- ☑ Web 服务器：Tomcat 9.0 及以上版本。

3.2.2 业务流程

启动项目后，员工信息管理系统的首页将被自动打开。在首页上，程序将分页显示企业员工的信息，并显示当前分页数和分页导航。在文本框中输入某一名员工的姓名后，单击"查询"按钮，程序将显示这名员工的信息；清除文本框中的员工姓名，再单击"查询"按钮，程序将跳转至首页。先单击"新增"按钮，输入新进员工的信息，再单击"保存"按钮，可完成新增员工信息的操作；先单击"修改"按钮，修改某一名员工的信息，再单击"更新"按钮，可完成修改员工信息的操作；先单击"删除"按钮，再单击"确定"按钮，可完成删除员工信息的操作；先勾选多名员工，再单击"批量删除"按钮，可完成批量删除员工信息的操作。在完成上述的每一个操作后，本项目都将跳转至首页。员工信息管理系统的业务流程如图 3.1 所示。

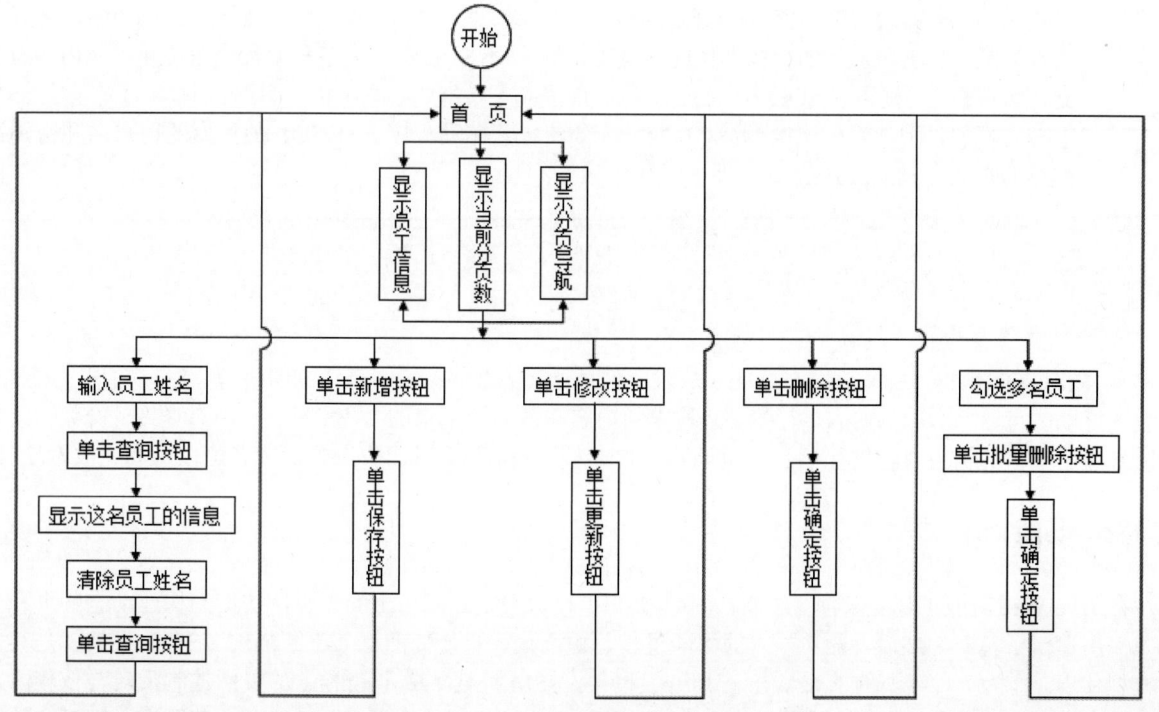

图 3.1　员工信息管理系统的业务流程图

3.2.3 功能结构

本项目旨在让读者理解 SSM 框架的实现原理。因此，在视觉上，除了首页，没有多元化、多彩的页面；在功能上，也仅限于对企业员工的信息执行增、删、改、查等操作。本项目的功能结构已经在章首页中给出。作为一个简易的员工信息管理系统，本项目实现的具体功能如下：

- ☑ 新增员工信息：把新进员工的信息添加到员工信息管理系统。
- ☑ 查询员工信息：既可以在首页上查看企业所有员工的信息，也可以根据姓名查询某一名员工的信息。
- ☑ 修改员工信息：修改某一名员工的邮箱、性别、所属部门等信息。
- ☑ 删除员工信息：删除某一名员工的姓名、邮箱、性别、所属部门等信息。
- ☑ 批量删除员工信息：删除多名员工的姓名、邮箱、性别、所属部门等信息。

3.3 技术准备

3.3.1 技术概览

- ☑ JSP 技术：JSP 全名为 java server pages，被译为 Java 服务器页面，它是一个简化的 Servlet 设计。JSP 技术有点类似 ASP 技术，它是在传统的网页 HTML 文件（后缀名为.html）中插入 Java 脚本程序（Scriptlet）和 JSP 指令标识（tag），从而形成 JSP 文件（后缀名为.jsp）。使用 JSP 技术开发的 Web 应用程序是可以跨平台的，既可以在 Windows 系统上运行，也可以在其他操作系统上运行。例如，本项目的首页就是使用 JSP 技术设计并完成的。
- ☑ MySQL 数据库：MySQL 是一个真正的多用户、多线程的 SQL 数据库服务器。SQL（结构化查询语言）是世界上最流行的、标准化的数据库语言。MySQL 不仅为多种编程语言提供了 API，而且支持多线程，还优化了 SQL 查询算法，极大地提高了性能、可扩展性、可用性。本项目是基于 SSM 框架实现的，因此用于连接 MySQL 数据库的路径和驱动程序、MySQL 数据库的用户名和密码都被存储在 dbconfig.properties 文件中。代码如下：

```
jdbc.jdbcUrl=jdbc:mysql://localhost:3306/ssm_crud?useUnicode=true&characterEncoding=utf-8&\
    useJDBCCompliantTimezoneShift=true&useLegacyDatetimeCode=false&serverTimezone=GMT%2B8
jdbc.driverClass=com.mysql.cj.jdbc.Driver
jdbc.user=root
jdbc.password=root
```

有关 MySQL 数据库的知识，在《Java 从入门到精通（第 7 版）》中有详细的讲解，对该知识不太熟悉的读者可以参考该书对应的内容。有关 JSP 技术的内容在本书第 1 章的"技术准备"中有详细的介绍，读者可以参考第 1 章的相关内容进行学习。下面将对 SSM 框架进行必要介绍，以确保读者可以顺利完成本项目。

3.3.2 Spring

在 Java Web 开发的历史长河中，Servlet 和 JSP 等技术曾是开疆拓土的先驱，它们为构建动态网页和交互式应用奠定了基石。然而，随着技术的不断进步，更高效、更便捷的框架不断涌现，SSM 框架就是其中的典型代表。作为一种经典的 Java Web 应用开发框架，SSM 框架集成了 Spring 的灵活控制反转与依赖注入机制、Spring MVC 的优雅 Web 请求处理能力，以及 MyBatis 的简洁数据库交互方式。若要深入理解和熟练运用 SSM 框架，首要任务是掌握其核心——Spring 框架。

Spring 是一个开源框架，它是于 2003 年兴起的一个轻量级的 Java 开发框架，由 Rod Johnson 在其著作 *Expert One-On-One J2EE Development and Design* 中阐述的部分理念和原型衍生而来。它是为了解决企业应用开发的复杂性而创建的。框架的主要优势之一就是其分层架构，这种架构允许使用者选择使用哪一个组件，同时为 J2EE 应用程序开发提供集成的框架。Spring 使用基本的 JavaBean 来完成以前只可能由 EJB 完成的事情。然而，Spring 的用途不仅限于服务器端的开发。从简单性、可测试性和松耦合的角度而言，任何 Java 应用都可以从 Spring 中受益。Spring 的核心是控制反转（IoC）和面向切面（AOP）。

简单来说，Spring 是一个分层的 JavaSE/EE full-stack（一站式）轻量级开源框架，其主要特点如下：

- ☑ 方便解耦，简化开发。通过 Spring 提供的 IoC 容器，可以将对象之间的依赖关系交由 Spring 进行控制，避免硬编码所造成的过度程序耦合。有了 Spring，用户不必再为单实例模式类、属性文件解析等这些很底层的需求编写代码，可以更专注于上层的应用。
- ☑ AOP 编程的支持。通过 Spring 提供的 AOP 功能，开发者可以方便地进行面向切面的编程，许多不容易用传统 OOP 实现的功能可以通过 AOP 轻松应付。
- ☑ 声明式事务的支持。在 Spring 中，开发者可以从单调烦闷的事务管理代码中解脱出来，通过声明式方式灵活地进行事务的管理，提高开发效率和质量。
- ☑ 方便程序的测试。可以用非容器依赖的编程方式进行几乎所有的测试工作。在 Spring 里，测试不再是昂贵的操作，而是随手可做的事情。例如，Spring 支持 Junit4，可以通过注解方便地测试 Spring 程序。
- ☑ 方便集成各种优秀框架。Spring 不仅不排斥各种优秀的开源框架，而且可以降低各种框架的使用难度，它提供了对各种优秀框架（如 Struts、Hibernate、Hessian、Quartz 等）的直接支持。
- ☑ 降低 Java EE API 的使用难度。Spring 对很多难用的 Java EE API（如 JDBC、JavaMail、远程调用等）提供了一个薄薄的封装层，通过 Spring 的简易封装，这些 Java EE API 的使用难度大为降低。

Spring 的源码设计精妙、结构清晰、匠心独运，处处体现着大师对 Java 设计模式的灵活运用以及在 Java 技术上的高深造诣。Spring 框架源码无疑是 Java 技术的最佳实践范例。如果想在短时间内迅速提高自己的 Java 技术水平和应用开发水平，学习和研究 Spring 源码将会使读者收到意想不到的效果。Spring 常用注解如表 3.1 所示。

表 3.1 Spring 常用注解

注　　解	说　　明
@Component	告诉 Spring，该类是一个实体类
@Repository	告诉 Spring，该类是一个持久层类
@Service	告诉 Spring，该类是一个业务层类
@Autowired	需要由 Spring 自动创建对象

原则上，Spring 不会关注每一层的类被标注的注解是否正规，意思就是如果实体类上用@Repository 或@Service 予以标注也是可以的，程序不会报错，只是不是很正规。

在对已经声明的某个类的对象使用@Autowired 标注以后，创建对象的操作将由 Spring 来完成。例如，如果想在 Service 层调用 DAO（data access object，数据访问对象）层里的方法，那么需要先声明、再创建 DAO 层的对象。如果在声明 DAO 层的对象时使用@Autowired 予以标注，就不需要使用 new 关键字创建 DAO 层的对象。

3.3.3　Spring MVC

Spring MVC 是一种基于 MVC（model-view-controller，模型-视图-控制器）的设计模式，它能够高效地处理用户请求与响应，同时提供了灵活的请求映射机制与视图渲染功能，使得 Web 应用开发更加便捷和高效。

顾名思义，MVC 可以被拆解为 3 个英文字母，即 M、V 和 C。需要说明的是，这 3 个英文字母各有所指：
- M，即 model，表示的是数据模型层。数据模型通常由 POJO（即 Java 对象）组成，其中封装了应用程序所需的数据。
- V，即 view，表示的是视图层。视图的作用是在浏览器呈现的界面上，展示数据模型中的各项数据。
- C，即 controller，表示的是控制层。控制层的作用是接收、处理、转发用户请求。

在 Web 项目的开发过程中，能够及时、准确地响应用户的请求是至关重要的。例如，用户在某个网页上，通过单击一个 URL 路径发送了一个请求；控制层获取了这个请求，先对其进行解析，待得到处理结果后，再把这个结果转发给视图层；在浏览器呈现的界面上，显示由控制层转发的处理结果。

凡事都有利弊，MVC 也不例外。MVC 的优点体现在以下几个方面：
- 多视图可以共享一个数据模型，从而提高代码的重用性。
- MVC 中的三个模块相互独立。
- 控制器提高了应用程序的灵活性和可配置性。
- 有利于软件工程化管理。

MVC 的缺点体现在如下的几个方面：
- 原理复杂。
- 增加了系统结构和实现过程的复杂性。
- 视图对模型数据的低效率访问。

Spring MVC 是一个典型的、教科书式的 MVC 设计模式。如图 3.2 所示，Spring MVC 的实现过程需要 Servlet、JSP 和 JavaBean 予以支持。其中，Servlet 相当于 controller，JSP 相当于 view，JavaBean 相当于 model。通过 Servlet、JSP 和 JavaBean，既可以实现数据模型层和视图层的代码分离，又可以将控制层单独划分出来，专门负责业务流程的控制、接收请求、创建所需的 JavaBean、将处理后的数据转发给视图层进行展示等操作。

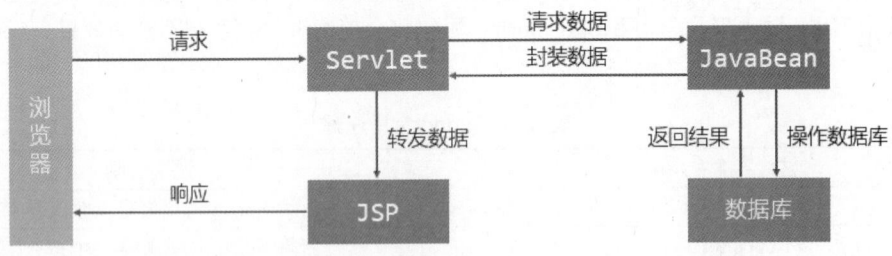

图 3.2　Spring MVC 的实现过程

Spring MVC 的主要特点如下：
- 拥有强大的灵活性、非入侵性和可配置性。
- 提供了一个前端控制器 DispatcherServlet，开发者无须额外开发控制器对象。
- 分工明确，包含控制器、验证器、命令对象、模型对象、处理程序映射、视图解析器等，每一个功能实现都由一个专门的对象负责完成。
- 可以自动绑定用户输入，并正确地转换数据类型。例如，Spring MVC 能自动解析字符串，并将其设置为模型的 int 或 float 类型的属性。
- 可以使用一个 key-value 的 Map 对象实现更加灵活的模型数据传输。
- 内置了常见的校验器，可以校验用户输入，如果校验不通过，则重定向回输入表单。输入校验是可选的，并且支持编程方式及声明方式。
- 支持国际化，支持根据用户区域显示多国语言，并且国际化的配置非常简单。

☑ 支持多种视图技术，最常见的有 JSP 技术以及其他技术，包括 Velocity 和 FreeMarker。

3.3.4 MyBatis

MyBatis 是一款优秀的持久层框架，它支持定制化 SQL、存储过程以及高级映射。MyBatis 几乎彻底消除了 JDBC 代码的复杂性，它能自动处理参数配置和结果集获取，极大地简化了与数据库的交互流程。MyBatis 可以使用简单的 XML 或注解来配置和映射原生信息，将接口和 Java 的 POJOs（Plain Old Java Objects，普通的 Java 对象）映射成数据库中的记录。

MyBatis 是一款半自动化的持久层框架。所谓半自动化，就是 MyBatis 不仅需要程序开发人员手动编写部分 SQL 语句，而且需要手动设置 SQL 语句与实体类的映射关系。在目前比较流行的项目架构设计中，不会让 Service 服务直接与数据库进行交互，而是需要通过持久层来获取数据实体。如图 3.3 所示，位于持久层的 MyBatis 可以将数据与实体类互相转换，MyBatis 既可以将一条数据封装成一个 Class 对象，也可以将多条数据封装成 List<Class>对象集合。

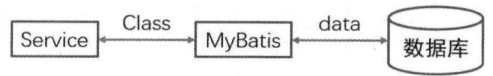

图 3.3 MyBatis 框架位于持久层

作为半自动化持久层框架，MyBatis 会将 SQL 语句中的字段与实体类的属性一一对应。不管是查询语句、添加语句、修改语句或者删除语句，MyBatis 的映射器（Mapper）都可以自动实现填充或者解析数据。如图 3.4 所示，根据一条查询语句，Mapper 就可以将查询结果交给实体类并创建包含这些数据的对象。

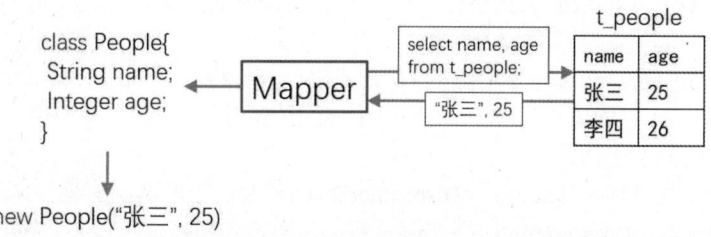

图 3.4 MyBatis 的映射器将 SQL 查询结果封装成类对象

映射器是 MyBatis 的核心功能，MyBatis 的绝大多数代码都是围绕着映射器展开的。

程序开发人员可以在映射器中编写待执行的 SQL 语句的同时，设定数据表与实体类之间的映射关系。如果不设定映射关系，则映射器会自动为同名的表字段与类属性建立映射关系。如图 3.5 所示，在数据表中有 4 个字段，在实体类中有 3 个属性，同名的表字段和类属性可以互相映射，不同的表字段和类属性就不可以了。自动映射不区分大小写，如果 People 类的属性名是 NAME，同样可以映射数据表中的表字段 name。

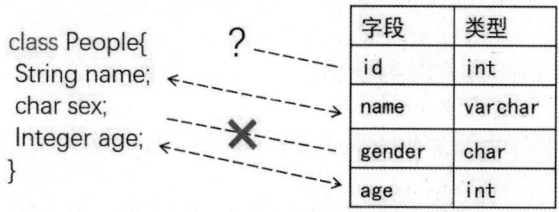

图 3.5 同名的表字段和类属性可以互相映射

创建 MyBatis 映射器应采用接口类型，接口名称以 Mapper 结尾。MyBatis 能够自动实现由程序开发人员定义的抽象方法。例如，创建员工类 Emp 的映射器 EmpMapper，让该映射器可以执行查询指定名称的员工、添加新员工、删除指定名称的员工等操作。

MyBatis 的主要特点如下：

- ☑ 简单易用：MyBatis 的配置和使用相对简单，学习门槛低。它采用了简洁的 SQL 映射语句，使得开发者可以更加专注于 SQL 编写和数据库操作。
- ☑ 灵活性强：MyBatis 支持灵活的 SQL 编写方式，开发者可以编写原生的 SQL 语句，充分发挥数据库的特性。同时，MyBatis 也支持动态 SQL、参数映射和结果集映射等高级特性，便于开发者进行复杂的数据库操作。
- ☑ 性能优秀：MyBatis 使用 JDBC 进行数据库访问，通过优化 SQL 语句、缓存查询结果和预编译等技术手段，提升了数据库访问的性能。此外，MyBatis 还支持二级缓存和本地缓存，进一步提升了系统的性能。
- ☑ 与 Spring 无缝集成：MyBatis 与 Spring 框架能够无缝集成，通过 Spring 容器来管理 MyBatis 的 SqlSessionFactory 和事务，为开发者提供了一种简洁高效的数据库操作方式，同时能够充分利用 Spring 框架的各种便利特性。

3.3.5 SSM 框架

SSM 框架是基于 MVC 设计模式的 Java Web 框架，它融合了 Spring、Spring MVC 和 MyBatis 三大技术。其中，Spring 作为核心容器，提供了强大的依赖注入和事务管理功能，从而实现了业务逻辑层的高效管理；Spring MVC 以其优雅的设计，专门负责 Web 层的请求处理与响应，确保了用户界面的流畅交互；MyBatis 作为持久层框架，简洁而高效地处理与数据库的交互，提供了灵活的 SQL 映射和数据库操作。这三者协同工作，使得 SSM 框架具备了清晰的层次结构、强大的功能和灵活的配置，成为 Java Web 开发的经典工具。

SSM 框架基本原理如下：

- ☑ 用户发起请求：用户在浏览器中输入 URL 或进行其他操作，向服务器发送请求。
- ☑ DispatcherServlet 拦截请求：DispatcherServlet 是 Spring MVC 的核心组件，它拦截所有请求并负责处理请求的分发工作。
- ☑ 处理器映射器与处理器适配器：DispatcherServlet 通过处理器映射器（Handler Mapping）将请求与对应的处理器（Controller）建立映射关系，并通过处理器适配器（Handler Adapter）调用相应的处理器进行处理。
- ☑ 处理器处理请求：处理器根据请求的具体内容进行相应的处理，如获取请求参数、调用业务逻辑、访问数据库等。
- ☑ 模型与视图生成：处理器处理完请求后，将处理结果封装到模型（model）中，并选择合适的视图（view）进行渲染。
- ☑ 响应生成与返回：视图通过模型生成具体的响应内容，并将响应返回给 DispatcherServlet。
- ☑ 将响应发送给用户：DispatcherServlet 将生成的响应发送回用户的浏览器，完成"请求-处理-响应"的整个过程。

SSM 框架基本原理的简易示意图如图 3.6 所示。

SSM 框架的主要特点如下：

- ☑ 模块化设计：SSM 框架将整个项目拆分成多个模块，每个模块承担不同的职责，使得项目结构清晰、易于管理和维护。
- ☑ 轻量级：SSM 框架是由 Spring、Spring MVC 和 MyBatis 构成的，不会增加过多的额外开销，适合在资源有限的环境下使用。
- ☑ 灵活性：SSM 框架采用了面向接口编程的思想，通过依赖注入和控制反转等机制，实现了各个模块之间的解耦合，使得系统更加灵活可拓展。

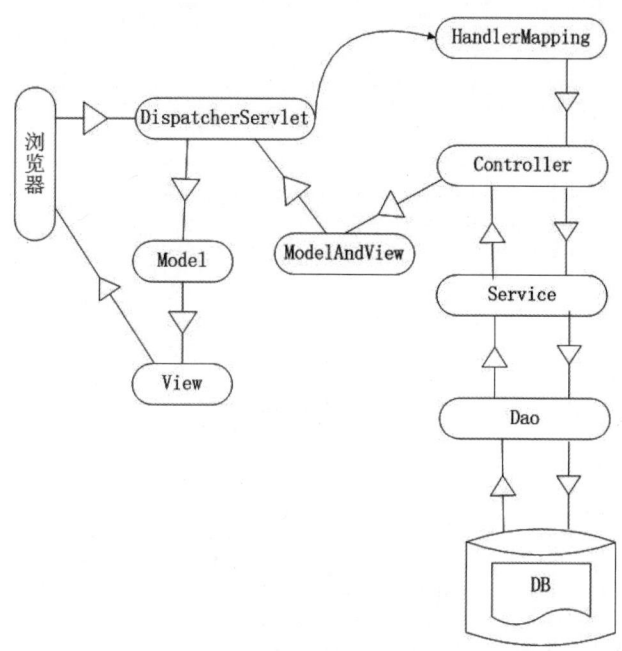

图 3.6　SSM 框架基本原理的简易示意图

☑ 易于学习和使用：SSM 框架文档丰富，社区活跃，大量的教程和案例可供参考，使得开发人员能够快速上手并快速构建项目。
☑ 优秀的生态系统：SSM 框架在 Java 开发领域有着庞大且活跃的生态系统，可以方便地集成其他优秀的开源组件和工具。

通过上述内容，读者已经大致熟悉了 SSM 框架的组成部分、基本原理和主要特点。下面将对本项目的 SSM 框架的搭建过程进行讲解。

本项目中的 applicationContext.xml 文件是 Spring 的配置文件，其中主要包含如下配置：

☑ 数据源配置：配置数据库的连接信息，包括用于连接 MySQL 数据库的路径和驱动程序、MySQL 数据库的用户名和密码等。
☑ 扫描器配置：配置需要自动扫描的包路径，使 Spring 能够自动发现和管理相应的组件和 Bean。
☑ 事务管理器配置：用于支持数据库事务操作。
☑ 其他模块配置：用于引入 MyBatis 的配置文件。

applicationContext.xml 文件中的代码如下：

```xml
<?xml version="1.0" encoding="UTF-8"?>
<beans xmlns="http://www.springframework.org/schema/beans"
    xmlns:xsi="http://www.w3.org/2001/XMLSchema-instance"
    xmlns:context="http://www.springframework.org/schema/context"
    xmlns:aop="http://www.springframework.org/schema/aop"
    xmlns:tx="http://www.springframework.org/schema/tx"
    xsi:schemaLocation="http://www.springframework.org/schema/aop
        http://www.springframework.org/schema/aop/spring-aop-4.3.xsd
        http://www.springframework.org/schema/beans http://www.springframework.org/schema/beans/spring-beans-3.2.xsd
        http://www.springframework.org/schema/tx http://www.springframework.org/schema/tx/spring-tx-4.3.xsd
        http://www.springframework.org/schema/context
        http://www.springframework.org/schema/context/spring-context-4.3.xsd">

    <context:component-scan base-package="com.test">
        <context:exclude-filter type="annotation"
            expression="org.springframework.stereotype.Controller" />
```

```xml
</context:component-scan>

<!--==================== 数据源配置 ==================-->
<context:property-placeholder location="classpath:dbconfig.properties" />
<bean id="pooledDataSource" class="com.mchange.v2.c3p0.ComboPooledDataSource">
    <property name="jdbcUrl" value="${jdbc.jdbcUrl}"></property>
    <property name="driverClass" value="${jdbc.driverClass}"></property>
    <property name="user" value="${jdbc.user}"></property>
    <property name="password" value="${jdbc.password}"></property>
</bean>

<!--==================== 引入MyBatis的配置文件 ================ -->
<bean id="sqlSessionFactory" class="org.mybatis.spring.SqlSessionFactoryBean">
    <!-- 指定mybatis全局配置文件的位置 -->
    <property name="configLocation" value="classpath:mybatis-config.xml"></property>
    <property name="dataSource" ref="pooledDataSource"></property>
    <!-- 指定mybatis，mapper文件的位置 -->
    <property name="mapperLocations" value="classpath:mapper/*.xml"></property>
</bean>

<!-- 配置扫描器，将mybatis接口的实现加入ioc容器中 -->
<bean class="org.mybatis.spring.mapper.MapperScannerConfigurer">
    <!--扫描所有dao接口的实现，加入ioc容器中 -->
    <property name="basePackage" value="com.test.crud.dao"></property>
</bean>

<!-- 配置一个可以批量执行的sqlSession -->
<bean id="sqlSession" class="org.mybatis.spring.SqlSessionTemplate">
    <constructor-arg name="sqlSessionFactory" ref="sqlSessionFactory"></constructor-arg>
    <constructor-arg name="executorType" value="BATCH"></constructor-arg>
</bean>

<!-- =============== 事务管理器配置 ==================-->
<bean id="transactionManager" class="org.springframework.jdbc.datasource.DataSourceTransactionManager">
    <!--控制住数据源 -->
    <property name="dataSource" ref="pooledDataSource"></property>
</bean>
<!--开启基于注解的事务，使用xml配置形式的事务（主要的都是使用配置式） -->
<aop:config>
    <!-- 切入点表达式 -->
    <aop:pointcut expression="execution(* com.test.crud.service..*(..))" id="txPoint"/>
    <!-- 配置事务增强 -->
    <aop:advisor advice-ref="txAdvice" pointcut-ref="txPoint"/>
</aop:config>

<!--配置事务增强，事务如何切入 -->
<tx:advice id="txAdvice" transaction-manager="transactionManager">
    <tx:attributes>
        <!-- 所有方法都是事务方法 -->
        <tx:method name="*"/>
        <!--以get开始的所有方法 -->
        <tx:method name="get*" read-only="true"/>
    </tx:attributes>
</tx:advice>
</beans>
```

本项目中的 dispatcherServlet-servlet.xml 文件是 Spring MVC 的配置文件，其中主要包含如下配置：

- ☑ 控制器配置：配置控制器（controller）的扫描路径，使得 Spring MVC 能够自动发现和管理相应的控制器。
- ☑ 视图解析器配置：配置视图解析器，用于将模型（model）渲染成具体的响应视图。

☑ 其他模块配置：如将 Spring MVC 不能处理的请求交给 tomcat 等。

dispatcherServlet-servlet.xml 文件中的代码如下：

```xml
<?xml version="1.0" encoding="UTF-8"?>
<beans xmlns="http://www.springframework.org/schema/beans"
    xmlns:xsi="http://www.w3.org/2001/XMLSchema-instance"
    xmlns:context="http://www.springframework.org/schema/context"
    xmlns:mvc="http://www.springframework.org/schema/mvc"
    xsi:schemaLocation="http://www.springframework.org/schema/mvc
        http://www.springframework.org/schema/mvc/spring-mvc-4.3.xsd
        http://www.springframework.org/schema/beans http://www.springframework.org/schema/beans/spring-beans-3.2.xsd
        http://www.springframework.org/schema/context
        http://www.springframework.org/schema/context/spring-context-4.3.xsd">

    <!--控制器配置 -->
    <context:component-scan base-package="com.test" use-default-filters="false">
        <!--只扫描控制器。 -->
        <context:include-filter type="annotation" expression="org.springframework.stereotype.Controller"/>
    </context:component-scan>

    <!--视图解析器配置 -->
    <bean class="org.springframework.web.servlet.view.InternalResourceViewResolver">
        <property name="prefix" value="/WEB-INF/views/"></property>
        <property name="suffix" value=".jsp"></property>
    </bean>

    <!--其他模块配置 -->
    <!-- 将 Spring MVC 不能处理的请求交给 tomcat -->
    <mvc:default-servlet-handler/>
    <!-- 能支持 Spring MVC 更高级的一些功能-->
    <mvc:annotation-driven/>
</beans>
```

本项目中的 mybatis-config.xml 文件是 MyBatis 的配置文件，其中主要包括开启驼峰命名、给实体类设置别名、配置分页插件等内容。

mybatis-config.xml 文件的代码如下：

```xml
<?xml version="1.0" encoding="UTF-8"?>
<!DOCTYPE configuration
  PUBLIC "-//mybatis.org//DTD Config 3.0//EN"
  "http://mybatis.org/dtd/mybatis-3-config.dtd">
<configuration>
    <settings>
        <!--开启驼峰命名 -->
        <setting name="mapUnderscoreToCamelCase" value="true"/>
    </settings>

    <typeAliases>
        <!--给实体类设置别名 -->
        <package name="com.test.crud.bean"/>
    </typeAliases>

    <plugins>
        <!--配置分页插件 -->
        <plugin interceptor="com.github.pagehelper.PageInterceptor">
            <!--分页参数合理化 -->
            <property name="reasonable" value="true"/>
        </plugin>
    </plugins>

</configuration>
```

3.4 数据库设计

3.4.1 数据库概述

本项目采用的数据库名称为 ssm_crud，数据库包含两张数据表，如表 3.2 所示。

表 3.2 员工信息管理系统的数据库结构

数据库名	表 名	表 说 明
ssm_crud	tbl_dept	部门信息表
	tbl_emp	员工信息表

3.4.2 数据表设计

tbl_dept 和 tbl_emp 两张表都采用了主键自增原则，下面将详细介绍这两张表的结构设计。

☑ tbl_dept（部门信息表）：主要用于存储企业部门的信息。该数据表的结构如表 3.3 所示。

表 3.3 tbl_dept 表结构

字 段 名 称	数 据 类 型	长 度	是 否 主 键	说 明
dept_id	INT		主键	部门的编号
dept_name	VARCHAR	255		部门的名称

☑ tbl_emp（员工信息表）：主要用于存储企业员工的信息。该数据表的结构如表 3.4 所示。

表 3.4 tbl_emp 表结构

字 段 名 称	数 据 类 型	长 度	是 否 主 键	说 明
emp_id	INT		主键	员工的编号
emp_name	VARCHAR	255		员工的姓名
gender	VARCHAR	255		员工的性别
email	VARCHAR	255		员工的邮箱
d_id	INT			部门的编号

3.5 实体类设计

实体类又称数据模型类，顾名思义，它是一种专门用于保存数据模型的类。每一个实体类都对应着一种数据模型，通常会将类的属性与数据表的字段相对应。虽然实体类的属性都是私有的，但是通过每一个属性的 Getter/Setter 方法，外部类能够获取或修改实体类的某一个属性值。实体类通常都会提供无参构造方法，并根据具体情况确定是否提供有参构造方法。

员工信息管理系统有两个实体类，它们分别是部门类和员工类。其中，部门类对应的是 com.test.crud.bean

包下的 Department.java 文件。在部门类中，有两个私有的属性，即 deptId（部门的编号）和 deptName（部门的名称）。这两个属性与 tbl_dept（部门信息表）中两个字段（dept_id 和 dept_name）相对应。为了方便外部类访问这两个私有属性，需要为它们添加 Getter/Setter 方法。Department 类的代码如下：

```java
package com.test.crud.bean;

public class Department {
    private Integer deptId;                                    //部门的编号
    private String deptName;                                   //部门的名称
    //无参构造方法
    public Department() {
        super();
    }
    //有参构造方法
    public Department(Integer deptId, String deptName) {
        super();
        this.deptId = deptId;
        this.deptName = deptName;
    }
    //添加 Getter/Setter 方法
    public Integer getDeptId() {
        return deptId;
    }

    public void setDeptId(Integer deptId) {
        this.deptId = deptId;
    }

    public String getDeptName() {
        return deptName;
    }

    public void setDeptName(String deptName) {
        this.deptName = deptName == null ? null : deptName.trim();
    }
}
```

员工类对应的是 com.test.crud.bean 包下的 Employee.java 文件。在员工类中，有 5 个私有的属性，即 empId（员工的编号）、empName（员工的姓名）、gender（员工的性别）、email（员工的邮箱）和 dId（部门的编号）。这 5 个属性与 tbl_emp（员工信息表）中的 5 个字段（emp_id、emp_name、gender、email 和 d_id）相对应。为了方便外部类访问这 5 个私有属性，需要为它们添加 Getter/Setter 方法。Employee 类的代码如下：

```java
public class Employee {
    private Integer empId;                                     //员工的编号
    @Pattern(regexp="(^[\u2E80-\u9FFF]{2,5})"
        ,message="用户名必须是2-5位的中文")
    private String empName;                                    //员工的姓名
    private String gender;                                     //员工的性别
    @Pattern(regexp="^([a-z0-9_\\.-]+)@([\\da-z\\.-]+)\\.([a-z\\.]{2,6})$",
        message="邮箱格式不正确")
    private String email;                                      //员工的邮箱
    private Integer dId;                                       //部门的编号

    private Department department;                             //部门类的对象

    @Override
    public String toString() {                                  //把员工信息转为字符串
        return "Employee [empId=" + empId + ", empName=" + empName
```

```java
            + ", gender=" + gender + ", email=" + email + ", dId=" + dId
            + "]";
}
//无参构造方法
public Employee() {
    super();
}
//有参构造方法
public Employee(Integer empId, String empName, String gender, String email, Integer dId) {
    super();
    this.empId = empId;
    this.empName = empName;
    this.gender = gender;
    this.email = email;
    this.dId = dId;
}
//添加 Getter/Setter 方法
public Department getDepartment() {
    return department;
}

public void setDepartment(Department department) {
    this.department = department;
}

public Integer getEmpId() {
    return empId;
}

public void setEmpId(Integer empId) {
    this.empId = empId;
}

public String getEmpName() {
    return empName;
}

public void setEmpName(String empName) {
    this.empName = empName == null ? null : empName.trim();
}

public String getGender() {
    return gender;
}

public void setGender(String gender) {
    this.gender = gender == null ? null : gender.trim();
}

public String getEmail() {
    return email;
}

public void setEmail(String email) {
    this.email = email == null ? null : email.trim();
}

public Integer getdId() {
    return dId;
}

public void setdId(Integer dId) {
```

```
        this.dId = dId;
    }
}
```

在上述代码中，对 empName（员工的姓名）和 email（员工的邮箱）这两个私有属性使用了 @Pattern 注解，其目的是使用正则表达式对这两个私有属性进行验证：验证员工的姓名是不是 2~5 位的中文；验证员工的邮箱是不是符合邮箱的格式。如果员工的姓名或者员工的邮箱没有通过验证，则本项目就会给出"用户名必须是 2~5 位的中文"或者"邮箱格式不正确"的提示信息。

此外，在上述代码中，有一个 Department 类型的私有属性 department，用以表示部门类的对象；还重写了一个 toString()方法，用以把员工信息转为字符串。

3.6 工具类设计

在实际开发过程中，需要编写很多个控制器。虽然这些控制器中的方法各不相同，但是这些控制器的作用都是先让后端处理由前端发送的请求，再把由后端返回的结果传递给前端。程序开发人员习惯把由后端返回的所有结果都统一封装成一个类，并把这个类称作"通用返回类"。

员工信息管理系统的通用返回类对应的是 com.test.crud.bean 包下的 Msg.java 文件。在 Msg 类（响应信息类）中，包含了 3 个私有的属性。其中，code 表示状态码；msg 表示提示信息；extend 表示一个 Map 集合的对象，该对象用于存储返回给浏览器的数据。为了方便外部类访问这 3 个私有的属性，需要为它们添加 Getter/Setter 方法。

此外，在 Msg 类中，还包含了 3 个方法，即 success()方法、fail()方法和 add()方法。success()和 fail()这两个方法用于统一管理返回浏览器的数据。其中，success()方法用于管理程序运行正常时返回浏览器的数据；fail()方法用于管理程序发生异常时返回浏览器的数据。add()方法用于把返回浏览器的数据存储在 Map 集合的对象里。Msg 类的代码如下：

```
public class Msg {
    private int code;                          //状态码：100（成功）、200（失败）
    private String msg;                        //提示信息

    //一个 Map 集合的对象，用于存储返回给浏览器的数据
    private Map<String, Object> extend = new HashMap<String, Object>();

    public static Msg success(){               //管理程序运行正常时返回给浏览器的数据
        Msg result = new Msg();
        result.setCode(100);
        result.setMsg("处理成功！");
        return result;
    }

    public static Msg fail(){                  //管理程序发生异常时返回给浏览器的数据
        Msg result = new Msg();
        result.setCode(200);
        result.setMsg("处理失败！");
        return result;
    }

    public Msg add(String key,Object value){   //把返回给浏览器的数据存储在 Map 集合的对象里
        this.getExtend().put(key, value);
        return this;
    }
```

```java
//添加 Getter/Setter 方法
public int getCode() {
    return code;
}

public void setCode(int code) {
    this.code = code;
}

public String getMsg() {
    return msg;
}

public void setMsg(String msg) {
    this.msg = msg;
}

public Map<String, Object> getExtend() {
    return extend;
}

public void setExtend(Map<String, Object> extend) {
    this.extend = extend;
}
}
```

3.7 Mapper 接口和 Example 类设计

Mapper 接口和 Example 类，即 MyBatis 的 Mapper 接口和 Example 类。MyBatis 的 Mapper 接口和 Example 类是相辅相成的。本项目包含两个 MyBatis 的 Mapper 接口（DepartmentMapper 和 EmployeeMapper）和两个 MyBatis 的 Example 类（DepartmentExample 和 EmployeeExample）。下面将分别介绍上述 Mapper 接口和 Example 类的设计过程。

3.7.1 Mapper 接口设计

MyBatis 的 Mapper 接口是一种用于定义数据库操作方法的接口。通过 Mapper 接口，能够将 SQL 语句与接口的方法绑定，使得在调用接口方法时能够执行对应的 SQL 语句。Mapper 接口提供了许多方法，其中主要的方法及其说明如表 3.5 所示。

表 3.5 Mapper 接口中的主要方法及其说明

方　　法	说　　明
int countByExample(UserExample example) throws SQLException	按条件计数
int deleteByPrimaryKey(Integer id) throws SQLException	按主键删除
int deleteByExample(UserExample example) throws SQLException	按条件查询
String/Integer insert(User record) throws SQLException	插入数据（返回值为 ID）
User selectByPrimaryKey(Integer id) throws SQLException	按主键查询
List selectByExample(UserExample example) throws SQLException	按条件查询
List selectByExampleWithBLOGs(UserExample example) throws SQLException	按条件查询（含 BLOB 字段）

续表

方法	说明
int updateByPrimaryKey(User record) throws SQLException	按主键更新
int updateByPrimaryKeySelective(User record) throws SQLException	按主键更新值非 null 的字段
int updateByExample(User record, UserExample example) throws SQLException	按条件更新
int updateByExampleSelective(User record, UserExample example) throws SQLException	按条件更新值非 null 的字段

DepartmentMapper 接口对应的是 com.test.crud.dao 包下的 DepartmentMapper.java 文件。在 DepartmentMapper 接口中，除了表 3.5 中介绍的方法，还包含一个用于选择性插入数据（如果数据表中有多个字段，那么可以只插入某一个字段的数据）的 insertSelective()方法。DepartmentMapper 接口的代码如下：

```java
public interface DepartmentMapper {
    long countByExample(DepartmentExample example);
    int deleteByExample(DepartmentExample example);
    int deleteByPrimaryKey(Integer deptId);
    int insert(Department record);
    int insertSelective(Department record);                              //选择性插入数据
    List<Department> selectByExample(DepartmentExample example);
    Department selectByPrimaryKey(Integer deptId);
    int updateByExampleSelective
        (@Param("record") Department record, @Param("example") DepartmentExample example);
    int updateByExample(@Param("record") Department record, @Param("example") DepartmentExample example);
    int updateByPrimaryKeySelective(Department record);
    int updateByPrimaryKey(Department record);
}
```

Mapper 接口能够将 SQL 语句与接口的方法绑定，进而在调用接口方法时能够执行对应的 SQL 语句。SSM 框架把 SQL 语句写在了 xml 文件中。在本项目中，与 DepartmentMapper 接口中的方法绑定的 SQL 语句被写在了 DepartmentMapper.xml 文件中。需要特别注意的是，在 DepartmentMapper.xml 文件中定义的查询（<select></select>），其 id 的值必须与 DepartmentMapper 接口中方法的名称保持一致，否则将无法进行查询。DepartmentMapper.xml 文件中代码如下：

```xml
<?xml version="1.0" encoding="UTF-8"?>
<!DOCTYPE mapper PUBLIC "-//mybatis.org//DTD Mapper 3.0//EN" "http://mybatis.org/dtd/mybatis-3-mapper.dtd">
<mapper namespace="com.test.crud.dao.DepartmentMapper">
  <resultMap id="BaseResultMap" type="com.test.crud.bean.Department">
    <id column="dept_id" jdbcType="INTEGER" property="deptId" />
    <result column="dept_name" jdbcType="VARCHAR" property="deptName" />
  </resultMap>
  <sql id="Example_Where_Clause">
    <where>
      <foreach collection="oredCriteria" item="criteria" separator="or">
        <if test="criteria.valid">
          <trim prefix="(" prefixOverrides="and" suffix=")">
            <foreach collection="criteria.criteria" item="criterion">
              <choose>
                <when test="criterion.noValue">
                  and ${criterion.condition}
                </when>
                <when test="criterion.singleValue">
                  and ${criterion.condition} #{criterion.value}
                </when>
                <when test="criterion.betweenValue">
                  and ${criterion.condition} #{criterion.value} and #{criterion.secondValue}
                </when>
                <when test="criterion.listValue">
                  and ${criterion.condition}
```

```xml
                <foreach close=")" collection="criterion.value" item="listItem" open="(" separator=",">
                  #{listItem}
                </foreach>
              </when>
            </choose>
          </foreach>
        </trim>
      </if>
    </foreach>
  </where>
</sql>
<sql id="Update_By_Example_Where_Clause">
  <where>
    <foreach collection="example.oredCriteria" item="criteria" separator="or">
      <if test="criteria.valid">
        <trim prefix="(" prefixOverrides="and" suffix=")">
          <foreach collection="criteria.criteria" item="criterion">
            <choose>
              <when test="criterion.noValue">
                and ${criterion.condition}
              </when>
              <when test="criterion.singleValue">
                and ${criterion.condition} #{criterion.value}
              </when>
              <when test="criterion.betweenValue">
                and ${criterion.condition} #{criterion.value} and #{criterion.secondValue}
              </when>
              <when test="criterion.listValue">
                and ${criterion.condition}
                <foreach close=")" collection="criterion.value" item="listItem" open="(" separator=",">
                  #{listItem}
                </foreach>
              </when>
            </choose>
          </foreach>
        </trim>
      </if>
    </foreach>
  </where>
</sql>
<sql id="Base_Column_List">
  dept_id, dept_name
</sql>
<select id="selectByExample" parameterType="com.test.crud.bean.DepartmentExample" resultMap="BaseResultMap">
  select
  <if test="distinct">
    distinct
  </if>
  <include refid="Base_Column_List" />
  from tbl_dept
  <if test="_parameter != null">
    <include refid="Example_Where_Clause" />
  </if>
  <if test="orderByClause != null">
    order by ${orderByClause}
  </if>
</select>
<select id="selectByPrimaryKey" parameterType="java.lang.Integer" resultMap="BaseResultMap">
  select
  <include refid="Base_Column_List" />
  from tbl_dept
  where dept_id = #{deptId,jdbcType=INTEGER}
```

```xml
    </select>
    <delete id="deleteByPrimaryKey" parameterType="java.lang.Integer">
      delete from tbl_dept
      where dept_id = #{deptId,jdbcType=INTEGER}
    </delete>
    <delete id="deleteByExample" parameterType="com.test.crud.bean.DepartmentExample">
      delete from tbl_dept
      <if test="_parameter != null">
        <include refid="Example_Where_Clause" />
      </if>
    </delete>
    <insert id="insert" parameterType="com.test.crud.bean.Department">
      insert into tbl_dept (dept_id, dept_name)
      values (#{deptId,jdbcType=INTEGER}, #{deptName,jdbcType=VARCHAR})
    </insert>
    <insert id="insertSelective" parameterType="com.test.crud.bean.Department">
      insert into tbl_dept
      <trim prefix="(" suffix=")" suffixOverrides=",">
        <if test="deptId != null">
          dept_id,
        </if>
        <if test="deptName != null">
          dept_name,
        </if>
      </trim>
      <trim prefix="values (" suffix=")" suffixOverrides=",">
        <if test="deptId != null">
          #{deptId,jdbcType=INTEGER},
        </if>
        <if test="deptName != null">
          #{deptName,jdbcType=VARCHAR},
        </if>
      </trim>
    </insert>
    <select id="countByExample" parameterType="com.test.crud.bean.DepartmentExample" resultType="java.lang.Long">
      select count(*) from tbl_dept
      <if test="_parameter != null">
        <include refid="Example_Where_Clause" />
      </if>
    </select>
    <update id="updateByExampleSelective" parameterType="map">
      update tbl_dept
      <set>
        <if test="record.deptId != null">
          dept_id = #{record.deptId,jdbcType=INTEGER},
        </if>
        <if test="record.deptName != null">
          dept_name = #{record.deptName,jdbcType=VARCHAR},
        </if>
      </set>
      <if test="_parameter != null">
        <include refid="Update_By_Example_Where_Clause" />
      </if>
    </update>
    <update id="updateByExample" parameterType="map">
      update tbl_dept
      set dept_id = #{record.deptId,jdbcType=INTEGER},
        dept_name = #{record.deptName,jdbcType=VARCHAR}
      <if test="_parameter != null">
        <include refid="Update_By_Example_Where_Clause" />
      </if>
    </update>
```

```xml
<update id="updateByPrimaryKeySelective" parameterType="com.test.crud.bean.Department">
  update tbl_dept
  <set>
    <if test="deptName != null">
      dept_name = #{deptName,jdbcType=VARCHAR},
    </if>
  </set>
  where dept_id = #{deptId,jdbcType=INTEGER}
</update>
<update id="updateByPrimaryKey" parameterType="com.test.crud.bean.Department">
  update tbl_dept
  set dept_name = #{deptName,jdbcType=VARCHAR}
  where dept_id = #{deptId,jdbcType=INTEGER}
</update>
</mapper>
```

说明

#{}表示一个占位符,使用占位符可以防止 sql 注入。${}可以将 parameterType 传入的内容拼接在 sql 中,不能防止 sql 注入。

EmployeeMapper 接口对应的是 com.test.crud.dao 包下的 EmployeeMapper.java 文件。在 EmployeeMapper 接口中,除了表 3.5 中介绍的方法,还包含一个用于选择性插入数据的 insertSelective()方法和一个用于按主键查询(包括"部门"字段)的 selectByPrimaryKeyWithDept()方法。EmployeeMapper 接口的代码如下:

```java
public interface EmployeeMapper {
    long countByExample(EmployeeExample example);
    int deleteByExample(EmployeeExample example);
    int deleteByPrimaryKey(Integer empId);
    int insert(Employee record);
    int insertSelective(Employee record);
    List<Employee> selectByExample(EmployeeExample example);
    Employee selectByPrimaryKey(Integer empId);
    List<Employee> selectByExampleWithDept(EmployeeExample example);
    Employee selectByPrimaryKeyWithDept(Integer empId);
    int updateByExampleSelective(@Param("record") Employee record, @Param("example") EmployeeExample example);
    int updateByExample(@Param("record") Employee record, @Param("example") EmployeeExample example);
    int updateByPrimaryKeySelective(Employee record);
    int updateByPrimaryKey(Employee record);
}
```

在本项目中,与 EmployeeMapper 接口中的方法绑定的 SQL 语句被写在了 EmployeeMapper.xml 文件中。EmployeeMapper.xml 文件中代码如下:

```xml
<?xml version="1.0" encoding="UTF-8"?>
<!DOCTYPE mapper PUBLIC "-//mybatis.org//DTD Mapper 3.0//EN" "http://mybatis.org/dtd/mybatis-3-mapper.dtd">
<mapper namespace="com.test.crud.dao.EmployeeMapper">
  <resultMap id="BaseResultMap" type="com.test.crud.bean.Employee">
    <id column="emp_id" jdbcType="INTEGER" property="empId" />
    <result column="emp_name" jdbcType="VARCHAR" property="empName" />
    <result column="gender" jdbcType="CHAR" property="gender" />
    <result column="email" jdbcType="VARCHAR" property="email" />
    <result column="d_id" jdbcType="INTEGER" property="dId" />
  </resultMap>
  <resultMap type="com.test.crud.bean.Employee" id="WithDeptResultMap">
    <id column="emp_id" jdbcType="INTEGER" property="empId" />
    <result column="emp_name" jdbcType="VARCHAR" property="empName" />
    <result column="gender" jdbcType="CHAR" property="gender" />
    <result column="email" jdbcType="VARCHAR" property="email" />
```

```xml
<result column="d_id" jdbcType="INTEGER" property="dId" />
<!-- 指定联合查询出的部门字段的封装 -->
<association property="department" javaType="com.test.crud.bean.Department">
    <id column="dept_id" property="deptId"/>
    <result column="dept_name" property="deptName"/>
</association>
</resultMap>
<sql id="Example_Where_Clause">
    <where>
        <foreach collection="oredCriteria" item="criteria" separator="or">
            <if test="criteria.valid">
                <trim prefix="(" prefixOverrides="and" suffix=")">
                    <foreach collection="criteria.criteria" item="criterion">
                        <choose>
                            <when test="criterion.noValue">
                                and ${criterion.condition}
                            </when>
                            <when test="criterion.singleValue">
                                and ${criterion.condition} #{criterion.value}
                            </when>
                            <when test="criterion.betweenValue">
                                and ${criterion.condition} #{criterion.value} and #{criterion.secondValue}
                            </when>
                            <when test="criterion.listValue">
                                and ${criterion.condition}
                                <foreach close=")" collection="criterion.value" item="listItem" open="(" separator=",">
                                    #{listItem}
                                </foreach>
                            </when>
                        </choose>
                    </foreach>
                </trim>
            </if>
        </foreach>
    </where>
</sql>
<sql id="Update_By_Example_Where_Clause">
    <where>
        <foreach collection="example.oredCriteria" item="criteria" separator="or">
            <if test="criteria.valid">
                <trim prefix="(" prefixOverrides="and" suffix=")">
                    <foreach collection="criteria.criteria" item="criterion">
                        <choose>
                            <when test="criterion.noValue">
                                and ${criterion.condition}
                            </when>
                            <when test="criterion.singleValue">
                                and ${criterion.condition} #{criterion.value}
                            </when>
                            <when test="criterion.betweenValue">
                                and ${criterion.condition} #{criterion.value} and #{criterion.secondValue}
                            </when>
                            <when test="criterion.listValue">
                                and ${criterion.condition}
                                <foreach close=")" collection="criterion.value" item="listItem" open="(" separator=",">
                                    #{listItem}
                                </foreach>
                            </when>
                        </choose>
                    </foreach>
                </trim>
            </if>
```

```xml
      </foreach>
    </where>
</sql>
<sql id="Base_Column_List">
  emp_id, emp_name, gender, email, d_id
</sql>
<sql id="WithDept_Column_List">
  e.emp_id, e.emp_name, e.gender, e.email, e.d_id,d.dept_id,d.dept_name
</sql>
<!--    List<Employee> selectByExampleWithDept(EmployeeExample example);
    Employee selectByPrimaryKeyWithDept(Integer empId);
  -->
<!-- 查询员工同时带部门信息 -->
<select id="selectByExampleWithDept" resultMap="WithDeptResultMap">
    select
     <if test="distinct">
       distinct
     </if>
     <include refid="WithDept_Column_List" />
    FROM tbl_emp e
    left join tbl_dept d on e.`d_id`=d.`dept_id`
     <if test="_parameter != null">
       <include refid="Example_Where_Clause" />
     </if>
     <if test="orderByClause != null">
       order by ${orderByClause}
     </if>
</select>
<select id="selectByPrimaryKeyWithDept" resultMap="WithDeptResultMap">
    select
  <include refid="WithDept_Column_List" />
   FROM tbl_emp e
 left join tbl_dept d on e.`d_id`=d.`dept_id`
   where emp_id = #{empId,jdbcType=INTEGER}
</select>

<!-- 查询员工不带部门信息 -->
<select id="selectByExample" parameterType="com.test.crud.bean.EmployeeExample" resultMap="BaseResultMap">
    select
    <if test="distinct">
      distinct
    </if>
    <include refid="Base_Column_List" />
    from tbl_emp
    <if test="_parameter != null">
      <include refid="Example_Where_Clause" />
    </if>
    <if test="orderByClause != null">
      order by ${orderByClause}
    </if>
</select>
<select id="selectByPrimaryKey" parameterType="java.lang.Integer" resultMap="BaseResultMap">
    select
    <include refid="Base_Column_List" />
    from tbl_emp
    where emp_id = #{empId,jdbcType=INTEGER}
</select>
<delete id="deleteByPrimaryKey" parameterType="java.lang.Integer">
    delete from tbl_emp
    where emp_id = #{empId,jdbcType=INTEGER}
</delete>
<delete id="deleteByExample" parameterType="com.test.crud.bean.EmployeeExample">
```

```xml
      delete from tbl_emp
      <if test="_parameter != null">
        <include refid="Example_Where_Clause" />
      </if>
  </delete>
  <insert id="insert" parameterType="com.test.crud.bean.Employee">
    insert into tbl_emp (emp_id, emp_name, gender,
      email, d_id)
    values (#{empId,jdbcType=INTEGER}, #{empName,jdbcType=VARCHAR}, #{gender,jdbcType=CHAR},
      #{email,jdbcType=VARCHAR}, #{dId,jdbcType=INTEGER})
  </insert>
  <insert id="insertSelective" parameterType="com.test.crud.bean.Employee">
    insert into tbl_emp
    <trim prefix="(" suffix=")" suffixOverrides=",">
      <if test="empId != null">
        emp_id,
      </if>
      <if test="empName != null">
        emp_name,
      </if>
      <if test="gender != null">
        gender,
      </if>
      <if test="email != null">
        email,
      </if>
      <if test="dId != null">
        d_id,
      </if>
    </trim>
    <trim prefix="values (" suffix=")" suffixOverrides=",">
      <if test="empId != null">
        #{empId,jdbcType=INTEGER},
      </if>
      <if test="empName != null">
        #{empName,jdbcType=VARCHAR},
      </if>
      <if test="gender != null">
        #{gender,jdbcType=CHAR},
      </if>
      <if test="email != null">
        #{email,jdbcType=VARCHAR},
      </if>
      <if test="dId != null">
        #{dId,jdbcType=INTEGER},
      </if>
    </trim>
  </insert>
  <select id="countByExample" parameterType="com.test.crud.bean.EmployeeExample" resultType="java.lang.Long">
    select count(*) from tbl_emp
    <if test="_parameter != null">
      <include refid="Example_Where_Clause" />
    </if>
  </select>
  <update id="updateByExampleSelective" parameterType="map">
    update tbl_emp
    <set>
      <if test="record.empId != null">
        emp_id = #{record.empId,jdbcType=INTEGER},
      </if>
      <if test="record.empName != null">
        emp_name = #{record.empName,jdbcType=VARCHAR},
```

```xml
      </if>
      <if test="record.gender != null">
        gender = #{record.gender,jdbcType=CHAR},
      </if>
      <if test="record.email != null">
        email = #{record.email,jdbcType=VARCHAR},
      </if>
      <if test="record.dId != null">
        d_id = #{record.dId,jdbcType=INTEGER},
      </if>
    </set>
    <if test="_parameter != null">
      <include refid="Update_By_Example_Where_Clause" />
    </if>
  </update>
  <update id="updateByExample" parameterType="map">
    update tbl_emp
    set emp_id = #{record.empId,jdbcType=INTEGER},
      emp_name = #{record.empName,jdbcType=VARCHAR},
      gender = #{record.gender,jdbcType=CHAR},
      email = #{record.email,jdbcType=VARCHAR},
      d_id = #{record.dId,jdbcType=INTEGER}
    <if test="_parameter != null">
      <include refid="Update_By_Example_Where_Clause" />
    </if>
  </update>
  <update id="updateByPrimaryKeySelective" parameterType="com.test.crud.bean.Employee">
    update tbl_emp
    <set>
      <if test="empName != null">
        emp_name = #{empName,jdbcType=VARCHAR},
      </if>
      <if test="gender != null">
        gender = #{gender,jdbcType=CHAR},
      </if>
      <if test="email != null">
        email = #{email,jdbcType=VARCHAR},
      </if>
      <if test="dId != null">
        d_id = #{dId,jdbcType=INTEGER},
      </if>
    </set>
    where emp_id = #{empId,jdbcType=INTEGER}
  </update>
  <update id="updateByPrimaryKey" parameterType="com.test.crud.bean.Employee">
    update tbl_emp
    set emp_name = #{empName,jdbcType=VARCHAR},
      gender = #{gender,jdbcType=CHAR},
      email = #{email,jdbcType=VARCHAR},
      d_id = #{dId,jdbcType=INTEGER}
    where emp_id = #{empId,jdbcType=INTEGER}
  </update>
</mapper>
```

3.7.2 Example 类设计

MyBatis 的 Example 类用于构建查询条件，它的作用相当于 SQL 语句中 WHERE 子句后面的部分，通过它可以方便地生成 SQL 查询语句。Example 类也提供了许多方法，其中主要的方法及其说明如表 3.6 所示。

表 3.6　Example 类中的主要方法及其说明

方　　法	说　　明
example.setOrderByClause("字段名 ASC")	添加升序排列条件，DESC 为降序
example.setDistinct(false)	去除重复，boolean 型，true 为选择不重复的记录
example.and(Criteria criteria)	为 example 添加 criteria 查询条件，关系为与
example.or(Criteria criteria)	为 example 添加 criteria 查询条件，关系为或
criteria.andXxxIsNull	添加字段 xxx 为 null 的条件
criteria.andXxxIsNotNull	添加字段 xxx 不为 null 的条件
criteria.andXxxEqualTo(value)	添加 xxx 字段等于 value 条件
criteria.andXxxNotEqualTo(value)	添加 xxx 字段不等于 value 条件
criteria.andXxxGreaterThan(value)	添加 xxx 字段大于 value 条件
criteria.andXxxGreaterThanOrEqualTo(value)	添加 xxx 字段大于等于 value 条件
criteria.andXxxLessThan(value)	添加 xxx 字段小于 value 条件
criteria.andXxxLessThanOrEqualTo(value)	添加 xxx 字段小于等于 value 条件
criteria.andXxxIn(List<?>)	添加 xxx 字段值在 List<?>条件
criteria.andXxxNotIn(List<?>)	添加 xxx 字段值不在 List<?>条件
criteria.andXxxLike("%"+value+"%")	添加 xxx 字段值为 value 的模糊查询条件
criteria.andXxxNotLike("%"+value+"%")	添加 xxx 字段值不为 value 的模糊查询条件
criteria.andXxxBetween(value1,value2)	添加 xxx 字段值在 value1 和 value2 之间条件
criteria.andXxxNotBetween(value1,value2)	添加 xxx 字段值不在 value1 和 value2 之间条件

每一个实体类都对应一个 Example 类。本项目有两个实体类，它们分别是 Department 类（部门类）和 Employee 类（员工类），因此与这两个实体类对应的 Example 类分别是 DepartmentExample 类和 EmployeeExample 类。DepartmentExample 类和 EmployeeExample 类的基本原理是相同的，除了包含表 3.6 中的方法，还包含 3 个受保护的属性 orderByClause、distinct 和 oredCriteria。这 3 个属性表示的含义如下：

```
/**
*升序还是降序，可以连接多个。因为只会有一个，所以直接在这里连接了
* eg: 字段+空格+asc(desc)
*/
protected String orderByClause;

/**
*去除重复，因为只会有一个，所以直接在这里设置
*
* true 是选择不重复记录
*/
protected boolean distinct;

/**
*自定义查询条件
* Criteria 的集合，集合中的对象由 or 连接
*/
protected List<Criteria> oredCriteria;
```

在 DepartmentExample 类和 EmployeeExample 类中，虽然都有各自的无参构造方法，但是这两个类的构造方法的作用是相同的，即实例化 oredCriteria 对象。以 DepartmentExample 类为例，其无参构造方法的定义如下：

```
/**
*将 oredCriteria 中的条件按照 or 连接起来
```

```
*/
public DepartmentExample() {
    oredCriteria = new ArrayList<Criteria>();
}
```

> **说明**
>
> DepartmentExample 类和 EmployeeExample 类还包含许多表 3.6 中的方法，读者可通过本项目的源码查看每个方法的具体实现。

3.8　查询员工信息模块设计

本项目既可以在首页上查看企业所有员工的信息，也可以根据姓名查询某一名员工的信息。在首页上，本项目将分页显示企业员工的序号、姓名、性别、邮箱和所属部门等信息，并显示当前分页数、分页导航和功能按钮。首页的效果如图 3.7 所示。

图 3.7　员工信息管理系统首页的效果图

在首页的左上角，有一个文本框。在文本框中，企业的人事工作人员输入某一名员工的姓名后（例如"小李"），单击"查询"按钮，程序会查询与当前姓名相匹配的员工的序号、姓名、性别、邮箱和所属部门等信息，并予以显示。根据姓名查询某一名员工的信息的效果如图 3.8 所示。

图 3.8　根据姓名查询某一名员工的信息的效果图

3.8.1 控制器类设计

com.test.crud.controller 包下的 EmployeeController 类为员工控制器类。在基于 SSM 框架的 Java Web 应用程序中，把被@Controller 注解标注的类称作控制器类。控制器负责处理用户请求并生成相应的响应。控制器根据用户请求的具体内容进行相应的业务逻辑处理。

对于本项目而言，不论是查看企业所有员工的信息，还是查询某一名员工的信息，都是以员工的姓名作为依据。只不过，当查看企业所有员工的信息时，依据的是企业所有员工的姓名；当查询某一名员工的信息时，依据的是这名员工的姓名。因此，要想同时实现上述两种查询，只需在员工控制器类中定义一个 getEmps() 方法即可。getEmps()方法的代码如下：

```java
/**
*查询员工的信息
*/
// @RequestMapping("/emps")
public String getEmps(
        @RequestParam(value = "pn", defaultValue = "1") Integer pn,
        Model model,String empName) {
    //引入 PageHelper 分页插件
    PageHelper.startPage(pn, 5);
    //根据员工姓名查询员工的信息
    List<Employee> emps = employeeService.getAll(empName);
    //使用 PageInfo 类对象存储查询后的结果
    PageInfo page = new PageInfo(emps, 5);
    //封装分页信息
    model.addAttribute("pageInfo", page);
    return "list";
}
```

3.8.2 服务类设计

com.test.crud.service 包下的 EmployeeService 类为员工服务类。在基于 SSM 框架的 Java Web 应用程序中，把被@Service 注解标注的类称作服务类。服务类即服务层，通常包含应用程序的业务逻辑和处理。服务层负责业务规则、数据处理和逻辑操作，通常调用 DAO 层来执行数据库操作。

在 getEmps()方法中，包含一个 getAll()方法，该方法是在员工服务类中予以定义的。getAll()方法包含一个 String 类型的参数，即表示员工姓名的 empName。getAll()方法具有返回值，返回值是一个用于存储员工对象的 List 类型的对象。getAll()方法的代码如下：

```java
/**
*根据员工姓名查询员工的信息
*/
public List<Employee> getAll(String empName) {
    EmployeeExample example = new EmployeeExample();
    if(!StringUtils.isEmpty(empName)) {
        example.createCriteria().andEmpNameLike("%" + empName + "%");
    }
    return employeeMapper.selectByExampleWithDept(example);
}
```

3.8.3 DAO 层设计

com.test.crud.dao 包下的 EmployeeMapper 接口为员工数据访问对象。员工数据访问对象即 DAO 层，负责与数据库进行交互。DAO 层提供了对数据的持久化操作,包括查询、新增、修改和删除数据等。

在 getAll()方法中,包含一个 selectByExampleWithDept()方法,该方法是在 EmployeeMapper 接口中予以定义的。selectByExampleWithDept() 方法包含一个参数,即 EmployeeExample 类的对象。selectByExampleWithDept()方法具有返回值,返回值是一个用于存储员工对象的 List 类型的对象。selectByExampleWithDept()方法的代码如下:

```
List<Employee> selectByExampleWithDept(EmployeeExample example);
```

说明

在本项目中,与 EmployeeMapper 接口中的方法绑定的 SQL 语句被写在了 EmployeeMapper.xml 文件中。因此,读者可以在 EmployeeMapper.xml 文件中查找与 selectByExampleWithDept()方法绑定的 SQL 语句。

3.9 新增员工信息模块设计

新增员工信息模块的作用是把新进员工的信息添加到员工信息管理系统。企业的人事工作人员单击首页右上角的"新增"按钮(如图 3.7 所示)后,程序会弹出"新增员工信息"对话框。在"新增员工信息"对话框中,企业的人事工作人员根据实际情况把新进员工的姓名、邮箱、性别和部门等信息填写完毕后,单击"保存"按钮,即可完成新增员工信息的操作。"新增员工信息"对话框的效果如图 3.9 所示。

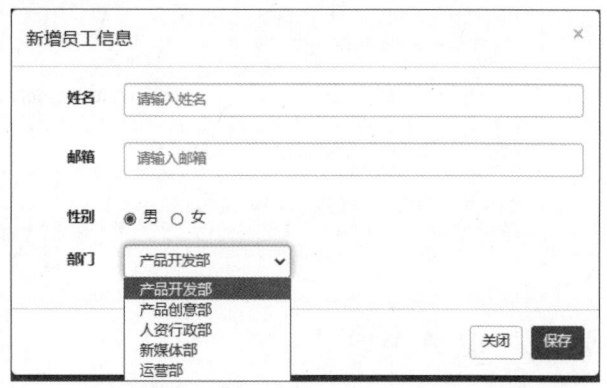

3.9.1 控制器类设计

图 3.9 "新增员工信息"对话框的效果图

在员工控制器类(EmployeeController 类)中,定义一个用于保存新进员工信息的方法,即 saveEmp()方法。该方法会对在"新增员工信息"对话框中填写完毕的姓名和邮箱进行校验。如果对姓名或者邮箱的格式校验失败,就显示错误信息;如果对姓名或者邮箱的格式校验成功,就把新进员工的信息保存在数据库中。saveEmp()方法的代码如下:

```java
/**
 * 保存新进员工信息
 */
@RequestMapping(value="/emp",method=RequestMethod.POST)
@ResponseBody
public Msg saveEmp(@Valid Employee employee,BindingResult result){
    if(result.hasErrors()){
        //校验失败,显示错误信息
        Map<String, Object> map = new HashMap();
        List<FieldError> errors = result.getFieldErrors();
        for (FieldError fieldError : errors) {
            System.out.println("错误的字段名:"+fieldError.getField());
            System.out.println("错误信息:"+fieldError.getDefaultMessage());
            map.put(fieldError.getField(), fieldError.getDefaultMessage());
        }
        return Msg.fail().add("errorFields", map);
```

```
    }else{
        employeeService.saveEmp(employee);
        return Msg.success();
    }
}
```

3.9.2 服务类设计

在 saveEmp()方法中，包含了另一个的 saveEmp()方法，只不过前者是在员工控制器类（EmployeeController 类）中予以定义的，而后者是在员工服务类（EmployeeService 类）中予以定义的。在员工服务类中予以定义的 saveEmp()方法用于保存员工信息，该方法没有返回值，但包含一个参数，即 Employee 类的对象。saveEmp()方法的代码如下：

```
/**
 * 保存员工信息
 */
public void saveEmp(Employee employee) {
    // TODO Auto-generated method stub
    employeeMapper.insertSelective(employee);
}
```

3.9.3 DAO 层设计

在 saveEmp()方法中，包含一个 insertSelective()方法，该方法是在 EmployeeMapper 接口（DAO 层）中予以定义的。insertSelective()方法包含一个参数，即 Employee 类的对象。insertSelective()方法具有返回值，返回值的类型是 int 型。insertSelective()方法的代码如下：

```
int insertSelective(Employee record);
```

说明

读者可以在 EmployeeMapper.xml 文件中查找与 insertSelective()方法绑定的 SQL 语句。

3.10 修改员工信息

修改员工信息模块的作用是修改某一名员工的邮箱、性别或者部门等信息。首页不仅显示了每一名员工的信息，还显示了用于操作每一条员工信息的功能按钮。企业的人事工作人员单击某一条员工信息后面的"修改"按钮（如图 3.7 所示）后，程序会弹出"修改员工信息"对话框。在"修改员工信息"对话框中，企业的人事工作人员根据实际情况对员工的邮箱、性别或部门等信息进行修改，修改完毕后单击"更新"按钮，即可完成修改员工信息的操作。"修改员工信息"对话框的效果如图 3.10 所示。

图 3.10 "修改员工信息"对话框的效果图

3.10.1 控制器类设计

在员工控制器类（EmployeeController 类）中，定义一个用于保存修改后的员工信息的方法，即 saveEmp() 方法。saveEmp() 方法的代码如下：

```java
/**
 * 保存修改后的员工信息
 */
@ResponseBody
@RequestMapping(value="/emp/{empId}",method=RequestMethod.PUT)
public Msg saveEmp(Employee employee, HttpServletRequest request){
    System.out.println("请求体中的值："+request.getParameter("gender"));
    System.out.println("将要更新的员工数据："+employee);
    employeeService.updateEmp(employee);
    return Msg.success();
}
```

虽然员工控制器类有两个 saveEmp() 方法，但是这两个方法同名不同参，且作用不同。第 3.9.1 节中的 saveEmp() 方法用于保存新进员工信息，而本节的 saveEmp() 方法用于保存修改后的员工信息。

3.10.2 服务类设计

在 saveEmp() 方法中，包含一个 updateEmp() 方法，该方法是在员工服务类（EmployeeService 类）中予以定义的，用于修改员工信息。updateEmp() 方法没有返回值，但包含一个参数，即 Employee 类的对象。updateEmp() 方法的代码如下：

```java
/**
 * 修改员工信息
 */
public void updateEmp(Employee employee) {
    // TODO Auto-generated method stub
    employeeMapper.updateByPrimaryKeySelective(employee);
}
```

3.10.3 DAO 层设计

在 updateEmp() 方法中，包含一个 updateByPrimaryKeySelective() 方法，该方法是在 EmployeeMapper 接口（DAO 层）中予以定义的。updateByPrimaryKeySelective() 方法包含一个参数，即 Employee 类的对象。updateByPrimaryKeySelective() 方法具有返回值，返回值的类型是 int 型。该方法的代码如下：

```java
int updateByPrimaryKeySelective(Employee record);
```

读者可以在 EmployeeMapper.xml 文件中查找与 updateByPrimaryKeySelective() 方法绑定的 SQL 语句。

3.11 删除员工信息

删除员工信息模块的作用是删除某一名员工的姓名、邮箱、性别、部门等信息。在首页上，用于操作每一条员工信息的功能按钮不仅有"修改"按钮，还有"删除"按钮。企业的人事工作人员单击某一条员工信息后面的"删除"按钮后，程序会弹出提示对话框。在提示对话框中，企业的人事工作人员只需单击"确定"按钮，即可完成删除某一名员工信息的操作。删除员工信息模块的效果如图3.11所示。

图 3.11 删除员工信息模块的效果图

3.11.1 控制器类设计

在员工控制器类（EmployeeController 类）中，定义一个用于删除员工信息的方法，即 deleteEmp()方法。该方法既可以删除某一名员工的信息，又可以批量删除多名员工的信息。在 deleteEmp()方法中，用于删除某一名员工的信息的代码如下：

```
/**
 * 删除员工信息
 */
@ResponseBody
@RequestMapping(value="/emp/{ids}",method=RequestMethod.DELETE)
public Msg deleteEmp(@PathVariable("ids")String ids){
    if(ids.contains("-")){
        //省略批量删除员工信息的代码
    }else{
        Integer id = Integer.parseInt(ids);
        employeeService.deleteEmp(id);
    }
    return Msg.success();
}
```

3.11.2 服务类设计

在 deleteEmp()方法（用于删除某一名员工的信息）的代码中，包含一个同名、不同参的 deleteEmp()方法，只不过后者是在员工服务类（EmployeeService 类）中予以定义的，用于根据 id 删除员工信息。deleteEmp(Integer id)方法没有返回值，但包含一个参数，即 Integer 类的对象，代码如下：

```
/**
 * 根据 id 删除员工信息
 */
public void deleteEmp(Integer id) {
    // TODO Auto-generated method stub
    employeeMapper.deleteByPrimaryKey(id);
}
```

3.11.3　DAO 层设计

在 deleteEmp()方法中，包含一个 deleteByPrimaryKey()方法，该方法是在 EmployeeMapper 接口（DAO 层）中予以定义的。deleteByPrimaryKey()方法包含一个参数，即 Integer 类的对象。deleteByPrimaryKey()方法具有返回值，返回值的类型是 int 型。deleteByPrimaryKey()方法的代码如下：

```
int deleteByPrimaryKey(Integer empId);
```

说明

读者可以在 EmployeeMapper.xml 文件中查找与 deleteByPrimaryKey()方法绑定的 SQL 语句。

3.12　批量删除员工信息

批量删除员工信息模块的作用是删除多名员工的姓名、邮箱、性别、部门等信息。在首页上，每一条员工信息的前面都有一个用于勾选的复选框。通过这些复选框，企业的人事工作人员能够同时勾选多条员工信息，而后单击首页右上角的"批量删除"按钮，程序会弹出提示对话框。在提示对话框中，企业的人事工作人员只需单击"确定"按钮，即可完成批量删除员工信息的操作。批量删除员工信息模块的效果如图 3.12 所示。

图 3.12　批量删除员工信息模块的效果图

3.12.1 控制器类设计

员工控制器类（EmployeeController 类）中的 deleteEmp()方法既可以删除某一名员工的信息，又可以批量删除多名员工的信息。deleteEmp()方法用于批量删除多名员工的信息的代码如下：

```
/**
 * 删除员工信息
 */
@ResponseBody
@RequestMapping(value="/emp/{ids}",method=RequestMethod.DELETE)
public Msg deleteEmp(@PathVariable("ids")String ids){
    //批量删除
    if(ids.contains("-")){
        List<Integer> del_ids = new ArrayList();
        String[] str_ids = ids.split("-");
        //组装 id 的集合
        for (String string : str_ids) {
            del_ids.add(Integer.parseInt(string));
        }
        employeeService.deleteBatch(del_ids);
    }else{
        //省略删除某一条员工信息的代码
    }
    return Msg.success();
}
```

3.12.2 服务类设计

在 deleteEmp()方法（用于批量删除多名员工的信息）的代码中，包含一个 deleteBatch()方法，该方法是在员工服务类（EmployeeService 类）中予以定义的，用于批量删除员工的信息。deleteBatch()方法没有返回值，但包含一个参数，即一个用于存储 Integer 类对象的 List 类型的对象。deleteBatch()方法的代码如下：

```
/**
 * 批量删除员工信息
 */
public void deleteBatch(List<Integer> ids) {
    // TODO Auto-generated method stub
    EmployeeExample example = new EmployeeExample();
    EmployeeExample.Criteria criteria = example.createCriteria();
    //delete from xxx where emp_id in(1,2,3)
    criteria.andEmpIdIn(ids);
    employeeMapper.deleteByExample(example);
}
```

3.12.3 DAO 层设计

在 deleteBatch()方法中，包含一个 deleteByExample()方法，该方法是在 EmployeeMapper 接口（DAO 层）中予以定义的。deleteByExample()方法包含一个参数，即 Integer 类的对象。deleteByExample()方法具有返回值，返回值的类型是 int 型。deleteByExample()方法的代码如下：

```
int deleteByExample(EmployeeExample example);
```

说明

读者可以在 EmployeeMapper.xml 文件中查找与 deleteByExample()方法绑定的 SQL 语句。

3.13　项目运行

通过前述步骤，设计并完成了"员工信息管理系统"项目的开发。下面运行本项目，以检验我们的开发成果。如图3.13所示，在 IntelliJ IDEA 中，单击▶快捷图标，即可运行本项目。

图 3.13　IntelliJ IDEA 的快捷图标

成功运行本项目，程序会自动打开如图 3.14 所示的"员工信息管理系统"首页。首页不仅分页显示了企业员工的信息，而且显示了当前分页数和分页导航，还显示了用于操作员工信息的功能按钮。通过这些功能按钮，企业的人事工作人员既能够查询所有员工的信息或者根据姓名查询某一名员工的信息，又能够对员工信息执行新增、修改、删除、批量删除等操作。这样，我们就成功地检验了本项目的运行。

图 3.14　成功运行项目后进入首页

本项目使用了 SSM 这个经典框架。SSM 框架是由 Spring、Spring MVC 和 MyBatis 整合而成的。其中，Spring 就像是整个项目中装配 JavaBean 的大工厂，在配置文件中可以使用特定的参数去调用实体类的构造方法来实例化对象。Spring MVC 在项目中拦截用户请求，它的 DispatcherServlet 将用户请求通过 HandlerMapping 去匹配 Controller，Controller 就是与用户请求相匹配的具体操作。MyBatis 是对 JDBC 的封装，它让数据库底层操作变得透明。MyBatis 通过配置文件关联各实体类的 Mapper 文件，Mapper 文件中绑定了每个类对数据库所需进行的 SQL 语句映射。

3.14　源码下载

虽然本章详细地讲解了如何编码实现"员工信息管理系统"项目的各个功能，但给出的代码都是代码片段，而非源码。为了方便读者学习，本书提供了完整的项目源码，扫描右侧二维码即可下载。

源码下载

第 4 章
好生活个人账本

——SSM + JSP + MySQL

随着个人收入的增加和生活水平的提高，人们的消费观念和消费模式也在逐渐发生变化。在现代社会中，越来越多的人开始注重理性消费和个人发展。一方面，他们会更加审慎地管理自己的财务，注重资金的合理使用，以确保每一分钱都用在刀刃上，避免盲目消费和浪费；另一方面，在保证基本生活需求得到满足的同时，他们也愿意在娱乐活动、社会交际、教育培训等方面进行适当的投入，进而提升个人的生活质量，促进个人的全面发展。因此，记账已经成为当下人们居家过日子的一项重要内容。本章将使用 Java Web 开发中的 SSM 框架、JSP 和 MySQL 数据库等关键技术开发一个简单的好生活个人账本。

项目微视频

本项目的核心功能及实现技术如下：

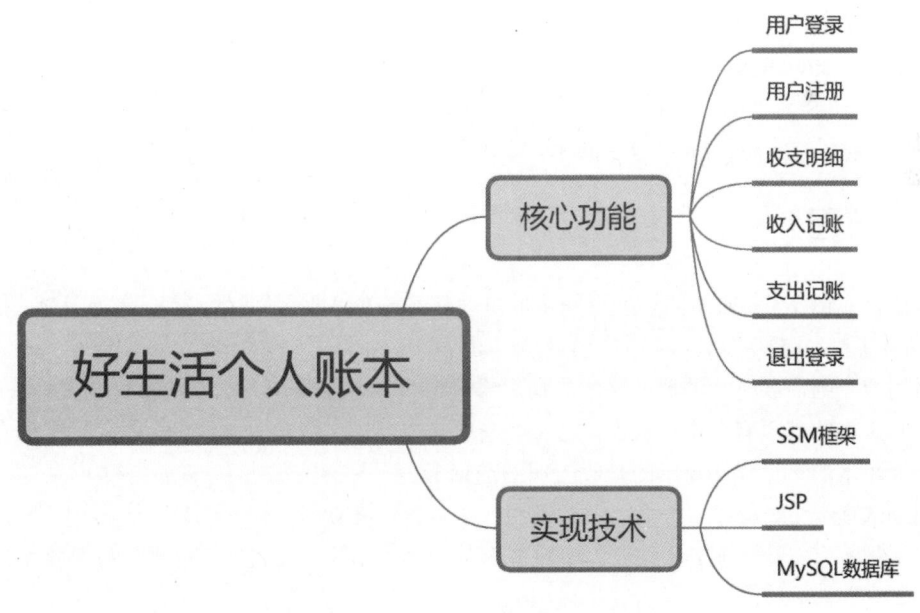

4.1 开发背景

在日常生活中，人们通过记账不仅可以管理资金，还可以用活资金。人们通过处理收入与支出的关系，当好家理好财，这样既能够提高生活水平，又能够根据自身状况适当地投资未来，均衡眼前利益和长远利益。以往居家过日子，人们在记账时经常用到的工具是纸和笔。随着计算机软件在各行各业、各个领域的不断深入，记账类软件层出不穷，这不仅大幅提高了人们记账的便利性，而且降低了纸张的使用率，进而对环境保护起到了促进作用。好生活个人账本是基于对个人收入和支出进行管理的网络平台，主要实现了查看收支明细、收入记账、支出记账等功能。

好生活个人账本将实现以下目标：
- ☑ 页面简洁大方、功能清晰明确、操作简单方便；
- ☑ 支持用户登录和用户注册；
- ☑ 用户可以查看或删除以往的收入项和支出项；
- ☑ 用户可以根据自身状况添加收入类型（如"工资收入"）和支出类型（如"购物支出"）；
- ☑ 用户可以记录新增的收入项和支出项。

4.2 系统设计

4.2.1 开发环境

本项目的开发及运行环境如下：
- ☑ 操作系统：推荐 Windows 10、11 及以上，兼容 Windows 7（SP1）。
- ☑ 开发工具：IntelliJ IDEA。
- ☑ 开发语言：Java EE。
- ☑ 数据库：MySQL 8.0。
- ☑ Web 服务器：Tomcat 9.0 及以上版本。

4.2.2 业务流程

启动项目后，好生活个人账本的登录页面将被自动打开。在登录页面上，当用户输入正确的用户名（mr）和密码（mrsoft）后，页面将跳转到好生活个人账本的首页。

如果是新用户，则需要进行注册，注册成功后将直接进入好生活个人账本的首页。

好生活个人账本的首页有 3 个选项卡，这 3 个选项卡的名称依次是收支明细、收入记账和支出记账。其中，好生活个人账本的首页默认显示的是收支明细选项卡。

在收支明细选项卡上，程序将分页显示当前用户以往的收支明细（即收入项和支出项）。其中，每一条收支明细的末尾处都有一个"删除"按钮，用户单击某一个"删除"按钮后，将删除与当前"删除"按钮对应的收支明细。

在收入记账选项卡上，用户输入收入备注、收入金额、收入日期、收入类型等信息后，单击"提交收入信息"按钮，即可完成新增收入项的操作，首页将显示收支明细选项卡。其中，用户可以通过单击"添加收入类型"按钮新增收入类型。

在支出记账选项卡上，用户输入支出备注、支出金额、支出日期、支出类型等信息后，单击"提交支出信息"按钮，即可完成新增支出项的操作，首页将显示收支明细选项卡。其中，用户可以通过单击"添加支出类型"按钮新增支出类型。

在首页上，用户单击"退出登录"超链接，即可退出好生活个人账本的首页，页面将跳转到好生活个人账本的登录页面。

好生活个人账本的业务流程如图 4.1 所示。

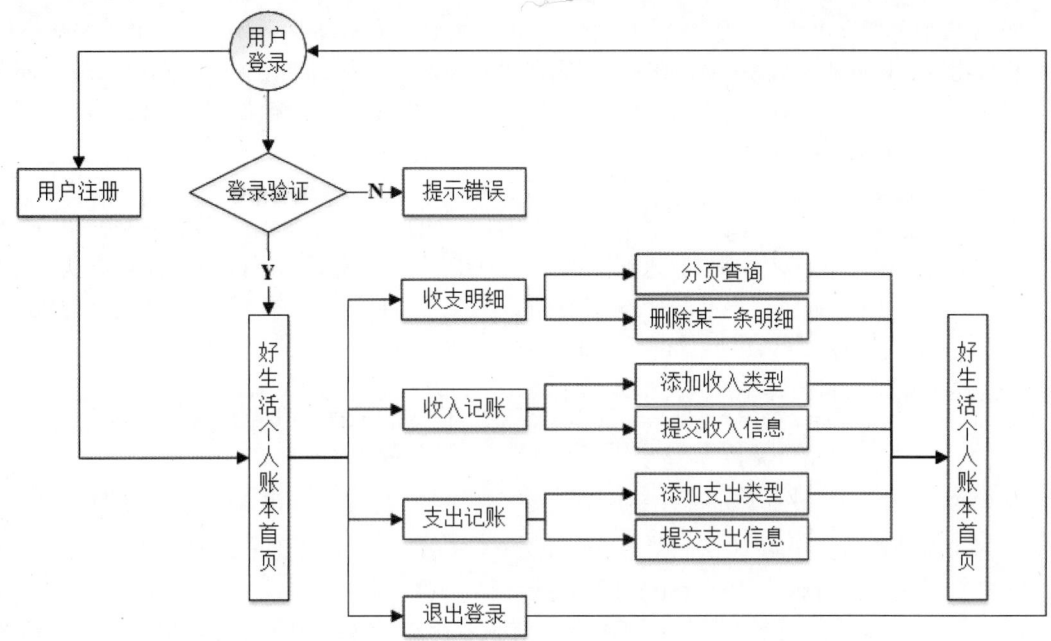

图 4.1 好生活个人账本的业务流程

4.2.3 功能结构

本项目的功能结构已经在章首页中给出。作为基于对个人收入和支出进行管理的网络平台，本项目实现的具体功能如下：

- ☑ 用户登录：用户输入正确的用户名和密码，单击"登录"按钮，页面将跳转到好生活个人账本的首页。
- ☑ 用户注册：新用户输入用户名、密码、邮箱、手机号、性别等信息，单击"注册"按钮，页面将跳转到好生活个人账本的首页。
- ☑ 收支明细：分页显示当前用户以往的收支明细，并支持删除某一条明细。
- ☑ 收入记账：新增收入类型和收入项。
- ☑ 支出记账：新增支出类型和支出项。
- ☑ 退出登录：退出好生活个人账本，页面将跳转到好生活个人账本的登录页面。

4.3 技术准备

4.3.1 技术概览

- ☑ SSM 框架：经典的 SSM 框架由 Spring、Spring MVC 和 MyBatis 组成。Spring 是一个轻量级的控制反转（IoC）和面向切面（AOP）的容器框架。Spring MVC 基于 MVC（model-view-controller：模型-视图-控制器）设计模式，用于处理用户请求和响应，并提供了灵活的请求映射和视图渲染功能。MyBatis 是一个良好的可持续性框架，支持普通 SQL 查询，同时允许对存储过程的高级映射进行数据的优化处理。
- ☑ JSP 技术：JSP 的工作原理是，首先通过客户端的浏览器，以超文本（HTML）形式通过表单（FORM）

向 Web 服务器提出请求。服务器收到客户端的请求后，由 Web 服务器上的 JSP 引擎把 JSP 代码、相关组件、Java 脚本程序以及 HTML 代码转化为 Servlet 代码。然后由 JSP 引擎调用 Web 服务器的 Java 编译器对 Servlet 代码进行编译。最后由 Java 虚拟机执行编译文件，并把客户端的请求结果以标准 HTML 页面的形式返回。

☑ MySQL 数据库：MySQL 是一款开源的、比较通用的关系型数据库系统，其应用领域非常广，目前在开源界备受关注。该数据库管理系统可以非常方便地从官方网站下载并安装使用。在安装该系统的过程中，可以灵活地对其进行配置。MySQL 具有 SQL 语句解析功能和 SQL 执行计划分析功能，其性能足够满足用户进行业务的访问和处理。本项目用于连接 MySQL 数据库的驱动程序和路径、MySQL 数据库的用户名和密码都被存储在 db.properties 文件中。代码如下：

```
jdbc.driver=com.mysql.cj.jdbc.Driver
jdbc.url=jdbc:mysql://localhost:3306/ssm_gerencaiwu_sys?useSSL=false&serverTimezone=Asia/Shanghai\
   &useUnicode=true&characterEncoding=utf-8&allowPublicKeyRetrieval=true
jdbc.username=root
jdbc.password=root
```

有关 MySQL 数据库的知识在《Java 从入门到精通（第 7 版）》中有详细的讲解，对这些知识不太熟悉的读者可以参考该书对应的内容。有关 JSP 技术的内容在本书第 1 章的"技术准备"中有详细的介绍，读者可以参考第 1 章的相关内容进行学习。有关 SSM 框架的内容在本书第 3 章的"技术准备"中有详细的介绍，读者可以参考第 3 章的相关内容进行学习。

在上文中提到了 Spring 的控制反转（IoC）和面向切面（AOP），下面将对其进行介绍，以确保读者可以更好地理解 SSM 框架。

4.3.2　Spring IoC

在传统的 Java 应用中，一个类想要调用另一个类中的属性或方法，通常会先在其代码中通过 new Object() 的方式将后者的对象创建出来，然后才能实现属性或方法的调用。为了方便理解和描述，我们可以将前者称为"调用者"，将后者称为"被调用者"。也就是说，调用者掌握着被调用者对象创建的控制权。

但在 Spring 框架中，Java 对象创建权由 IoC 容器掌握着。使用 IoC 容器创建对象的大致步骤如下：

（1）开发人员通过 XML 配置文件、注解、Java 配置类等方式，对 Java 对象进行定义，如在 XML 配置文件中使用<bean>标签、在 Java 类上使用@Component 注解等。

（2）Spring 框架启动时，IoC 容器会自动根据对象定义，将这些对象创建并管理起来。这些被 IoC 容器创建并管理的对象被称为 Spring Bean。

（3）当开发人员想要使用某个 Bean 时，可以直接从 IoC 容器中获取（例如，通过 ApplicationContext 的 getBean()方法），而不需要手动通过代码（如 new Obejct()的方式）创建。

IoC 容器带来的最大改变不是代码层面的，而是从思想层面上发生了"主从换位"的改变。原本调用者是主动的一方，它想要使用什么资源就会主动出击，自己创建。但在 Spring 框架中，IoC 容器掌握着主动权，调用者则变成了被动的一方，被动地等待 IoC 容器创建它所需要的对象（Bean）。

这个过程在职责层面发生了控制权的反转，把原本调用者通过代码实现的对象的创建，反转给 IoC 容器来帮忙实现，因此开发人员将这个过程称为 Spring 的"控制反转"。

在了解了 IoC 容器之后，还需要了解另外一个非常重要的概念——依赖注入。

依赖注入（denpendency injection，DI）是 Martin Fowler 于 2004 年在对"控制反转"进行解释时提出的。Martin Fowler 认为"控制反转"一词很晦涩，无法让人很直接地理解"到底是哪里反转了"，因此他建议使用"依赖注入"来代替"控制反转"。

在面向对象中，对象和对象之间存在一种叫作"依赖"的关系。简单来说，依赖关系就是在一个对象中需要用到另外一个对象,即对象中存在一个属性,该属性是另外一个类的对象。

例如,在一个 B 类中,有一个 A 类的对象 a,那么就可以说 B 类的对象依赖于对象 a。依赖注入就是基于这种"依赖关系"而产生的。

控制反转的核心思想就是由 Spring 负责对象的创建。在对象创建过程中,Spring 会自动根据依赖关系,将它依赖的对象注入当前对象中,这就是所谓的"依赖注入"。

依赖注入本质上是 Spring Bean 属性注入的一种,只不过这个属性是一个对象属性而已。

在 Java 程序开发过程中,系统中的各个对象之间、各个模块之间、软件系统和硬件系统之间,或多或少都存在一定的耦合关系。

若一个系统的耦合度过高,那么就会造成难以维护的问题,但完全没有耦合的代码几乎无法完成任何工作,这是由于几乎所有的功能都需要代码之间相互协作、相互依赖才能完成。因此开发人员在进行程序设计时,所秉承的思想一般都是在不影响系统功能的前提下,最大限度地降低耦合度。

IoC 容器底层通过工厂模式、Java 的反射机制、XML 解析等技术,将代码的耦合度降低到最低限度,其主要步骤如下:

(1)在配置文件(如 Bean.xml)中,对各个对象以及它们之间的依赖关系进行配置。
(2)开发人员可以把 IoC 容器当作一个工厂,这个工厂的产品就是 Spring Bean。
(3)容器启动时会加载并解析这些配置文件,得到对象的基本信息以及它们之间的依赖关系。
(4)IoC 容器利用 Java 的反射机制,根据类名生成相应的对象(即 Spring Bean),并根据依赖关系将这个对象注入依赖它的对象中。

由于对象的基本信息、对象之间的依赖关系都是在配置文件中定义的,并没有在代码中紧密耦合,因此即使对象发生改变,开发人员也只需要在配置文件中进行修改即可,而无须对 Java 代码进行修改,这就是 Spring IoC 容器实现解耦的原理。

IoC 思想是基于 IoC 容器实现的,IoC 容器底层其实就是一个 Bean 工厂。Spring 框架提供了两种不同类型 IoC 容器,它们分别是 BeanFactory 和 ApplicationContext。

1)BeanFactory

BeanFactory 是 IoC 容器的基本实现,也是 Spring 框架提供的最简单的 IoC 容器,它提供了 IoC 容器最基本的功能,由 org.springframework.beans.factory.BeanFactory 接口定义。

BeanFactory 采用懒加载(lazy-load)机制,容器在加载配置文件时并不会立刻创建 Java 对象,只有程序中获取(使用)这个对象时才会创建。

例如,使用 BeanFactory 获取上下文对象。代码如下:

```
package com.mrsoft;
import org.springframework.context.ApplicationContext;
import org.springframework.context.support.ClassPathXmlApplicationContext;

public class Main {
    public static void main(String[] args) {
        BeanFactory context = new ClassPathXmlApplicationContext("Beans.xml");
        HelloWorld obj = (HelloWorld) context.getBean("helloWorld");
        obj.getMessage();
    }
}
```

2)ApplicationContext

ApplicationContext 是 BeanFactory 接口的子接口,是对 BeanFactory 的扩展。ApplicationContext 在 BeanFactory 的基础上增加了许多企业级的功能,如 AOP(面向切面编程)、国际化、事务支持等。

ApplicationContext 接口有两个常用的实现类,具体如表 4.1 所示。

表 4.1　ApplicationContext 接口的两个常用的实现类

实 现 类	描　　述	示 例 代 码
ClassPathXmlApplicationContext	加载类路径 ClassPath 下指定的 XML 配置文件，并完成 ApplicationContext 的实例化工作	ApplicationContext applicationContext = new ClassPathXmlApplicationContext (String configLocation);
FileSystemXmlApplicationContext	加载指定的文件系统路径中指定的 XML 配置文件，并完成 ApplicationContext 的实例化工作	ApplicationContext applicationContext = new FileSystemXmlApplicationContext (String configLocation);

例如，使用 FileSystemXmlApplicationContext 实现类获取上下文对象。代码如下：

```
package com.mrsoft;
import org.springframework.beans.factory.BeanFactory;
import org.springframework.context.support.FileSystemXmlApplicationContext;

public class Main {
    public static void main(String[] args) {
        BeanFactory context = new FileSystemXmlApplicationContext
        ("D:/eclipse_J2EE_2024/eclipse/eclipse-workspace/SpringHelloWord/src/Beans.xml");
        HelloWorld obj = (HelloWorld) context.getBean("helloWorld");
        obj.getMessage();
    }
}
```

4.3.3　Spring AOP

AOP 是通过横向的抽取机制实现的。AOP 会把应用程序中的一些非业务的通用功能抽取出来，对其进行单独的维护，并通过 XML 配置文件，把已经抽取出来的这些非业务的通用功能按照指定的方式应用在应用程序中。

Spring 框架的一个关键组件是 AOP 框架。通常情况下，AOP 框架又称作 AOP 实现。为了更好地应用 AOP 技术，AOP 联盟应运而生。AOP 联盟定义了一套用于规范 AOP 框架的底层 API，因为底层 API 的表现形式主要是接口，所以各个 AOP 框架都是这些接口的具体实现。因此，凭借着底层 API，使得各个 AOP 框架可以相互移植。

为了更加高效地使用 AOP 技术，AOP 提供了 7 个术语。这 7 个术语及其说明如表 4.2 所示。

表 4.2　AOP 提供的 7 个术语及其说明

术 语 名 称	说　　明
Joinpoint（连接点）	AOP 的核心概念，指的是程序执行期间明确定义的某个位置，如类初始化前、后，类的某个方法被调用前、后，方法抛出异常后等。Spring 只支持类的某个方法被调用前、后和方法抛出异常后的连接点
Pointcut（切入点）	又称切点，如果把连接点当作数据库中的记录，那么切点就是查找该记录的查询条件。切点指要对哪些 Joinpoint 进行拦截，即被拦截的连接点
Advice（通知）	为拦截到的 Joinpoint 添加一些特殊的功能，即对切点增强的内容
Target（目标）	指需要进行增强的目标对象，通常也称作被通知（advised）对象
Weaving（织入）	指把增强代码应用到目标对象上，生成代理对象的过程
Proxy（代理）	一个类被 AOP 织入后生成了一个结果类，它是融合了原类和增强逻辑的代理类，即生成的代理对象
Aspect（切面）	由切点（Pointcut）和通知（Advice）组成

在表 4.2 中，Advice 被直译为"通知"，也有程序开发人员将其翻译为"增强处理"。Advice 有 5 种通知类型，这 5 种通知类型及其说明如表 4.3 所示。

表 4.3　Advice 的 5 种通知类型及其说明

通 知 类 型	说　　明
before（前置通知）	在目标方法被调用前，执行通知方法
after（后置通知）	在目标方法被调用后（不论是抛出异常还是执行成功），执行通知方法
after-returning（返回后通知）	在目标方法被成功地执行后，执行通知方法
after-throwing（抛出异常通知）	在目标方法抛出异常后，执行通知方法
around（环绕通知）	通知方法会将目标方法封装起来。也就是说，在目标方法被调用前、后，均可执行通知方法

AOP 被分为两个不同的类型：动态 AOP 和静态 AOP。

动态 AOP 的织入过程是在应用程序运行时动态执行的。最具代表性的动态 AOP 框架是 Spring AOP，它会为所有被通知的对象创建代理对象，并通过代理对象对原对象进行增强。

与静态 AOP 相比较，动态 AOP 的性能通常较差，但随着技术的不断发展，它的性能也在稳步提升。动态 AOP 的优点是它可以轻松地对应用程序的所有切面进行修改，无须对主程序代码进行重新编译。

静态 AOP 是通过修改或者扩展应用程序的 Java 代码实现织入过程的。最具代表性的静态 AOP 框架是 AspectJ。

与动态 AOP 相比较，静态 AOP 的性能更好。但是，它也有一个明显的缺点，即对切面的任何修改都需要重新编译整个应用程序。

在 Spring 框架中使用 AOP 主要有以下几个优势：

- ☑ 提供声明式企业服务，这种服务是声明式事务管理。
- ☑ 允许用户实现自定义切面。在某些不适合使用面向对象编程的场景中，使用面向切面编程来实现。
- ☑ 对业务逻辑的各个部分进行隔离，降低了业务逻辑各个部分间的耦合度。这样，既可以提高程序的可重用性，又可以提高开发效率。

AOP 联盟为通知（Advice）定义了一个 org.aopalliance.aop.Interface.Advice 接口。Spring AOP 根据通知（Advice）织入目标类方法的连接点位置，为 Advice 接口提供了 5 个子接口，这 5 个子接口及其说明如表 4.4 所示。

表 4.4　Advice 接口中的 6 个子接口及其说明

通知类型	子　接　口	说　　明
前置通知	org.springframework.aop.MethodBeforeAdvice	在执行目标方法前，实施增强
后置通知	org.springframework.aop.AfterReturningAdvice	在执行目标方法后，实施增强
环绕通知	org.aopalliance.intercept.MethodInterceptor	在执行目标方法前、后，实施增强
异常通知	org.springframework.aop.ThrowsAdvice	在方法抛出异常后实施增强
引入通知	org.springframework.aop.IntroductionInterceptor	在目标类中添加一些新的方法和属性

Spring 框架使用 org.springframework.aop.Advisor 接口表示切面的概念，实现对通知（Adivce）和连接点（Joinpoint）的管理。在 Spring AOP 中，切面可以分为 3 类：一般切面、切点切面和引介切面，这 3 类切面及其说明如表 4.5 所示。

表 4.5　Spring AOP 中的 3 类切面及其说明

切面类型	子接口	说明
一般切面	org.springframework.aop.Advisor	Spring AOP 默认的切面类型。由于 Advisor 接口仅包含一个 Advice（通知）类型的属性，且没有定义 PointCut（切点），因此它表示一个不带切点的简单切面。这样的切面会对目标对象（Target）中的所有方法进行拦截并织入增强代码。由于这个切面太过宽泛，因此一般不会直接使用
切点切面	org.springframework.aop.PointcutAdvisor	Advisor 的子接口，用来表示带切点的切面，该接口在 Advisor 的基础上还维护了一个 PointCut（切点）类型的属性。使用它，可以通过包名、类名、方法名等信息更加灵活地定义切面中的切点，提供更具有适用性的切面
引介切面	org.springframework.aop.IntroductionAdvisor	Advisor 的子接口，用来表示引介切面。引介切面是对应引介增强的特殊的切面，它应用于类层面上，所以引介切面适用 ClassFilter 进行定义

说明　在这 3 种类型的切面中，一般切面和切点切面更加常用。

4.4　数据库设计

4.4.1　数据库概述

本项目采用的数据库包含 3 张数据表，如表 4.6 所示。

表 4.6　好生活个人账本的数据库结构

表名	表说明
user	用户信息表
shouzhi_category	收支类型表
shouzhi_record	收支明细表

4.4.2　数据表设计

表 4.6 中的 3 张表都采用了主键自增原则，下面将详细介绍这 3 张表的结构设计。

☑　user（用户信息表）：主要用于存储用户注册时输入的信息。该数据表的结构如表 4.7 所示。

表 4.7　user 表结构

字段名称	数据类型	长度	是否主键	说明
uid	INT		主键	用户的编号
username	VARCHAR	20		用户的名称
password	VARCHAR	20		用户的密码
sex	VARCHAR	5		用户的性别
email	VARCHAR	20		用户的邮箱
phone	VARCHAR	11		用户的手机号码

- shouzhi_category（收支类型表）：主要用于存储收入类型和支出类型。该数据表的结构如表 4.8 所示。

表 4.8 shouzhi_category 表结构

字 段 名 称	数 据 类 型	长　度	是 否 主 键	说　明
szcid	INT		主键	收支类型的编号
parent_category	VARCHAR	20		收支类型的父类型（"收入"或者"支出"）
son_category	VARCHAR	20		收支类型的子类型（如"工资收入"或者"购物支出"）

- shouzhi_record（收支明细表）：主要用于存储收入信息和支出信息。该数据表的结构如表 4.9 所示。

表 4.9 shouzhi_record 表结构

字 段 名 称	数 据 类 型	长　度	是 否 主 键	说　明
szrid	INT		主键	收支明细的编号
szr_num	INT			收入或者支出的金额
user_id	INT			用户的编号
szr_date	VARCHAR	20		收支明细的日期
szr_comment	VARCHAR	100		收入明细的备注内容
shouzhiCategory	VARCHAR	50		收支类型对象
shouzhi_category_id	INT			收支类型的编号

4.5 SSM 框架的主要配置文件

4.5.1 Spring 的配置文件

本项目中的 applicationContext.xml 文件是 Spring 的配置文件，其中主要包含如下配置：
- 数据源配置：配置数据库的连接信息。
- 扫描器配置：配置需要自动扫描的包路径。
- 事务管理器配置：用于支持数据库事务操作。
- 其他模块配置：用于引入 MyBatis 的配置文件。

applicationContext.xml 文件中的代码如下：

```xml
<?xml version="1.0" encoding="UTF-8"?>
<beans xmlns="http://www.springframework.org/schema/beans"
    xmlns:context="http://www.springframework.org/schema/context" xmlns:p="http://www.springframework.org/schema/p"
    xmlns:aop="http://www.springframework.org/schema/aop" xmlns:tx="http://www.springframework.org/schema/tx"
    xmlns:xsi="http://www.w3.org/2001/XMLSchema-instance"
    xsi:schemaLocation="http://www.springframework.org/schema/beans
       http://www.springframework.org/schema/beans/spring-beans-4.0.xsd
    http://www.springframework.org/schema/context http://www.springframework.org/schema/context/spring-context-4.0.xsd
    http://www.springframework.org/schema/aop http://www.springframework.org/schema/aop/spring-aop-4.0.xsd
       http://www.springframework.org/schema/tx http://www.springframework.org/schema/tx/spring-tx-4.0.xsd
    http://www.springframework.org/schema/util http://www.springframework.org/schema/util/spring-util-4.0.xsd">

    <!-- 加载 properties 配置文件 -->
    <context:property-placeholder location="classpath:db.properties"/>
```

```xml
<!-- 配置数据源 -->
<!-- dbcp 数据库连接池 -->
<bean id="dataSource" class="org.apache.commons.dbcp.BasicDataSource"
    destroy-method="close">
    <property name="driverClassName" value="${jdbc.driver}" />
    <property name="url" value="${jdbc.url}" />
    <property name="username" value="${jdbc.username}" />
    <property name="password" value="${jdbc.password}" />
    <property name="maxActive" value="10" />
    <property name="maxIdle" value="5" />
</bean>

<!-- 配置 sqlSessionFactory -->
<bean id="sqlSessionFactory" class="org.mybatis.spring.SqlSessionFactoryBean">
    <!-- 配置数据源 -->
    <property name="dataSource" ref="dataSource"></property>
    <!-- 引入 MyBatis 的配置文件 -->
    <property name="configLocation" value="classpath:SqlMapConfig.xml"></property>
</bean>

<!-- 使用扫描器配置 mapper -->
<bean class="org.mybatis.spring.mapper.MapperScannerConfigurer">
    <!-- 指定扫描的包。每个 mapper 代理对象的 id 就是类名,首字母小写 -->
    <property name="basePackage" value="cn.zhku.jsj144.zk.financialManage.mapper"></property>
</bean>

<!-- 使用扫描器扫描 service 包 -->
<context:component-scan base-package="cn.zhku.jsj144.zk.financialManage.service"/>

<!-- 事务管理器 -->
<bean id="transactionManager" class="org.springframework.jdbc.datasource.DataSourceTransactionManager">
    <!-- 数据源 -->
    <property name="dataSource" ref="dataSource"></property>
</bean>

<!-- 开启事务注解 -->
<tx:annotation-driven transaction-manager="transactionManager"/>
</beans>
```

4.5.2 Spring MVC 的配置文件

本项目中的 springmvc.xml 文件是 Spring MVC 的配置文件,其中主要包含如下配置:

- ☑ 控制器配置:配置控制器(Controller)的扫描路径,使得 Spring MVC 能够自动发现和管理相应的控制器。
- ☑ 视图解析器配置:配置视图解析器,用于将模型(model)渲染成具体的响应视图。
- ☑ 其他模块配置:如配置注解驱动等。

springmvc.xml 文件中的代码如下:

```xml
<?xml version="1.0" encoding="UTF-8"?>
<beans xmlns="http://www.springframework.org/schema/beans"
    xmlns:xsi="http://www.w3.org/2001/XMLSchema-instance" xmlns:p="http://www.springframework.org/schema/p"
    xmlns:context="http://www.springframework.org/schema/context"
    xmlns:mvc="http://www.springframework.org/schema/mvc"
    xsi:schemaLocation="http://www.springframework.org/schema/beans
        http://www.springframework.org/schema/beans/spring-beans-4.0.xsd
        http://www.springframework.org/schema/mvc
        http://www.springframework.org/schema/mvc/spring-mvc-4.0.xsd
        http://www.springframework.org/schema/context
```

```xml
             http://www.springframework.org/schema/context/spring-context-4.0.xsd">

    <!-- 配置 controller 的扫描路径 -->
    <context:component-scan base-package="cn.zhku.jsj144.zk.financialManage.controller" />

    <!-- 配置注解驱动 -->
    <mvc:annotation-driven />

    <!-- 配置视图解析器 -->
    <mvc:interceptors>
        <mvc:interceptor>
            <mvc:mapping path="/**"/>
            <bean class="cn.zhku.jsj144.zk.financialManage.interceptor.LoginInterceptor">
                <property name="exceptUrls">
                    <list>
                        <value>/user/**</value>
                        <value>/index.jsp</value>
                        <value>/regist.jsp </value>
                        <value>/userManage/**</value>
                        <value>/categoryManage/**</value>
                        <value>/newsManage/**</value>
                    </list>
                </property>
            </bean>
        </mvc:interceptor>
    </mvc:interceptors>
</beans>
```

4.5.3　MyBatis 的配置文件

本项目中的 SqlMapConfig.xml 文件是 MyBatis 的配置文件，其作用主要是给实体类设置别名。SqlMapConfig.xml 文件的代码如下：

```xml
<?xml version="1.0" encoding="UTF-8"?>
<!DOCTYPE configuration
PUBLIC "-//mybatis.org//DTD Config 3.0//EN"
"http://mybatis.org/dtd/mybatis-3-config.dtd">
<configuration>
    <typeAliases>
        <!-- 给实体类设置别名，别名为类名 -->
        <package name="cn.zhku.jsj144.zk.financialManage"/>
    </typeAliases>
</configuration>
```

4.6　登录拦截器设计

在本项目的 Spring MVC 的配置文件中，包含一个用于配置视图解析器的代码段。在这个代码段中，包含一个登录拦截器。下面将对这个登录拦截器进行介绍。

在 Java 语言中，拦截器（Interceptor）是 Spring MVC 框架中对请求进行拦截和处理的组件，可以实现权限验证、日志记录、异常处理等功能。在 HttpServletRequest 到达 Controller 之前，可以根据需要检查 HttpServletRequest，也可以进行修改；在 HttpServletResponse 返回之前，可以根据需要检查 HttpServletResponse，也可以进行修改。

通过实现 HandlerInterceptor 接口（属于 org.springframework.web.servlet 包），并且实现 preHandle（前置）、postHandle（后置）和 afterCompletion（完成后）方法，即可自定义拦截器。自定义拦截器的步骤如下：

（1）自定义一个实现了 Interceptor 接口的类，或者继承抽象类 AbstractInterceptor。
（2）在配置文件中注册定义的拦截器。
（3）根据需要在请求中引用上述定义的拦截器。或者，为了方便可以将拦截器定义为默认的拦截器，这样在不加特殊说明的情况下，所有请求都会被这个拦截器拦截。

本项目的登录拦截器对应的是 LoginInterceptor.java 文件，是一个自定义拦截器，实现了 HandlerInterceptor 接口。在登录拦截器中，虽然重写了 preHandle()、postHandle()、afterCompletion()方法，但是只实现了 preHandle()方法。preHandle()方法在处理请求之前被执行，其返回值为 true 或者 false。如果 preHandle()方法返回 true，那么登录拦截器不会拦截请求，会处理请求。如果 preHandle()方法返回 false，那么登录拦截器会拦截请求，不会处理请求。preHandle()方法的代码如下：

```java
@Override
public boolean preHandle(HttpServletRequest request,
        HttpServletResponse response, Object handler) throws Exception {
    //请求资源路径
    String requestUri = request.getRequestURI();
    if (requestUri.startsWith(request.getContextPath())) {
        requestUri = requestUri.substring(
                request.getContextPath().length(), requestUri.length());
    }
    //放行 exceptUrls 中配置的 url
    for (String url : exceptUrls) {
        if (url.endsWith("/**")) {
            if (requestUri.startsWith(url.substring(0, url.length() - 3))) {
                return true;                                    //放行
            }
        }
        else if (requestUri.startsWith(url)) {                  //与放行资源匹配
            return true;                                        //放行
        }
    }
    //拦截用户请求，判断用户是否登录
    HttpSession session = request.getSession();
    User user = (User) session.getAttribute("user");
    if (user != null) {
        return true;//如果用户已经登录。放行
    }
    //如果用户未登录，跳转到登录页面
    response.sendRedirect(request.getContextPath() + "/index.jsp");   //登录页面
    return false;                                               //拦截
}
```

4.7 实体类设计

实体类又称数据模型类，每一个实体类都对应着一种数据模型。在第 4.4.2 节中，已经介绍了本项目的 3 张数据表。下面将分别介绍与这 3 张数据表对应的实体类。

4.7.1 用户类

用户类对应的是 User.java 文件。在用户类中，有 6 个私有的属性，即 uid（用户的编号）、username（用

户的名称)、password (用户的密码)、sex (用户的性别)、email (用户的邮箱) 和 phone (用户的手机号码)。不难发现,这 6 个属性与 user (用户信息表) 中 6 个字段是一一对应的。为了方便外部类访问这 6 个私有属性,需要为它们添加 Getter/Setter 方法。User 类的代码如下:

```java
public class User{
    private int uid;                        //用户的编号
    private String username;                //用户的名称
    private String password;                //用户的密码
    private String sex;                     //用户的性别
    private String email;                   //用户的邮箱
    private String phone;                   //用户的手机号码

    public int getUid() {
        return uid;
    }
    public void setUid(int uid) {
        this.uid = uid;
    }
    public String getUsername() {
        return username;
    }
    public void setUsername(String username) {
        this.username = username;
    }
    public String getPassword() {
        return password;
    }
    public void setPassword(String password) {
        this.password = password;
    }
    public String getSex() {
        return sex;
    }
    public void setSex(String sex) {
        this.sex = sex;
    }
    public String getEmail() {
        return email;
    }
    public void setEmail(String email) {
        this.email = email;
    }
    public String getPhone() {
        return phone;
    }
    public void setPhone(String phone) {
        this.phone = phone;
    }
}
```

4.7.2 收支类型类

收支类型类对应的是 ShouzhiCategory.java 文件。在收支类型类中,有 3 个私有的属性,即 szcid (收支类型的编号)、parent_category (收支类型的父类型) 和 son_category (收支类型的子类型)。这 3 个属性与 shouzhi_category (收支类型表) 中 3 个字段也是一一对应的。为了方便外部类访问这 3 个私有属性,需要为它们添加 Getter/Setter 方法。ShouzhiCategory 类的代码如下:

```java
public class ShouzhiCategory {
```

```java
    private int szcid;                                  //收支类型的编号
    private String parent_category;                     //收支类型的父类型（"收入"或者"支出"）
    private String son_category;                        //收支类型的子类型（如"工资收入"或者"购物支出"）

    public int getSzcid() {
        return szcid;
    }
    public void setSzcid(int szcid) {
        this.szcid = szcid;
    }
    public String getParent_category() {
        return parent_category;
    }
    public void setParent_category(String parent_category) {
        this.parent_category = parent_category;
    }
    public String getSon_category() {
        return son_category;
    }
    public void setSon_category(String son_category) {
        this.son_category = son_category;
    }
}
```

4.7.3 收支明细类

收支明细类对应的是 ShouzhiRecord.java 文件。在收支明细类中，有 6 个私有的属性，即 szrid（收支明细的编号）、szr_num（收入或者支出的金额）、szr_date（收支明细的日期）、szr_comment（收支明细的备注内容）、shouzhiCategory（收支类型对象）和 user_id（用户的编号）。然而，shouzhi_record（收支明细表）有 7 个字段，比收支明细类的属性多出一个字段，即 shouzhi_category_id（收支类型的编号）。这是因为 shouzhi_record（收支明细表）中的 shouzhi_category_id（收支类型的编号）可以从收支明细类的 shouzhiCategory（收支类型对象）中获取。为了方便外部类访问这 6 个私有属性，需要为它们添加 Getter/Setter 方法。ShouzhiRecord 类的代码如下：

```java
public class ShouzhiRecord {
    private int szrid;                                  //收支明细的编号
    private int szr_num;                                //收入或者支出的金额
    private String szr_date;                            //收支明细的日期
    private String szr_comment;                         //收支明细的备注内容
    private ShouzhiCategory shouzhiCategory;            //收支类型对象
    private int user_id;                                //用户的编号

    public int getSzrid() {
        return szrid;
    }
    public void setSzrid(int szrid) {
        this.szrid = szrid;
    }
    public int getSzr_num() {
        return szr_num;
    }
    public void setSzr_num(int szr_num) {
        this.szr_num = szr_num;
    }
    public String getSzr_date() {
        return szr_date;
    }
    public void setSzr_date(String szr_date) {
```

```
        this.szr_date = szr_date;
    }
    public String getSzr_comment() {
        return szr_comment;
    }
    public void setSzr_comment(String szr_comment) {
        this.szr_comment = szr_comment;
    }
    public int getUser_id() {
        return user_id;
    }
    public void setUser_id(int user_id) {
        this.user_id = user_id;
    }
    public ShouzhiCategory getShouzhiCategory() {
        return shouzhiCategory;
    }
    public void setShouzhiCategory(ShouzhiCategory shouzhiCategory) {
        this.shouzhiCategory = shouzhiCategory;
    }
}
```

4.8 Mapper 接口设计

通过 MyBatis 的 Mapper 接口，能够将 SQL 语句与接口的方法绑定，使得在调用接口方法时能够执行对应的 SQL 语句。下面将分别介绍本项目中主要的 Mapper 接口。

4.8.1 UserMapper 接口

UserMapper 接口对应的是 UserMapper.java 文件。在 UserMapper 接口中有 4 个方法，它们分别是 queryUserByUser()、queryUserByUsername()、updatePasswordByUsername()和 insertUser()方法。有关上述 4 个方法的说明，读者可参照 UserMapper 接口的代码，代码如下：

```
public interface UserMapper {
    User queryUserByUser(User user);                          //通过用户名和密码查询用户是否存在
    User queryUserByUsername(String username);                //通过用户名查询用户是否存在
    void updatePasswordByUsername(User user);                 //修改密码（找回密码）
    void insertUser(User user);                               //用户注册（添加用户）
}
```

MyBatis 的 Mapper 接口能够将 SQL 语句与接口的方法绑定，SSM 框架把 SQL 语句写在了 xml 文件中。在本项目中，与 UserMapper 接口中的方法绑定的 SQL 语句被写在了 UserMapper.xml 文件中。UserMapper.xml 文件中的代码如下：

```xml
<?xml version="1.0" encoding="UTF-8"?>
<!DOCTYPE mapper
PUBLIC "-//mybatis.org//DTD Mapper 3.0//EN"
"http://mybatis.org/dtd/mybatis-3-mapper.dtd">
<mapper namespace="cn.zhku.jsj144.zk.financialManage.mapper.UserMapper">

    <!-- 通过用户名和密码查询用户是否存在-->
    <select id="queryUserByUser" parameterType="user" resultType="user">
        select * from user
        where username=#{username} and password=#{password}
```

```xml
    </select>

    <!--通过用户名查询用户是否存在 -->
    <select id="queryUserByUsername" parameterType="String" resultType="user">
        select * from user
        where username=#{username}
    </select>

    <!-- 修改密码（找回密码）-->
    <update id="updatePasswordByUsername" parameterType="user">
        update user
        set password=#{password}
        where username=#{username}
    </update>

    <!-- 用户注册（添加用户）-->
    <insert id="insertUser">
        <!-- 自增主键返回 -->
        <selectKey keyProperty="uid" order="AFTER" resultType="int">
            select last_insert_id()
        </selectKey>
        insert into
        user(username,password,sex,email,phone)
        values(#{username},#{password},#{sex},#{email},#{phone})
    </insert>
</mapper>
```

4.8.2 ShouzhiCategoryMapper 接口

ShouzhiCategoryMapper 接口对应的是 ShouzhiCategoryMapper.java 文件。在 ShouzhiCategoryMapper 接口中有 4 个方法，它们分别是 findSonCategoryByParent()方法、findCategoryBySonCategory()方法、findShouzhiCategoryByParent()方法和 addShouzhiCategory()方法。有关上述 4 个方法的说明，读者可参照 ShouzhiCategoryMapper 接口的代码，代码如下：

```java
public interface ShouzhiCategoryMapper {
    //通过收支父类型，查询出该父类型的所有子类型
    List<String> findSonCategoryByParent(String parent_category);
    //根据收支子类型，获得收支类型对象
    ShouzhiCategory findCategoryBySonCategory(String son_category);
    //查询收入或者支出子类型
    List<ShouzhiCategory> findShouzhiCategoryByParent(String parent_category);
    //添加收支子类型
    void addShouzhiCategory(ShouzhiCategory shouzhiCategory);
}
```

在本项目中，ShouzhiCategoryMapper.xml 文件含有与 ShouzhiCategoryMapper 接口中的方法绑定的 SQL 语句。ShouzhiCategoryMapper.xml 文件中的代码如下：

```xml
<?xml version="1.0" encoding="UTF-8"?>
<!DOCTYPE mapper
PUBLIC "-//mybatis.org//DTD Mapper 3.0//EN"
"http://mybatis.org/dtd/mybatis-3-mapper.dtd">
<mapper namespace="cn.zhku.jsj144.zk.financialManage.mapper.ShouzhiCategoryMapper">

    <!-- 通过收支父类型，查询出该父类型的所有子类型 -->
    <select id="findSonCategoryByParent" parameterType="String" resultType="String">
        select son_category from shouzhi_category where parent_category=#{parent_category}
    </select>
```

```xml
<!-- 根据收支子类型，获得收支类型对象 -->
<select id="findCategoryBySonCategory" parameterType="String" resultType="ShouzhiCategory">
    select * from shouzhi_category where son_category=#{son_category}
</select>

<!-- 查询收入或者支出子类型 -->
<select id="findShouzhiCategoryByParent" parameterType="String" resultType="ShouzhiCategory">
    select * from shouzhi_category where parent_category=#{parent_category}
</select>

<!-- 添加收支子类型 -->
<insert id="addShouzhiCategory" parameterType="ShouzhiCategory">
    insert into shouzhi_category(parent_category,son_category)
    values(#{parent_category},#{son_category})
</insert>
</mapper>
```

4.8.3 ShouzhiRecordMapper 接口

ShouzhiRecordMapper 接口对应的是 ShouzhiRecordMapper.java 文件。在 ShouzhiRecordMapper 接口中有 5 个方法，它们分别是 findShouzhiRecordCount()方法、findCurrenPageRecordList()方法、deleteOneShouzhiRecord() 方法、addShouzhiRecord()方法和 findParentCategoryById()方法。有关上述 5 个方法的说明，读者可参照 ShouzhiRecordMapper 接口的代码，代码如下：

```java
public interface ShouzhiRecordMapper {
    int findShouzhiRecordCount(Map<String, Object> map);             //查询用户的收支明细记录数
    //查询用户的收支明细记录
    List<ShouzhiRecord> findCurrenPageRecordList(ShouzhiRecordQueryVo queryVo);
    //删除用户的收支信息（一条）
    void deleteOneShouzhiRecord(int id);
    //添加收支信息（收入 或者 支出）
    void addShouzhiRecord(ShouzhiRecord shouzhiRecord);
    //通过收支类型的编号确认是收入 还是 支出
    String findParentCategoryById(int szcid);
}
```

在本项目中，ShouzhiRecordMapper.xml 文件含有与 ShouzhiRecordMapper 接口中的方法绑定的 SQL 语句。ShouzhiRecordMapper.xml 文件中的代码如下：

```xml
<?xml version="1.0" encoding="UTF-8"?>
<!DOCTYPE mapper
PUBLIC "-//mybatis.org//DTD Mapper 3.0//EN"
"http://mybatis.org/dtd/mybatis-3-mapper.dtd">
<mapper namespace="cn.zhku.jsj144.zk.financialManage.mapper.ShouzhiRecordMapper">

<!-- 查询用户的收支明细记录数-->
<select id="findShouzhiRecordCount" parameterType="java.util.Map" resultType="int">
    select count(*) from shouzhi_record
    where user_id=#{user.uid}
    <if test="szr_date != null and szr_date!='' >
        and szr_date like '${szr_date}%'
    </if>
    <if test="szr_comment != null and szr_comment!=''">
        and szr_comment like '%${szr_comment}%'
    </if>
</select>

<resultMap type="ShouzhiRecord" id="ShouzhiRecordResultMap">
    <id property="szrid" column="szrid"/>
```

```xml
        <result property="szr_num" column="szr_num"/>
        <result property="szr_date" column="szr_date"/>
        <result property="szr_comment" column="szr_comment"/>
        <result property="user_id" column="user_id"/>
        <!-- 一对一关系 -->
        <association property="shouzhiCategory" javaType="ShouzhiCategory">
            <id property="szcid" column="szcid"/>
            <result property="parent_category" column="parent_category"/>
            <result property="son_category" column="son_category"/>
        </association>
    </resultMap>

    <select id="findCurrenPageRecordList"    parameterType="ShouzhiRecordQueryVo"
     resultMap="ShouzhiRecordResultMap">
      select
      rec.szrid,
      rec.szr_num,
      rec.szr_date,
      rec.szr_comment,
      rec.user_id,
      cat.szcid,
      cat.parent_category,
      cat.son_category
      from shouzhi_record    rec
      left join shouzhi_category cat
      on rec.shouzhi_category_id=cat.szcid
      where rec.user_id=#{uid}
      <if test="szr_date != null and szr_date!=''">
          and rec.szr_date like '${szr_date}%'
      </if>
      <if test="szr_comment != null and szr_comment!= '''>
          and rec.szr_comment like '%${szr_comment}%'
      </if>
      order by rec.szr_date desc
      limit #{startPosition},#{pageRecord}
    </select>

    <!-- 删除用户的收支信息（一条） -->
    <delete id="deleteOneShouzhiRecord" parameterType="int">
        delete from shouzhi_record
        where szrid=#{id}
    </delete>

    <!-- 添加收支记录信息（收入 或者 支出） -->
    <insert id="addShouzhiRecord" parameterType="ShouzhiRecord">
        insert into shouzhi_record(szr_num,szr_date,szr_comment,shouzhi_category_id,user_id)
        values(#{szr_num},#{szr_date},#{szr_comment},#{shouzhiCategory.szcid},#{user_id})
    </insert>

    <!-- 通过收支类型的编号确认是收入还是支出 -->
    <select id="findParentCategoryById" parameterType="int" resultType="String">
        select parent_category from shouzhi_category
        where szcid=#{szcid}
    </select>
</mapper>
```

4.9 用户登录模块设计

已经完成注册的用户可以在好生活个人账本的登录页面上直接输入用户名和密码，而后单击"登录"按钮进行登录。程序会校验用户输入的用户名和密码是否正确，如果输入的用户名和密码是正确的，那么页面会跳转到好生活个人账本的首页；如果输入的用户名和密码是错误的，那么在登录页面上会显示"用户名或者密码输入错误，请重新输入"的提示信息。此外，用户可以通过单击"重置"按钮，清空已经输入的用户名或者密码。用户登录模块的效果如图 4.2 所示。

4.9.1 用户控制器类设计

本项目的 UserController 类为用户控制器类，被@Controller 注解标注。在 UserController 类中，定义了一个用于表示用户登录的 login()方法。该方法首先获取用户在登录页面上输入的用户名和密码，然后判断输入的用户名和密码是否存在于 user（用户信息表）中。如果输入的用户名和密码存在于 user 中，那么页面会跳转到好生活个人账本的首页；如果输入的用户名不存在于 user 中，那么在登录页面上会显示"用户名或者密码输入错误，请重新输入"的提示信息。login()方法的代码如下：

图 4.2　用户登录模块的效果图

```
@RequestMapping("login.action")
public String login(User user,HttpServletRequest request){
    System.out.println("用户名和密码： "+user.getUsername()+"::"+user.getPassword());
    //获得用户名和密码，判断是否存在
    User findUser=userService.queryUserByUser(user);

    if(findUser!=null){
        System.out.println("查找的用户名和密码： "+findUser.getUsername()+"::"+findUser.getPassword());
    }
    System.out.println("");
    //用户名和密码存在，将其保存到 session 中，页面跳转到好生活个人账本的首页
    if(findUser!=null){
        HttpSession session = request.getSession();
        session.setAttribute("user", findUser);
        //查询当前用户的收支明细
        return "redirect:/shouzhiRecord/findShouzhiRecord.action";
    }

    //用户名和密码不存在，就显示登录失败的信息
    String msg="用户名或者密码输入错误，重新输入";
    request.setAttribute("msg", msg);
    return "/index.jsp";                                    //跳转到登录页面
}
```

4.9.2 用户服务类设计

本项目的 UserService 接口为用户服务接口。在 login()方法中，调用了一个抽象方法，即 queryUserByUser()方法，该方法是在 UserService 接口中予以定义的。queryUserByUser()方法有一个参数，即 User 类型的对象，表示用户类的对象。queryUserByUser()方法具有返回值，返回值是一个 User 类型的对象。queryUserByUser()方法的代码如下：

```
User queryUserByUser(User user);                    //通过用户名和密码查询用户是否存在
```

本项目的 UserServiceImpl 类是 UserService 接口的实现类，被@Service 注解标注，表示用户服务类。在 UserServiceImpl 类中，重写了 UserService 接口中的 queryUserByUser()方法。该方法的作用是调用 DAO 层

（数据访问对象）来执行数据库操作。重写后的 queryUserByUser()方法的代码如下：

```
@Override
public User queryUserByUser(User user) {
    return userMapper.queryUserByUser(user);
}
```

4.9.3 用户 DAO 层设计

本项目的 UserMapper 接口为用户数据访问对象，即用户 DAO 层。在重写后的 queryUserByUser()方法中，有一个同名且同参的 queryUserByUser()方法，该同名且同参的方法是在 UserMapper 接口中予以定义的。该同名且同参的方法有一个参数，即用户类的对象。该同名且同参的方法具有返回值，返回值是一个 User 类型的对象。UserMapper 接口中的 queryUserByUser()方法的代码如下：

```
User queryUserByUser(User user);                    //通过用户名和密码查询用户是否存在
```

> 在本项目中，与 UserMapper 接口中的方法绑定的 SQL 语句被写在了 UserMapper.xml 文件中。因此，读者可以在 UserMapper.xml 文件中查找与 queryUserByUser()方法绑定的 SQL 语句。

4.10 用户注册模块设计

未注册的用户不能登录到好生活个人账本的首页，须在好生活个人账本的注册页面完成注册操作。在注册页面上，用户须根据提示依次输入用户名、密码、邮箱、手机号、性别等信息。其中，程序会对用户输入的邮箱和手机号进行格式校验。用户只有输入格式正确的邮箱和手机号，才能单击"注册"按钮，以完成注册操作，而后页面会直接跳转到好生活个人账本的首页。用户注册模块的效果如图 4.3 所示。

4.10.1 用户控制器类设计

在 UserController 类（用户控制器类）中，定义了一个用于表示用户注册的 regist()方法。用户在注册页面上输入用户名、密码、邮箱、手机号、性别等信息后，regist()方法首先获取其中的用户名，然后判断该用户名是否存在于 user（用户信息表）中。如果该用户名存在于 user 中，那么注册页面会显示"当前用户已经存在，请重新输入用户名"的提示信息；如果该用户名不存在于 user 中，那么当前用户输入的信息将被添加到 user 中，以完成注册操作，而后页面会直接跳转到好生活个人账本的首页。regist()方法的代码如下：

图 4.3 用户注册模块的效果图

```
@RequestMapping("regist.action")
public String regist(User user,String repassword,HttpServletRequest request){
    //通过用户名查询用户是否存在
    User findUser=userService.queryUserByUsername(user.getUsername());
    if(findUser!=null){
```

```
                //用户名已存在，给出提示信息
                request.setAttribute("msg", "当前用户已经存在，请重新输入用户名");
                request.setAttribute("user", user);                    //保存原来的输入数据
                request.setAttribute("repassword", repassword);
                return"/regist.jsp";                                    //json 格式
        }

        //添加用户
        userService.insertUser(user);
        //保存登录信息
        HttpSession session = request.getSession();
        session.setAttribute("user", user);
        //直接跳转到首页
        return "redirect:/shouzhiRecord/findShouzhiRecord.action";
}
```

4.10.2　用户服务类设计

在 regist()方法中，调用了一个抽象方法，即 queryUserByUsername()方法，该方法是在 UserService 接口（用户服务接口）中予以定义的。queryUserByUsername()方法有一个参数，即 String 类型的 username，表示用户名。queryUserByUser()方法具有返回值，返回值是一个 User 类型的对象，即用户类的对象。queryUserByUsername()方法的代码如下：

```
User queryUserByUsername(String username);                //通过用户名查询用户是否存在
```

在 UserServiceImpl 类（用户服务类）中，重写了 UserService 接口中的 queryUserByUsername()方法。该方法的作用是调用 DAO 层（数据访问对象）来执行数据库操作。重写后的 queryUserByUsername()方法的代码如下：

```
@Override
public User queryUserByUsername(String username) {
    return userMapper.queryUserByUsername(username);
}
```

4.10.3　用户 DAO 层设计

在重写后的 queryUserByUsername()方法中，有一个同名且同参的 queryUserByUsername()方法，该同名且同参的方法是在 UserMapper 接口（即用户 DAO 层）中予以定义的。该同名且同参的方法包含一个参数，即 String 类型的 username。该同名且同参的方法具有返回值，返回值是一个 User 类型的对象。UserMapper 接口中的 queryUserByUsername()方法的代码如下：

```
User queryUserByUsername(String username);                //通过用户名查询用户是否存在
```

 说明

读者可以在 UserMapper.xml 文件中查找与 queryUserByUsername()方法绑定的 SQL 语句。

4.11　收支明细模块设计

用户在登录好生活个人账本的首页后，会发现好生活个人账本的首页有 3 个选项卡，这 3 个选项卡的

名称依次是收支明细、收入记账和支出记账。其中，好生活个人账本的首页默认显示的是收支明细选项卡。在收支明细选项卡上，分页显示了当前用户的收支数据明细，每一条收支明细都包括收支备注、金额、收支日期、收支类型、操作等内容。当前用户可单击"删除"按钮，删除某一条收支明细。收支明细模块的效果如图 4.4 所示。

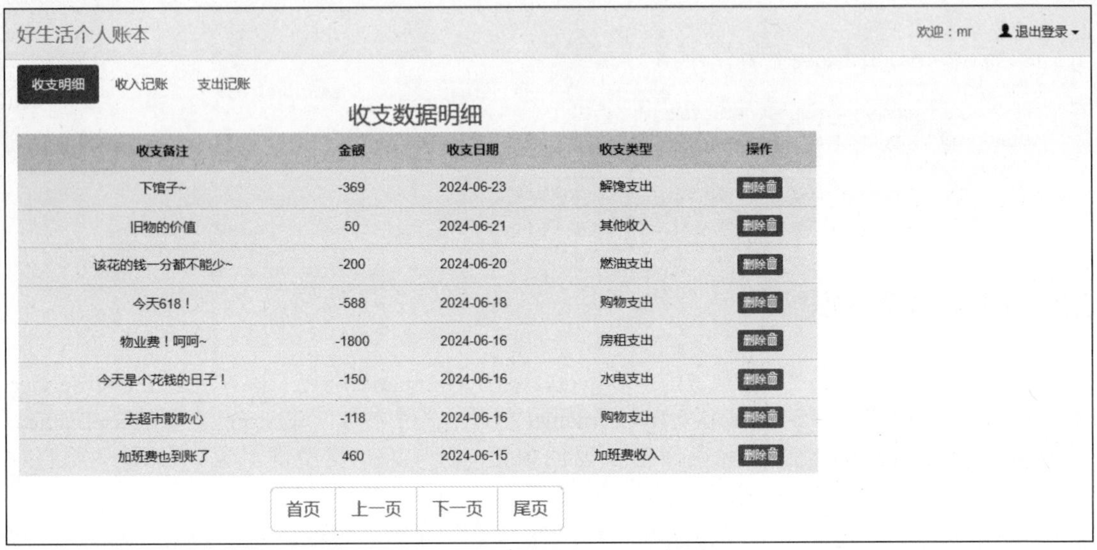

图 4.4 收支明细模块的效果图

4.11.1 收支明细控制器类设计

本项目的 ShouzhiRecordController 类为收支明细控制器类，被 @Controller 注解标注。在 ShouzhiRecordController 类中定义了两个方法：一个是用于表示查找并分页显示收支数据明细的 findShouzhiRecord()方法；另一个是用于表示删除某一条收支数据信息的 deleteOneShouzhiRecord()方法。findShouzhiRecord()方法和 deleteOneShouzhiRecord()方法的代码分别如下：

```
//查找并分页显示收支数据明细
@RequestMapping(value="findShouzhiRecord.action")
public String findShouzhiRecord(ShouzhiRecord shouzhiRecord,HttpServletRequest request)
        throws UnsupportedEncodingException{
    //获取当前页和用户名
    int currentPage=0;
    if(request.getParameter("currentPage")!=null){
        currentPage=Integer.parseInt((String) request.getParameter("currentPage"));
    }
    User user=(User) request.getSession().getAttribute("user");
    if(user==null){
        return "/index.jsp";                                    //登录页面
    }
    //保存查询结果
    if(shouzhiRecord!=null){
        if(shouzhiRecord.getSzr_date()!=null){
            request.setAttribute("date_condition", shouzhiRecord.getSzr_date());
        }
        if(shouzhiRecord.getSzr_comment()!=null){
            String com=new String((shouzhiRecord.getSzr_comment()).getBytes("ISO-8859-1"),"utf-8");
            request.setAttribute("comment_condition", com);
```

```
            shouzhiRecord.setSzr_comment(com);                    //重新赋值
        }
}

//查询账单明细
PageBean<ShouzhiRecord> pageBean= shouzhiRecordService.findShouzhiRecord(currentPage,user,shouzhiRecord);
//查询收入子类型
List<ShouzhiCategory> incomes=shouzhiCategoryService.findShouzhiCategoryByParent("收入");
request.setAttribute("incomes", incomes);
//查询支出子类型
List<ShouzhiCategory> spends=shouzhiCategoryService.findShouzhiCategoryByParent("支出");
request.setAttribute("spends", spends);

if(pageBean.getPageList().size()==0){                             //查询结果为null时,确保数据为空
        pageBean.setPageList(null);
}
request.setAttribute("pageBean", pageBean);                       //分页记录
return "/jsp/main.jsp";                                           //跳转到首页
}

//删除某一条收支数据信息
@RequestMapping("deleteOne.action")
@ResponseBody
public String deleteOneShouzhiRecord(int id){
        shouzhiRecordService.deleteOneShouzhiRecord(id);
        return "OK";
}
```

4.11.2 收支明细服务类设计

本项目的 ShouzhiRecordService 接口为收支明细服务接口。在 ShouzhiRecordController 类的 findShouzhiRecord()方法和 deleteOneShouzhiRecord()方法中，分别调用了一个抽象的、同名的方法，即 findShouzhiRecord()方法和 deleteOneShouzhiRecord()方法，只不过后者都是在 ShouzhiRecordService 接口中予以定义的。ShouzhiRecordService 接口的 findShouzhiRecord()方法和 deleteOneShouzhiRecord()方法的代码分别如下：

```
//查询账单明细
PageBean<ShouzhiRecord> findShouzhiRecord(int currentPage, User user, ShouzhiRecord shouzhiRecord);
//删除用户收支信息（一条）
void deleteOneShouzhiRecord(int id);
```

本项目的 ShouzhiRecordServiceImpl 类是 ShouzhiRecordService 接口的实现类，被@Service 注解标注，表示收支明细服务类。在 ShouzhiRecordServiceImpl 类中，分别重写了 ShouzhiRecordService 接口的 findShouzhiRecord()方法和 deleteOneShouzhiRecord()方法。这两个方法的作用都是调用 DAO 层（数据访问对象）来执行数据库操作。重写后的 findShouzhiRecord()方法和 deleteOneShouzhiRecord()方法的代码如下：

```
//查询账单明细
@Override
public PageBean<ShouzhiRecord> findShouzhiRecord(int currentPage, User user, ShouzhiRecord shouzhiRecord) {
        int pageRecord = 8;                                       //每页记录数
        int startPosition=0;                                      //开始位置
        //因为当前页 currentPage 比实际的当前页少 1
        //所以当 currentPage=0 时,当前页是第 1 页;当 currentPage=1 时,当前页是第 2 页
        if(currentPage!=0){
                //失败
                //startPosition = (currentPage - 1) * pageRecord;  //开始位置
```

```java
            startPosition = currentPage * pageRecord;          //开始位置
        }
        //总记录数
        int allRecord = 0;
        //通过查询，获得总记录数
        Map<String,Object> map=new HashMap<String,Object>();
        map.put("user", user);
        if(shouzhiRecord!=null){
            if(shouzhiRecord.getSzr_date()!=null&&shouzhiRecord.getSzr_date()!=""){
                map.put("szr_date", shouzhiRecord.getSzr_date());
            }
            if(shouzhiRecord.getSzr_comment()!=null&&shouzhiRecord.getSzr_comment()!=""){
                map.put("szr_comment", shouzhiRecord.getSzr_comment());
            }
        }
        allRecord=shouzhiRecordMapper.findShouzhiRecordCount(map);
        //总页数
        int allPage = allRecord / pageRecord;
        if (allRecord % pageRecord != 0)
            allPage = allPage + 1;
        ShouzhiRecordQueryVo queryVo=new ShouzhiRecordQueryVo();
        queryVo.setUid(user.getUid());
        queryVo.setStartPosition(startPosition);
        queryVo.setPageRecord(pageRecord);
        //保存查询结果
        if(shouzhiRecord!=null){
            if(shouzhiRecord.getSzr_date()!=null&&shouzhiRecord.getSzr_date()!=""){
                queryVo.setSzr_date(shouzhiRecord.getSzr_date());
            }
            if(shouzhiRecord.getSzr_comment()!=null&&shouzhiRecord.getSzr_comment()!=""){
                queryVo.setSzr_comment(shouzhiRecord.getSzr_comment());
            }
        }

        //收支明细
        List<ShouzhiRecord> pageList=shouzhiRecordMapper.findCurrenPageRecordList(queryVo);

        //分页查询
        PageBean<ShouzhiRecord> pageBean = new PageBean<ShouzhiRecord>();
        pageBean.setAllPage(allPage);
        pageBean.setAllRecord(allRecord);
        pageBean.setCurrentPage(currentPage);
        pageBean.setPageList(pageList);
        pageBean.setPageRecord(pageRecord);
        pageBean.setStartPosition(startPosition);

        return pageBean;
}

//删除某一条收支数据信息
@Override
public void deleteOneShouzhiRecord(int id) {
    shouzhiRecordMapper.deleteOneShouzhiRecord(id);
}
```

4.11.3 收支明细 DAO 层设计

本项目的 shouzhiRecordMapper 接口为收支明细数据访问对象，即收支明细 DAO 层。在重写后的 findShouzhiRecord()方法和 deleteOneShouzhiRecord()方法中，依次调用了 shouzhiRecordMapper 接口的 3 个

方法，即findShouzhiRecordCount()、findCurrenPageRecordList()和deleteOneShouzhiRecord()方法。这3个方法的代码分别如下：

```
//查询用户的收支明细记录数
int findShouzhiRecordCount(Map<String, Object> map);
//分页查询当前用户的当前页记录数
List<ShouzhiRecord> findCurrenPageRecordList(ShouzhiRecordQueryVo queryVo);
//删除用户收支信息（一条）
void deleteOneShouzhiRecord(int id);
```

说明

读者可以在ShouzhiRecordMapper.xml文件中查找与findShouzhiRecordCount()、findCurrenPageRecordList()和deleteOneShouzhiRecord()方法绑定的SQL语句。

4.12 收入记账模块设计

用户在登录好生活个人账本的首页后，单击收入记账选项卡，首页将显示收入记账选项卡的相关元素。在收入记账选项卡上，用户输入收入备注、收入金额、收入日期、收入类型等信息后，单击"提交收入信息"按钮，即可完成新增收入项的操作，页面会跳转至首页，并显示收支明细选项卡的相关元素。另外，用户可以通过单击"添加收入类型"按钮新增收入类型。收入记账模块的效果如图4.5所示。

图4.5 收入记账模块的效果图

4.12.1 收支明细控制器类设计

在 ShouzhiRecordController 类（收支明细控制器类）中，定义了一个用于表示新增收支明细的addShouzhiRecord()方法。该方法既可以执行新增收入项的操作，又可以执行新增支出项的操作。区别在于，如果执行的操作是新增收入项，那么在收支明细选项卡上显示的金额是正整数；如果执行的操作是新增支出项，那么在收支明细选项卡上显示的金额是负整数。因为在收入记账选项卡上执行的操作是新增收入项，所以addShouzhiRecord()方法的代码如下：

```
@RequestMapping("addShouzhiRecord.action")
```

```java
public String addShouzhiRecord(ShouzhiRecord shouzhiRecord,HttpServletRequest request) throws IOException{
    int szcid = shouzhiRecord.getShouzhiCategory().getSzcid();        //获取收支类型的编号
    String cat=shouzhiRecordService.findParentCategoryById(szcid);    //通过收支类型的编号，查询是收入 还是 支出
    if("支出".equals(cat)){
        //省略"如果是支出，金额为负数"的代码
    }
    else{                                                              //如果是收入
        int num=shouzhiRecord.getSzr_num();
        if(num<=0){
            shouzhiRecord.setSzr_num(-num);                            //金额为正数
        }
    }
    shouzhiRecordService.addShouzhiRecord(shouzhiRecord);
    return "redirect:/shouzhiRecord/findShouzhiRecord.action";
}
```

4.12.2　收支明细服务类设计

在 addShouzhiRecord()方法中，依次调用了两个抽象方法，即 findParentCategoryById()方法和 addShouzhiRecord()方法，它们都是在 ShouzhiRecordService 接口中予以定义的。ShouzhiRecordService 接口的 findParentCategoryById()方法和 addShouzhiRecord()方法的代码分别如下：

```java
//通过收支类型的编号，查询是收入 还是 支出
String findParentCategoryById(int szcid);
//新增收支明细（收入 或者 支出）
void addShouzhiRecord(ShouzhiRecord shouzhiRecord);
```

在 ShouzhiRecordServiceImpl 类（收支明细服务类）中，分别重写了 ShouzhiRecordService 接口的 findParentCategoryById()方法和 addShouzhiRecord()方法。这两个方法的作用都是调用 DAO 层（数据访问对象）来执行数据库操作。重写后的 findParentCategoryById()方法和 addShouzhiRecord()方法的代码分别如下：

```java
//通过收支类型的编号，查询是收入 还是 支出
@Override
public String findParentCategoryById(int szcid) {
    // TODO Auto-generated method stub
    return shouzhiRecordMapper.findParentCategoryById(szcid);
}
//新增收支明细（收入 或者 支出）
@Override
public void addShouzhiRecord(ShouzhiRecord shouzhiRecord) {
    shouzhiRecordMapper.addShouzhiRecord(shouzhiRecord);
}
```

4.12.3　收支明细 DAO 层设计

在重写后的 findParentCategoryById()方法和 addShouzhiRecord()方法中，依次调用了 shouzhiRecordMapper 接口（收支明细 DAO 层）的 findParentCategoryById()和 addShouzhiRecord()方法。这两个方法的代码分别如下：

```java
//通过收支类别 id 查询是收入 还是 支出
String findParentCategoryById(int szcid);
//添加收支记录信息（收入 或者 支出）
void addShouzhiRecord(ShouzhiRecord shouzhiRecord);
```

好生活个人账本　第4章

说明

读者可以在 ShouzhiRecordMapper.xml 文件中查找与 findParentCategoryById()和 addShouzhiRecord()方法绑定的 SQL 语句。

4.13　支出记账模块设计

用户在登录好生活个人账本的首页后，单击支出记账选项卡，首页将显示支出记账选项卡的相关元素。在支出记账选项卡上，用户输入支出备注、支出金额、支出日期、支出类型等信息后，单击"提交支出信息"按钮，即可完成新增支出项的操作，页面会跳转至首页，并显示收支明细选项卡的相关元素。另外，用户可以通过单击"添加支出类型"按钮新增支出类型。支出记账模块的效果如图 4.6 所示。

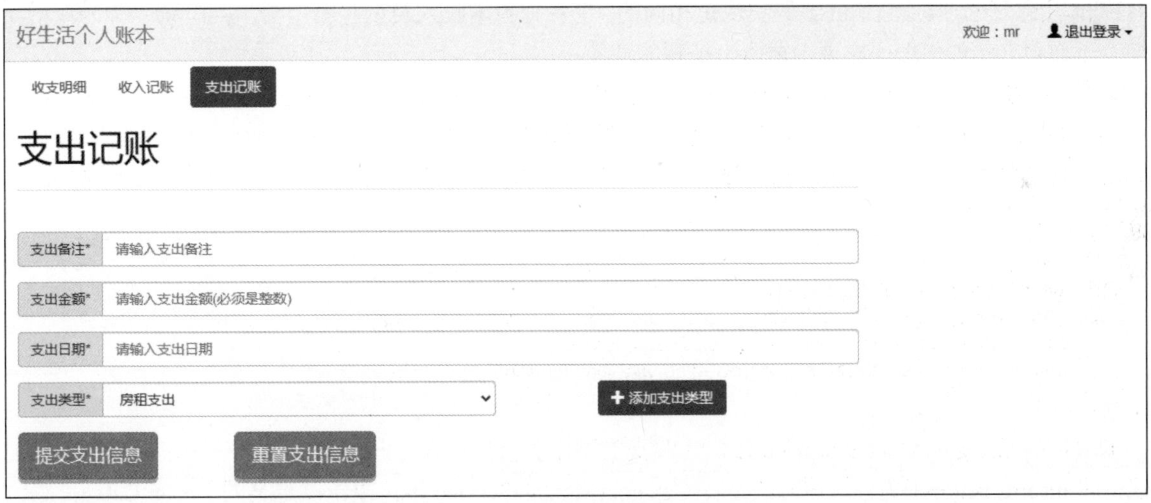

图 4.6　支出记账模块的效果图

4.13.1　收支明细控制器类设计

在 ShouzhiRecordController 类（收支明细控制器类）中，addShouzhiRecord()方法既可以执行新增收入项的操作，又可以执行新增支出项的操作。因为在支出记账选项卡上执行的操作是新增支出项，所以 addShouzhiRecord()方法的代码如下：

```
@RequestMapping("addShouzhiRecord.action")
public String addShouzhiRecord(ShouzhiRecord shouzhiRecord,HttpServletRequest request) throws IOException{
    int szcid = shouzhiRecord.getShouzhiCategory().getSzcid();      //获取收支类型的编号
    String cat=shouzhiRecordService.findParentCategoryById(szcid);  //通过收支类型的编号，查询是收入 还是 支出
    if("支出".equals(cat)){                                         //如果是支出
        int num=shouzhiRecord.getSzr_num();
        if(num>=0){
            shouzhiRecord.setSzr_num(-num);                         //金额为负数
        }
    }
    else{
```

121

```
            //省略"如果是收入,金额为正数"的代码
        }
        shouzhiRecordService.addShouzhiRecord(shouzhiRecord);
        return "redirect:/shouzhiRecord/findShouzhiRecord.action";
}
```

收入记账模块和支出记账模块除了在收支明细控制器类设计上有所不同,在收支明细服务类设计和收支明细 DAO 层设计上皆相同。因此,本节将不介绍支出记账模块的收支明细服务类和收支明细 DAO 层的设计过程。

4.13.2 其他功能模块设计

在图 4.9 中,有一个"添加收入类型"按钮,该按钮用于新增收入类型(如"加班费收入"等)。在图 4.10 中,有一个"添加支出类型"按钮,该按钮用于新增支出类型(如"水电支出"等)。虽然这两个按钮的名称和功能不同,但是它们的编码实现是相同的。笔者把新增收入类型和新增支出类型统一称为新增收支类型。下面将介绍如何编码实现上述两个按钮的功能。

本项目的 CategoryManageController 类为收支类型控制器类,被 @Controller 注解标注。在 CategoryManageController 类中,定义了一个用于表示新增收支类型的 addCategory()方法。程序调用该方法完成新增收支类型的操作后,页面仍将显示收入记账选项卡或者支出记账选项卡的相关元素。addCategory() 方法的代码如下:

```
//新增收支类型
@RequestMapping("/addCategory.action")
public String addCategory(ShouzhiCategory shouzhiCategory,Integer currentPage){         //添加后,仍旧返回当前页
    categoryManageService.insertCategory(shouzhiCategory);
    return "redirect:/categoryManage/findCategorys.action?currentPage="+currentPage;    //重定向
}
```

本项目的 CategoryManageService 接口为收支类型服务接口。在 addCategory()方法中,调用了一个抽象方法,即 insertCategory()方法,该方法是在 CategoryManageService 接口中予以定义的。insertCategory()方法有一个参数,即 ShouzhiCategory 类型的对象,表示收支类型对象。insertCategory()方法的代码如下:

```
//新增收支类型
void insertCategory(ShouzhiCategory shouzhiCategory);
```

本项目的 CategoryManageServiceImpl 类是 CategoryManageService 接口的实现类,被@Service 注解标注,表示收支类型服务类。在 CategoryManageServiceImpl 类中,重写了 CategoryManageService 接口中的 insertCategory()方法。该方法的作用是调用 DAO 层(数据访问对象)来执行数据库操作。重写后的 insertCategory()方法的代码如下:

```
//新增收支类型
@Override
public void insertCategory(ShouzhiCategory shouzhiCategory) {
    categoryManageMapper.insertCategory(shouzhiCategory);
}
```

本项目的 CategoryManageMapper 接口为收支类型数据访问对象,即收支类型 DAO 层。在重写后的 insertCategory()方法中,有一个同名且同参的 insertCategory()方法,该同名且同参的方法是在 CategoryManageMapper 接口中予以定义的。CategoryManageMapper 接口中的 insertCategory()方法的代码

如下：

```
//新增收支类型
void insertCategory(ShouzhiCategory shouzhiCategory);
```

说明
读者可以在 CategoryManageMapper.xml 文件中查找与 insertCategory()方法绑定的 SQL 语句。

4.14 退出登录模块设计

用户在登录好生活个人账本的首页后，可以单击"退出登录"超链接，退出好生活个人账本的首页，页面将跳转到好生活个人账本的登录页面。退出登录模块的效果如图 4.7 所示。

图 4.7 退出登录模块的效果图

在 UserController 类（用户控制器类）中，定义了一个用于表示用户退出登录的 logout()方法。用户在首页上单击"退出登录"超链接后，程序将删除 session 数据（即服务器端存储的用户会话数据），页面将跳转到好生活个人账本的登录页面。logout()方法的代码如下：

```
@RequestMapping("/logout.action")
public String logout(HttpServletRequest request){
    request.getSession().removeAttribute("user");          //删除 session 数据
    return "/index.jsp";
}
```

4.15 项 目 运 行

通过前述步骤，设计并完成了"好生活个人账本"项目的开发。下面运行本项目，以检验我们的开发成果。如图 4.8 所示，在 IntelliJ IDEA 中，单击█快捷图标，即可运行本项目。

图 4.8 IntelliJ IDEA 的快捷图标

成功运行本项目，程序会自动打开如图 4.9 所示的好生活个人账本的登录页面。新用户须事先完成注册操作。新用户成功注册或者老用户成功登录后，页面将跳转到好生活个人账本的首页。好生活个人账本的首页有 3 个选项卡：在收支明细选项卡上，程序将分页显示当前用户以往的收支明细（即收入项和支出项）；在收入记账选项卡上，当前用户可执行新增收入项的操作；在支出记账选项卡上，当前用户可完成新增支出项的操作；在首页上，用户单击"退出登录"超链接，即可退出好生活个人账本的首页，页面将跳转到好生活个人账本的登录页面。这样，我们就成功地检验了本项目的运行。

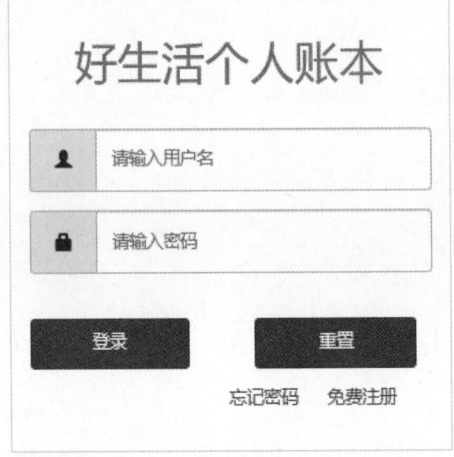

图 4.9　好生活个人账本的登录页面

本项目主要练习了如何开发 SSM 框架的 3 层内容，它们分别是 Controller 层（控制器类）、Service 层（服务类）和 DAO 层（数据访问对象）。其中，Controller 层主要负责具体的业务模块流程的控制，调用 Service 层提供的接口来控制业务流程，根据业务流程的不同会有不同的控制器；Service 层主要负责业务模块的逻辑应用设计，先设计接口，再设计实类。DAO 层主要负责与数据库进行交互，首先设计的是接口，然后可以在模块中进行接口的调用来进行数据业务的处理。

4.16　源码下载

虽然本章详细地讲解了如何编码实现"好生活个人账本"项目的各个功能，但给出的代码都是代码片段，而非源码。为了方便读者学习，本书提供了完整的项目源码，扫描右侧二维码即可下载。

第 5 章 嗨乐影评平台

——SSM + JSP + MySQL

影评平台是电影爱好者获取电影名称、电影类型、剧情介绍、演职人员介绍、电影评论等信息的在线平台。这些平台旨在以简洁清晰的页面增强用户体验,以平民化的视角解读电影,便于电影爱好者发表各类原创的观后感,分享关于电影的优秀评论。本章将使用 Java Web 开发中的 SSM 框架、JSP 和 MySQL 数据库等关键技术开发一个影评平台项目——嗨乐影评平台。

项目微视频

本项目的核心功能及实现技术如下:

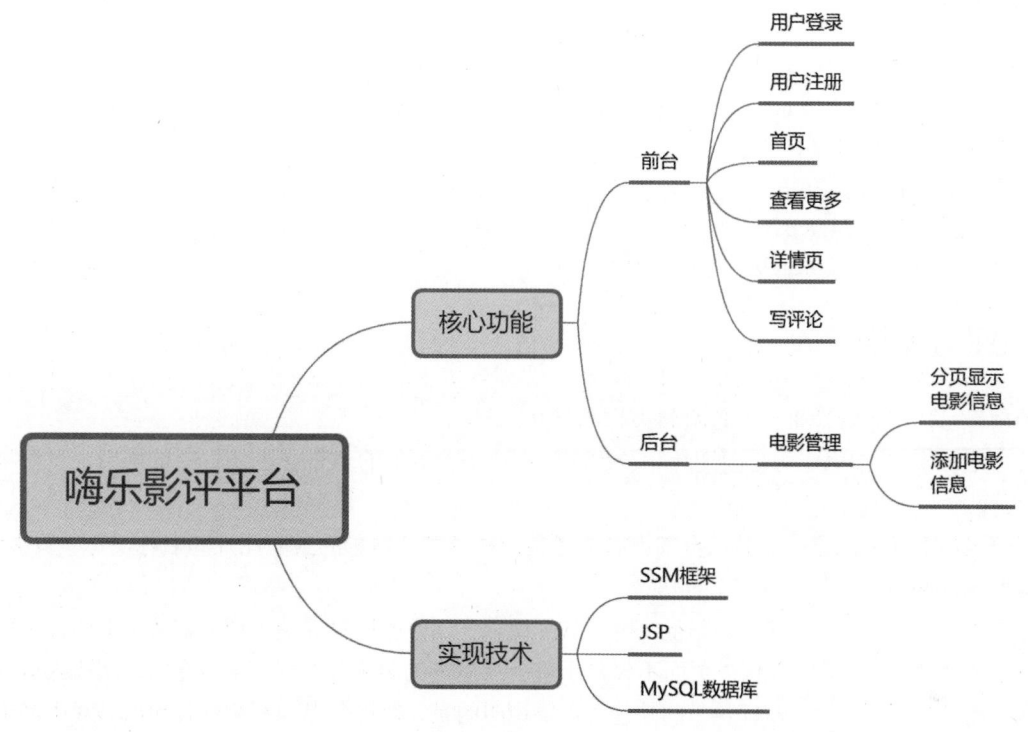

5.1 开发背景

电影是有内涵的,一部好的电影是值得回味的。电影评论的内容多种多样,任何一篇电影评论都是对一部电影进行总结的观后感,带有强烈的个人属性,代表观影者自己的思考与感受。一篇好的电影评论更是值得电影爱好者驻足。电影评论的目的在于分析、鉴定和评价蕴含在一部电影中的审美价值、认识价值、社会意义、镜头语言等内容。通过有理有据地分析和评价一部电影,既能够帮助电影爱好者更好地理解这部电影，又能够提高电影爱好者的鉴赏水平，还能够间接地促进电影艺术的繁荣和发展。嗨乐影评平台就是一个用于电影爱好者分享对电影的看法和感受的平台。

嗨乐影评平台将实现以下目标：
- ☑ 页面简洁清晰、操作简单方便；
- ☑ 支持用户登录和用户注册；
- ☑ 平台上的所有电影会按照电影类型（如剧情、冒险、科幻等）进行分类；
- ☑ 用户既可以查看平台上的所有电影，也可以查看某一部电影的名称、类型、剧情介绍、演职人员等信息和评论内容，还可以为某一部电影编写新的评论；
- ☑ 管理员可以对平台上的所有电影进行管理，如查询电影、添加电影等。

5.2 系统设计

5.2.1 开发环境

本项目的开发及运行环境如下：
- ☑ 操作系统：推荐 Windows 10、11 及以上，兼容 Windows 7（SP1）。
- ☑ 开发工具：IntelliJ IDEA。
- ☑ 开发语言：Java EE。
- ☑ 数据库：MySQL 8.0。
- ☑ Web 服务器：Tomcat 9.0 及以上版本。

5.2.2 业务流程

启动项目后，嗨乐影评平台的首页将被自动打开。

在首页上，有 5 部电影的封面图片和名称。用户单击首页上的"查看更多"超链接后，页面将跳转到"更多"页面。

在"更多"页面上，将分页显示全部电影或者某一种类型的电影的封面图片和名称，每一页显示 15 部电影的封面图片和名称。

不论在首页上还是在"更多"页面上，用户单击任意一部电影的封面图片或者名称，页面都可以跳转到当前电影的"详情"页面。"详情"页面不仅显示着当前电影的名称、类型、剧情介绍、演职人员等信息和评论内容，而且为用户提供"写评论"的功能。需要说明的是，在提交已经写完的评论时，对于尚未登录的用户，页面会跳转到"登录"页面，提示用户完成登录操作；对于尚未注册的用户，须先完成注册操作，再完成登录操作。

为了避免上述"先操作再登录"的情况，嗨乐影评平台在首页上提供了"用户登录"的功能。对于已经注册的用户，可以直接输入正确的用户名和密码，完成登录操作。

对于尚未注册的用户，应先注册，再登录。用户完成登录操作后，页面将跳转到首页。需要说明的是，如果用户输入的用户名是 admin、密码是 123456，那么程序会把当前用户视作嗨乐影评平台的管理员。嗨乐影评平台的管理员可以在"后台管理"页面对平台上的所有电影进行管理，如查询电影、添加电影等。

嗨乐影评平台的业务流程如图 5.1 所示。

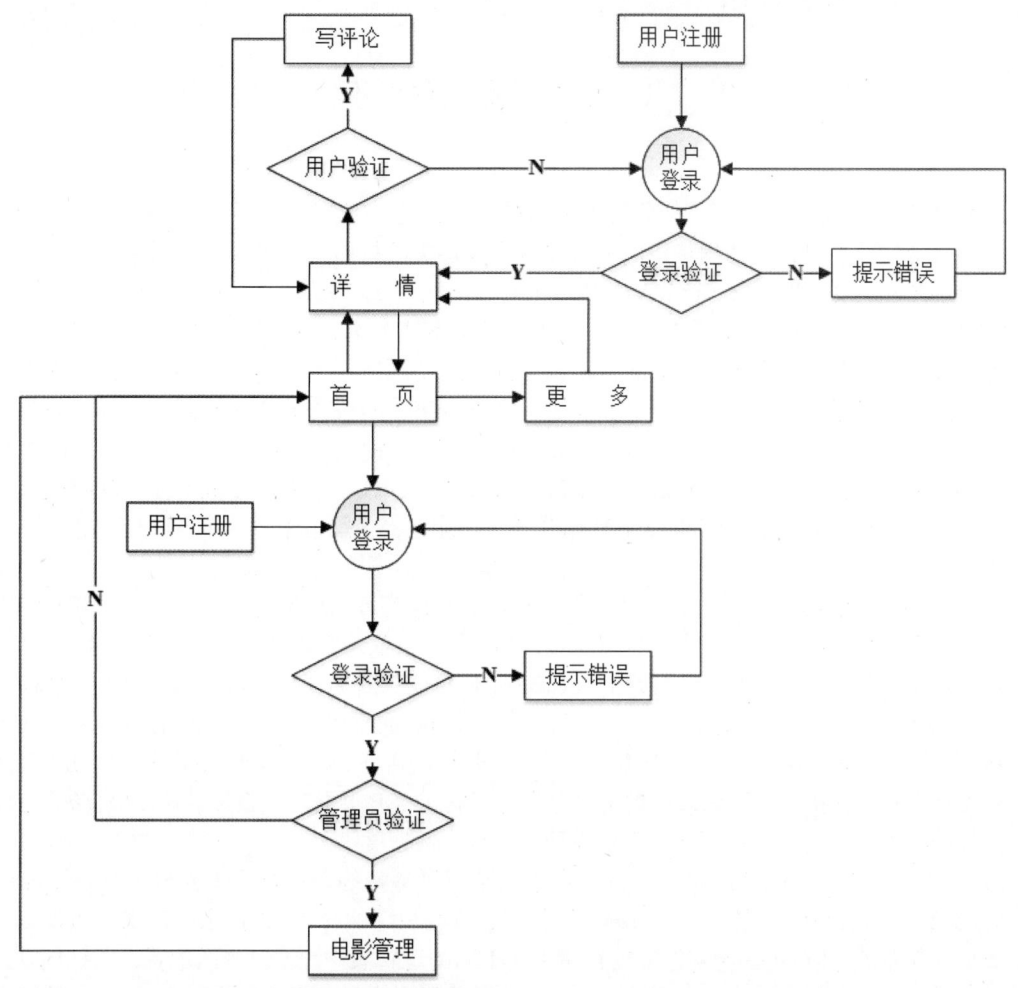

图 5.1　嗨乐影评平台的业务流程图

5.2.3　功能结构

本项目的功能结构已经在章首页中给出。作为用于电影爱好者分享对电影的观后感的平台，本项目实现的具体功能如下：

- ☑ 用户登录：支持普通用户和管理员的登录操作。
- ☑ 用户注册：支持新用户的注册操作。
- ☑ 首页：不仅显示 5 部电影的封面图片和名称，而且提供"查看更多"超链接，还提供了"用户登录"的功能。
- ☑ 更多：将分页显示全部电影或者某一种类型的电影的封面图片和名称，每一页显示 15 部电影的封面图片和名称。
- ☑ 详情：不仅显示当前电影的名称、类型、剧情介绍、演职人员等信息和评论内容，而且为用户提供"写评论"的功能。
- ☑ 写评论：用户编写并提交电影的观后感。
- ☑ 电影管理：对平台上的所有电影进行管理，如查询电影、添加电影等。

5.3 技术准备

- ☑ SSM 框架：Spring 是轻量级的、开源的 JavaEE 框架，它类似一个工厂，用于装配和生产 JavaBean，在配置文件中可以指定使用特定的参数去调用实体类的构造方法来实例化对象，它大大降低了企业应用开发的复杂度。Spring MVC 是 Spring 的一个后续产品，是 Spring 的一个子项目。Spring MVC 是 Spring 为表述层开发提供的一整套完备的解决方案。目前业界普遍选择了 Spring MVC 作为 Java EE 项目表述层开发的首选方案。表述层是三层架构之一，另外的两层架构分别为业务逻辑层和数据访问层。MyBatis 是一个支持定制化的 SQL、存储过程以及高级映射的优秀的持久层框架，它避免了几乎所有的通过 JDBC 代码获取数据库结果集的情况，可以使用简单的 XML 配置和注解，用于配置和原始对象映射，将接口和 Java 的 POJO（普通的 Java 对象）映射成数据库中的记录。MyBatis 将 SQL 语句和 Java 代码分开，功能边界更加清晰，一个专注于业务，一个专注于数据。
- ☑ JSP 技术：因为 JSP 是基于 Java 语言的，所以 JSP 也是可以跨平台的。采用 JSP 开发的项目，能够将业务层和视图层分离开来。也就是说，JSP 只负责显示数据即可，即使业务代码被修改，JSP 页面也不受影响。JSP 可以使用一个 JavaBean 封装业务处理代码或者作为一个数据处理模型，这个 JavaBean 既可以被重复使用，也可以将其应用到其他应用程序中。JSP 的本质是 Servlet，因此 JSP 具有 Servlet 的基础功能。当用户通过浏览器首次访问 JSP 页面时，服务器对 JSP 页面代码进行编译，并且仅执行一次编译；当用户再次访问同一 JSP 页面时，服务器直接执行编译过的代码，这样既可以节约服务器资源，又可以提升客户端访问速度。
- ☑ MySQL 数据库：MySQL 最大的优势在于它是开源免费的，可以极大减少项目建设的成本。把 MySQL 作为后端的数据库管理系统，不仅能够满足快速响应 Web 请求的需求，而且支持安全的数据存储服务。MySQL 能够将数据存储于在不同的逻辑表中，通过修改配置、字符集等方式，能够支持多种开发语言。本项目用于连接 MySQL 数据库的驱动程序和路径、MySQL 数据库的用户名和密码都被存储在 db.properties 文件中。代码如下：

```
jdbc.driver=com.mysql.cj.jdbc.Driver
jdbc.url=jdbc:mysql://localhost:3306/movie?useUnicode=true&characterEncoding=utf8
jdbc.username=root
jdbc.password=root
```

有关 MySQL 数据库的知识在《Java 从入门到精通（第 7 版）》中有详细的讲解，对这些知识不太熟悉的读者可以参考该书对应的内容。有关 JSP 技术的内容在本书第 1 章的"技术准备"中有详细的介绍，读者可以参考第 1 章的相关内容进行学习。有关 SSM 框架的内容在本书第 3 章和第 4 章均有介绍，读者可以参考这两章的相关内容进行学习。

5.4 数据库设计

5.4.1 数据库概述

本项目采用的数据库主要包含 5 张数据表，如表 5.1 所示。

表 5.1 嗨乐影评平台的数据库结构

表 名	表 说 明	表 名	表 说 明
t_comment	电影评论表	t_sort_movie	电影类型与电影关联表
t_movie	电影信息表	t_userinfo	用户信息表
t_sort	电影类型表		

5.4.2 数据表设计

表 5.1 中除了 t_userinfo（用户信息表），其余 4 张表都采用了主键自增原则，下面将详细介绍这 5 张表的结构设计。

- ☑ t_comment（电影评论表）：主要用于存储与每一条评论对应的编号、电影的编号、内容、账户名、时间等信息。该数据表的结构如表 5.2 所示。

表 5.2 t_comment 表结构

字 段 名 称	数据类型	长 度	是否主键	说 明
commentId	INT		主键	评论的编号
movieObj	INT			电影的编号
content	TEXT			评论的内容
userObj	VARCHAR	20		账户名
commentTime	VARCHAR	20		评论的时间

- ☑ t_movie（电影信息表）：主要用于存储电影的相关信息。该数据表的结构如表 5.3 所示。

表 5.3 t_movie 表结构

字 段 名 称	数据类型	长 度	是否主键	说 明
movieId	INT		主键	电影的编号
movieName	VARCHAR	20		电影的名称
moviePhoto	VARCHAR	20		电影的封面图片
director	VARCHAR	20		电影的导演
mainPerformer	VARCHAR			电影的演职人员
duration	VARCHAR	20		电影的时长
area	VARCHAR	20		电影的区域
releaseDate	VARCHAR	20		电影的上映时间
price	VARCHAR	20		电影的票价
hit	INT			电影的点击量
opera	TEXT			电影的剧情简介

- ☑ t_sort（电影类型表）：主要用于存储电影类型的编号和名称。该数据表的结构如表 5.4 所示。

表 5.4 t_sort 表结构

字 段 名 称	数据类型	长 度	是否主键	说 明
id	INT		主键	电影类型的编号
sorts	VARCHAR	10		电影类型的名称

- ☑ t_sort_movie（电影类型与电影关联表）：主要用于存储电影类型与电影关联的编号、电影的编号

和电影类型的编号。该数据表的结构如表 5.5 所示。

表 5.5 t_sort_movie 表结构

字 段 名 称	数 据 类 型	长　　度	是 否 主 键	说　　　明
id	INT		主键	电影类型与电影关联的编号
movieid	INT			电影的编号
sortsid	INT			电影类型的编号

☑ t_userinfo（用户信息表）：主要用于存储用户的相关信息。该数据表的结构如表 5.6 所示。

表 5.6 t_userinfo 表结构

字 段 名 称	数 据 类 型	长　　度	是 否 主 键	说　　　明
user_name	VARCHAR	20	主键	账户名
password	VARCHAR	255		密码
userPhoto	VARCHAR	60		用户的头像
userType	VARCHAR	20		用户的类型（ROLE_USER/ROLE_ADMIN）
regTime	VARCHAR	20		用户的注册时间

5.5 SSM 框架的主要配置文件

5.5.1 Spring 的配置文件

本项目中的 applicationContext.xml 文件是 Spring 的配置文件，其中主要包含如下配置：

☑ 扫描器配置：配置需要自动扫描的包路径。
☑ 数据源配置：配置数据库的连接信息。
☑ 事务管理器配置：用于支持数据库事务操作。
☑ 其他模块配置：配置 sqlSessionfactory 工厂和 AOP 增强。

applicationContext.xml 文件中的代码如下：

```xml
<?xml version="1.0" encoding="UTF-8"?>
<beans xmlns="http://www.springframework.org/schema/beans"
       xmlns:xsi="http://www.w3.org/2001/XMLSchema-instance"
       xmlns:context="http://www.springframework.org/schema/context"
       xmlns:aop="http://www.springframework.org/schema/aop"
       xmlns:tx="http://www.springframework.org/schema/tx"
       xsi:schemaLocation="http://www.springframework.org/schema/beans
       http://www.springframework.org/schema/beans/spring-beans.xsd
       http://www.springframework.org/schema/context
       http://www.springframework.org/schema/context/spring-context.xsd
       http://www.springframework.org/schema/aop
       http://www.springframework.org/schema/aop/spring-aop.xsd
       http://www.springframework.org/schema/tx
       http://www.springframework.org/schema/tx/spring-tx.xsd">

    <!--配置需要自动扫描的包路径-->
    <context:component-scan base-package="com.zxl">
        <context:exclude-filter type="annotation" expression="org.springframework.stereotype.Controller"/>
    </context:component-scan>
```

```xml
<!--配置数据库的连接信息-->
<context:property-placeholder location="classpath:db.properties"/>
<!--配置链接池-->
<bean id="dataSource" class="org.springframework.jdbc.datasource.DriverManagerDataSource">
    <property name="driverClassName" value="${jdbc.driver}"/>
    <property name="url" value="${jdbc.url}"/>
    <property name="username" value="${jdbc.username}"/>
    <property name="password" value="${jdbc.password}"/>
</bean>
<!--配置sqlSessionfactory工厂-->
<bean id="sqlSessionFactory" class="org.mybatis.spring.SqlSessionFactoryBean">
    <property name="dataSource" ref="dataSource"/>
    <!--取别名-->
    <property name="typeAliasesPackage" value="com.zxl.entity"/>
    <property name="configLocation" value="classpath:mybatis-config.xml" />
    <!--配置xml文件路径-->
    <property name="mapperLocations" value="classpath:mapper/*.xml"/>
    <!--配置分页插件-->
    <property name="plugins">
        <array>
            <bean class="com.github.pagehelper.PageInterceptor">
                <property name="properties">
                    <!--使用下面的方式配置参数，一行配置一个 -->
                    <props>
                        <prop key="helperDialect">mysql</prop>
                        <prop key="reasonable">true</prop>
                    </props>
                </property>
            </bean>
        </array>
    </property>
</bean>
<!--配置dao接口所在包-->
<bean id="mapperScannerConfigurer" class="org.mybatis.spring.mapper.MapperScannerConfigurer">
    <property name="basePackage" value="com.zxl.dao"/>
</bean>
<!--配置事务管理器-->
<bean id="dataSourceTransactionManager"
 class="org.springframework.jdbc.datasource.DataSourceTransactionManager">
    <property name="dataSource" ref="dataSource"/>
</bean>
<!--配置事务通知-->
<tx:advice id="txAdvice" transaction-manager="dataSourceTransactionManager">
    <tx:attributes>
        <tx:method name="find*" read-only="true"/>
        <tx:method name="*" isolation="DEFAULT"/>
    </tx:attributes>
</tx:advice>

<!--配置AOP增强-->
<aop:config>
    <aop:advisor advice-ref="txAdvice" pointcut="execution(* com.zxl.service.impl.*ServiceImpl.*(..))"/>
</aop:config>
</beans>
```

5.5.2 Spring MVC 的配置文件

本项目中的 springmvc.xml 文件是 Spring MVC 的配置文件，其中主要包含如下配置：

- ☑ 控制器配置：配置控制器（Controller）的扫描路径，使得 Spring MVC 能够自动发现和管理相应的控制器。
- ☑ 视图解析器配置：用于将模型（model）渲染成具体的响应视图。
- ☑ 静态资源配置：配置处理静态资源的映射路径，使得访问静态资源时能够正确加载。
- ☑ 其他模块配置：如配置注解驱动等。

springmvc.xml 文件中的代码如下：

```xml
<?xml version="1.0" encoding="UTF-8"?>
<beans xmlns="http://www.springframework.org/schema/beans"
    xmlns:mvc="http://www.springframework.org/schema/mvc"
    xmlns:context="http://www.springframework.org/schema/context"
    xmlns:xsi="http://www.w3.org/2001/XMLSchema-instance"
    xmlns:aop="http://www.springframework.org/schema/aop"
    xsi:schemaLocation="
        http://www.springframework.org/schema/beans
        http://www.springframework.org/schema/beans/spring-beans.xsd
        http://www.springframework.org/schema/mvc
        http://www.springframework.org/schema/mvc/spring-mvc.xsd
        http://www.springframework.org/schema/context
        http://www.springframework.org/schema/context/spring-context.xsd
        http://www.springframework.org/schema/aop
        http://www.springframework.org/schema/aop/spring-aop.xsd">
    <!--配置控制器的扫描路径-->
    <context:component-scan base-package="com.zxl">
        <context:include-filter type="annotation" expression="org.springframework.stereotype.Controller"/>
    </context:component-scan>
    <!--配置视图解析器-->
    <bean id="internalResourceViewResolver"
        class="org.springframework.web.servlet.view.InternalResourceViewResolver">
        <property name="prefix" value="/pages/"/>
        <property name="suffix" value=".jsp"/>
    </bean>
    <!--配置处理静态资源的映射路径-->
    <mvc:resources location="/static/" mapping="/static/**" />

    <mvc:annotation-driven/>

    <!--
        支持 AOP 的注解支持，AOP 底层使用代理技术
        JDK 动态代理，要求必须有接口
        cglib 代理，生成子类对象，proxy-target-class="true" 默认使用 cglib 的方式
    -->
    <aop:aspectj-autoproxy proxy-target-class="true"/>
    <!-- 配置文件上传解析器 --><!-- id 的值是固定的-->
    <bean id="multipartResolver" class="org.springframework.web.multipart.commons.CommonsMultipartResolver">
        <!-- 设置上传文件的最大尺寸为 5MB -->
        <property name="maxUploadSize" value="5242880"/>
    </bean>
</beans>
```

5.5.3 MyBatis 的配置文件

本项目中的 mybatis-config.xml 文件是 MyBatis 的配置文件。mybatis-config.xml 文件的代码如下：

```xml
<?xml version="1.0" encoding="UTF-8"?>
<!DOCTYPE configuration
    PUBLIC "-//mybatis.org//DTD Config 3.0//EN"
```

```
         "http://mybatis.org/dtd/mybatis-3-config.dtd">
<configuration>
</configuration>
```

5.6 实体类设计

在 Java 语言中，一个实体类对应一种数据模型。在第 5.4.2 节中，已经介绍了本项目的 5 张数据表。除了 t_sort_movie（电影类型与电影关联表），其余 4 张数据表都有对应的实体类。下面将分别介绍与这 4 张数据表对应的实体类。

5.6.1 电影评论类

电影评论类对应的是 TComment.java 文件。在电影评论类中，有 6 个私有的属性，即 commentid（评论的编号）、movieobj（电影的编号）、content（评论的内容）、userobj（账户名）、commenttime（评论的时间）和 tMovie（电影信息类对象）。不难发现，前 5 个属性与 t_comment（电影评论表）中的字段是一一对应的。为了方便外部类访问这 6 个私有属性，需要为它们添加 Getter/Setter 方法。TComment 类的代码如下：

```java
public class TComment implements Serializable {
    private static final long serialVersionUID = -67987860200491556L;

    private Integer commentid;                    //评论的编号
    private Integer movieobj;                     //电影的编号
    private String content;                       //评论的内容
    private String userobj;                       //账户名
    private String commenttime;                   //评论的时间
    private TMovie tMovie;                        //电影信息类对象

    public TMovie gettMovie() {
        return tMovie;
    }
    public void settMovie(TMovie tMovie) {
        this.tMovie = tMovie;
    }
    public Integer getCommentid() {
        return commentid;
    }
    public void setCommentid(Integer commentid) {
        this.commentid = commentid;
    }
    public Integer getMovieobj() {
        return movieobj;
    }
    public void setMovieobj(Integer movieobj) {
        this.movieobj = movieobj;
    }
    public String getContent() {
        return content;
    }
    public void setContent(String content) {
        this.content = content;
    }
    public String getUserobj() {
        return userobj;
```

```java
    }
    public void setUserobj(String userobj) {
        this.userobj = userobj;
    }
    public String getCommenttime() {
        return commenttime;
    }
    public void setCommenttime(String commenttime) {
        this.commenttime = commenttime;
    }
    @Override
    public String toString() {
        return "TComment{" +
                "commentid=" + commentid +
                ", movieobj=" + movieobj +
                ", content='" + content + '\'' +
                ", userobj='" + userobj + '\'' +
                ", commenttime='" + commenttime + '\'' +
                '}';
    }
}
```

5.6.2 电影信息类

电影信息类对应的是 TMovie.java 文件。在电影信息类中，有 13 个私有的属性。除了与 t_movie（电影信息表）中的字段对应的 11 个属性，其余两个私有的属性分别为 sorts（用于存储电影类型类对象的列表）和 sortid（电影类型的编号）。为了方便外部类访问这 13 个私有属性，需要为它们添加 Getter/Setter 方法。TMovie 类的代码如下：

```java
public class TMovie implements Serializable {
    private static final long serialVersionUID = 203638591866512926L;

    private Integer movieid;                    //电影的编号
    private String moviename;                   //电影的名称
    private String moviephoto;                  //电影的封面图片
    private String director;                    //电影的导演
    private String mainperformer;               //电影的演职人员
    private String duration;                    //电影的时长
    private String area;                        //电影的区域
    private String releasedate;                 //电影的上映时间
    private String price;                       //电影的票价
    private String opera;                       //电影的剧情简介
    private Integer hit;                        //电影的点击量
    private List<TSort> sorts;                  //用于存储电影类型类对象的列表
    private Integer [] sortid;                  //电影类型的编号

    public Integer[] getSortid() {
        return sortid;
    }
    public void setSortid(Integer[] sortid) {
        this.sortid = sortid;
    }
    public List<TSort> getSorts() {
        return sorts;
    }
    public void setSorts(List<TSort> sorts) {
        this.sorts = sorts;
    }
```

```java
public Integer getMovieid() {
    return movieid;
}
public void setMovieid(Integer movieid) {
    this.movieid = movieid;
}
public String getMoviename() {
    return moviename;
}
public void setMoviename(String moviename) {
    this.moviename = moviename;
}
public String getMoviephoto() {
    return moviephoto;
}
public void setMoviephoto(String moviephoto) {
    this.moviephoto = moviephoto;
}
public String getDirector() {
    return director;
}
public void setDirector(String director) {
    this.director = director;
}
public String getMainperformer() {
    return mainperformer;
}
public void setMainperformer(String mainperformer) {
    this.mainperformer = mainperformer;
}
public String getDuration() {
    return duration;
}
public void setDuration(String duration) {
    this.duration = duration;
}
public String getArea() {
    return area;
}
public void setArea(String area) {
    this.area = area;
}
public String getReleasedate() {
    return releasedate;
}
public void setReleasedate(String releasedate) {
    this.releasedate = releasedate;
}
public String getPrice() {
    return price;
}
public void setPrice(String price) {
    this.price = price;
}
public String getOpera() {
    return opera;
}
public void setOpera(String opera) {
    this.opera = opera;
}
public Integer getHit() {
    return hit;
```

```java
    public void setHit(Integer hit) {
        this.hit = hit;
    }
    @Override
    public String toString() {
        return "TMovie{" +
                "movieid=" + movieid +
                ", moviename='" + moviename + '\'' +
                ", moviephoto='" + moviephoto + '\'' +
                ", director='" + director + '\'' +
                ", mainperformer='" + mainperformer + '\'' +
                ", duration='" + duration + '\'' +
                ", area='" + area + '\'' +
                ", releasedate='" + releasedate + '\'' +
                ", price=" + price +
                ", opera='" + opera + '\'' +
                ", hit=" + hit +
                ", sorts=" + sorts +
                ", sortid=" + Arrays.toString(sortid) +
                '}';
    }
}
```

5.6.3 电影类型类

电影类型类对应的是 TSort.java 文件。在电影类型类中，有 3 个私有的属性。其中，id（电影类型的编号）和 sorts（电影类型的名称）是与 t_sort（电影类型表）中的字段一一对应的；hit 表示电影的点击量。为了方便外部类访问这 3 个私有属性，需要为它们添加 Getter/Setter 方法。TSort 类的代码如下：

```java
public class TSort implements Serializable {
    private static final long serialVersionUID = -43408998428089381L;

    private Integer id;                                    //电影类型的编号
    private String sorts;                                  //电影类型的名称
    private Integer hit;                                   //电影的点击量

    public Integer getHit() {
        return hit;
    }
    public void setHit(Integer hit) {
        this.hit = hit;
    }

    public Integer getId() {
        return id;
    }
    public void setId(Integer id) {
        this.id = id;
    }
    public String getSorts() {
        return sorts;
    }
    public void setSorts(String sorts) {
        this.sorts = sorts;
    }
    @Override
    public String toString() {
        return "TSort{" +
```

```
            "id=" + id +
            ", sorts='" + sorts + "\" +
            '}';
    }
}
```

5.6.4 用户信息类

用户信息类对应的是 TUserinfo.java 文件。在用户信息类中，有 6 个私有的属性。其中，usertype 和 usertypeStr 都表示用户的类型，只不过 usertype 的值是英文的，而 usertypeStr 的值是中文的。因此，在用户信息类中，定义了一个 getUsertypeStr()方法，用于把用户的类型由英文转为中文。其余 4 个属性与 t_userinfo（用户信息表）中的字段是一一对应的。为了方便外部类访问这 6 个私有属性，需要为它们添加 Getter/Setter 方法。TUserinfo 类的代码如下：

```java
public class TUserinfo implements Serializable {
    private static final long serialVersionUID = 590560556651666747L;

    private String userName;                    //账户名
    private String password;                    //密码
    private String userphoto;                   //用户的头像
    private String usertype;                    //用户的类型（ROLE_USER/ROLE_ADMIN）
    private String usertypeStr;                 //用户的类型（普通用户/管理员）
    private String regtime;                     //用户的注册时间

    public String getUsertypeStr() {
        if (this.usertype.equals("ROLE_ADMIN")){
            usertypeStr="管理员";
        }else if (this.usertype.equals("ROLE_USER")){
            usertypeStr="普通用户";
        }
        return usertypeStr;
    }

    public void setUsertypeStr(String usertypeStr) {
        this.usertypeStr = usertypeStr;
    }
    public String getUserName() {
        return userName;
    }
    public void setUserName(String userName) {
        this.userName = userName;
    }
    public String getPassword() {
        return password;
    }
    public void setPassword(String password) {
        this.password = password;
    }
    public String getUserphoto() {
        return userphoto;
    }
    public void setUserphoto(String userphoto) {
        this.userphoto = userphoto;
    }
    public String getUsertype() {
        return usertype;
    }
    public void setUsertype(String usertype) {
```

```
        this.usertype = usertype;
    }
    public String getRegtime() {
        return regtime;
    }
    public void setRegtime(String regtime) {
        this.regtime = regtime;
    }
    @Override
    public String toString() {
        return "TUserinfo{" +
                "userName='" + userName + '\'' +
                ", password='" + password + '\'' +
                ", userphoto='" + userphoto + '\'' +
                ", usertype='" + usertype + '\'' +
                ", regtime='" + regtime + '\'' +
                '}';
    }
}
```

5.7　Mapper 接口设计

MyBatis 的 Mapper 接口能够将 SQL 语句与其中的方法绑定，进而在调用 Mapper 接口中的方法时能够执行对应的 SQL 语句。下面将分别介绍本项目中的 Mapper 接口。

本项目中的 Mapper 接口都被命名为 XXxxDao。

5.7.1　TCommentDao 接口

TCommentDao 接口对应的是 TCommentDao.java 文件。在 TCommentDao 接口中，主要包含两个方法，它们分别是 queryAll()和 insert()方法。其中，queryAll()方法用于查询符合筛选条件的电影评论；insert()方法用于新增电影评论。TCommentDao 接口中的代码如下：

```
public interface TCommentDao {
    /**
     * 通过实体作为筛选条件查询
     *
     * @param tComment 实例对象
     * @return 对象列表
     */
    List<TComment> queryAll(TComment tComment);
    /**
     * 新增数据
     *
     * @param tComment 实例对象
     * @return 影响行数
     */
    int insert(TComment tComment);
}
```

SSM 框架把 SQL 语句写在了 xml 文件中。在本项目中，与 TCommentDao 接口中的方法绑定的 SQL 语句被写在了 TCommentDao.xml 文件中。TCommentDao.xml 文件中的代码如下：

```xml
<?xml version="1.0" encoding="UTF-8"?>
<!DOCTYPE mapper PUBLIC "-//mybatis.org//DTD Mapper 3.0//EN" "http://mybatis.org/dtd/mybatis-3-mapper.dtd">
<mapper namespace="com.zxl.dao.TCommentDao">

    <resultMap type="com.zxl.entity.TComment" id="TCommentMap">
        <result property="commentid" column="commentId" jdbcType="INTEGER"/>
        <result property="movieobj" column="movieObj" jdbcType="INTEGER"/>
        <result property="content" column="content" jdbcType="VARCHAR"/>
        <result property="userobj" column="userObj" jdbcType="VARCHAR"/>
        <result property="commenttime" column="commentTime" jdbcType="VARCHAR"/>
        <association property="tMovie" javaType="TMovie">
            <result property="moviename" column="movieName"/>
        </association>
    </resultMap>

    <!--通过实体作为筛选条件查询-->
    <select id="queryAll" resultMap="TCommentMap">
        select
        commentId, movieObj, content, userObj, commentTime
        from movie.t_comment
        <where>
            <if test="commentid != null">
                and commentId = #{commentid}
            </if>
            <if test="movieobj != null">
                and movieObj = #{movieobj}
            </if>
            <if test="content != null and content != ''">
                and content = #{content}
            </if>
            <if test="userobj != null and userobj != ''">
                and userObj = #{userobj}
            </if>
            <if test="commenttime != null and commenttime != ''">
                and commentTime = #{commenttime}
            </if>
        </where>
    </select>

    <!--新增数据-->
    <insert id="insert" keyProperty="commentid" useGeneratedKeys="true">
        insert into movie.t_comment(movieObj, content, userObj, commentTime)
        values (#{movieobj}, #{content}, #{userobj}, #{commenttime})
    </insert>
</mapper>
```

5.7.2　TMovieDao 接口

TMovieDao 接口对应的是 TMovieDao.java 文件。在 TMovieDao 接口中，主要包含了用于查询符合筛选条件的电影信息的方法和用于新增电影信息的方法。有关 TMovieDao 接口中各个方法的名称及其说明，读者可参照 TMovieDao 接口中的代码，代码如下：

```java
public interface TMovieDao {
    /**
     * 通过ID查询单条数据
     *
     * @param movieid 主键
```

```
     * @return 实例对象
     */
    TMovie queryById(Integer movieid);

    /**
     *
     * @return 所有数据
     */
    List<TMovie> findall();

    /**
     * 查询数据的总条数
     */
    Integer findCount();

    /**
     * 通过实体作为筛选条件查询
     *
     * @param tMovie 实例对象
     * @return 对象列表
     */
    List<TMovie> queryAll(TMovie tMovie);

    /**
     * 新增数据
     *
     * @param tMovie 实例对象
     * @return 影响行数
     */
    int insert(TMovie tMovie);

    /**
     * 通过电影类型来查询电影
     */
    List<TMovie> findBySort(Integer sid);

    List<TMovie> findByname(String name);

    List<TSort> countHit();

    List<TMovie> findSome(TMovie tMovie);

    /**
     * 通过类型来查询电影数据
     */
    List<TMovie> findmoviesBySort(Integer id);

    /**
     * 根据类型来计算count
     */
    Integer countBysort(Integer id);

}
```

在本项目中，TMovieDao.xml 文件含有与 TMovieDao 接口中各个方法绑定的 SQL 语句。TMovieDao.xml 文件中的代码如下：

```xml
<?xml version="1.0" encoding="UTF-8"?>
<!DOCTYPE mapper PUBLIC "-//mybatis.org//DTD Mapper 3.0//EN" "http://mybatis.org/dtd/mybatis-3-mapper.dtd">
<mapper namespace="com.zxl.dao.TMovieDao">

    <resultMap type="com.zxl.entity.TMovie" id="TMovieMap">
```

```xml
        <result property="movieid" column="movieId" jdbcType="INTEGER"/>
        <result property="moviename" column="movieName" jdbcType="VARCHAR"/>
        <result property="moviephoto" column="moviePhoto" jdbcType="VARCHAR"/>
        <result property="director" column="director" jdbcType="VARCHAR"/>
        <result property="mainperformer" column="mainPerformer" jdbcType="VARCHAR"/>
        <result property="duration" column="duration" jdbcType="VARCHAR"/>
        <result property="area" column="area" jdbcType="VARCHAR"/>
        <result property="releasedate" column="releaseDate" jdbcType="VARCHAR"/>
        <result property="price" column="price" jdbcType="OTHER"/>
        <result property="opera" column="opera" jdbcType="VARCHAR"/>
        <result property="hit" column="hit"/>
        <association property="tSchedule" javaType="TSchedule">
            <result property="scheduleid" column="scheduleId"/>
        </association>

        <collection property="sorts" ofType="TSort">
            <result property="sorts" column="sorts"/>
            <result property="id" column="sid"/>
        </collection>
</resultMap>

<resultMap type="com.zxl.entity.TMovie" id="baseMapper">
    <result property="movieid" column="movieId" jdbcType="INTEGER"/>
    <result property="moviename" column="movieName" jdbcType="VARCHAR"/>
    <result property="moviephoto" column="moviePhoto" jdbcType="VARCHAR"/>
    <result property="director" column="director" jdbcType="VARCHAR"/>
    <result property="mainperformer" column="mainPerformer" jdbcType="VARCHAR"/>
    <result property="duration" column="duration" jdbcType="VARCHAR"/>
    <result property="area" column="area" jdbcType="VARCHAR"/>
    <result property="releasedate" column="releaseDate" jdbcType="VARCHAR"/>
    <result property="price" column="price" jdbcType="OTHER"/>
    <result property="opera" column="opera" jdbcType="VARCHAR"/>
    <result property="hit" column="hit"/>
    <collection property="sorts" ofType="TSort" javaType="java.util.List"
                select="com.zxl.dao.TSortDao.findMS" column="movieId">
    </collection>
</resultMap>

<!--查询单个-->
<select id="queryById" resultMap="TMovieMap">
    SELECT m.movieId,
           m.movieName,
           m.moviePhoto,
           m.director,
           m.mainPerformer,
           m.duration,
           m.area,
           m.releaseDate,
           m.price,
           m.opera,
           m.hit,
           s.id sid,
           s.sorts
    FROM t_movie m
           JOIN t_sort_movie sm ON m.movieId = sm.movieid
           JOIN t_sort s ON sm.sortsid = s.id
    WHERE m.movieId = #{movieid}
</select>

<!--查询所有数据-->
<select id="findall" resultMap="baseMapper">
    SELECT movieId,
```

```xml
        movieName,
        moviePhoto,
        director,
        mainPerformer,
        duration,
        AREA,
        releaseDate,
        price,
        opera,
        hit
    FROM t_movie
    ORDER BY movieId DESC, releaseDate DESC
</select>

<!--查询数据的总条数-->
<select id="findCount" resultType="java.lang.Integer">
    SELECT COUNT(movieId)
    FROM t_movie
</select>

<!--通过实体作为筛选条件查询-->
<select id="queryAll" resultMap="TMovieMap">
    select
        movieId, movieName, moviePhoto, director, mainPerformer, duration, area, releaseDate, price, opera
    from movie.t_movie
        <where>
            <if test="moviename != null and moviename != ''">
                and movieName = #{moviename}
            </if>
            <if test="moviephoto != null and moviephoto != ''">
                and moviePhoto = #{moviephoto}
            </if>
            <if test="director != null and director != ''">
                and director = #{director}
            </if>
            <if test="mainperformer != null and mainperformer != ''">
                and mainPerformer = #{mainperformer}
            </if>
            <if test="duration != null and duration != ''">
                and duration = #{duration}
            </if>
            <if test="area != null and area != ''">
                and area = #{area}
            </if>
            <if test="releasedate != null and releasedate != ''">
                and releaseDate = #{releasedate}
            </if>
            <if test="price != null">
                and price = #{price}
            </if>
            <if test="opera != null and opera != ''">
                and opera = #{opera}
            </if>
        </where>
    order by m.movieId desc
</select>

<!--通过电影类型来查询电影-->
<select id="findBySort" resultMap="TMovieMap">
    select m.movieId,
           m.movieName,
           m.moviePhoto,
```

```xml
            m.director,
            m.mainPerformer,
            m.duration,
            m.area,
            m.releaseDate,
            m.price,
            m.opera,
            s.id sid,
            s.sorts
        from t_movie m
            join t_sort_movie sm on m.movieId = sm.movieid
            join t_sort s on sm.sortsid = s.id
        where s.id = #{sid}

</select>
<select id="countHit" resultType="TSort">
    SELECT ts.sorts, SUM(tm.hit) hit
    FROM t_movie tm
            JOIN t_sort_movie tms ON tm.movieId = tms.movieid
            JOIN t_sort ts ON ts.id = tms.sortsid
    GROUP BY ts.sorts
</select>

<select id="findSome" resultMap="TMovieMap" parameterType="TMovie">
    select
    movieId, movieName, moviePhoto, director, mainPerformer, duration, area, releaseDate, price, opera
    from movie.t_movie
        <where>
            <if test="moviename != null and moviename != ''">
                and movieName like #{moviename}
            </if>
            <if test="director != null and director != ''">
                and director like #{director}
            </if>
            <if test="mainperformer != null and mainperformer != ''">
                and mainPerformer like #{mainperformer}
            </if>
        </where>
</select>

<!--新增所有列-->
<insert id="insert" keyProperty="movieid" useGeneratedKeys="true">
    insert into movie.t_movie(movieName, moviePhoto, director, mainPerformer, duration, area, releaseDate, price, opera)
    values (#{moviename}, #{moviephoto}, #{director}, #{mainperformer}, #{duration}, #{area}, #{releasedate}, #{price}, #{opera})
</insert>

<!--根据名字模糊查询-->
<select id="findByname" resultMap="TMovieMap">
    SELECT m.movieId,
            m.movieName,
            m.moviePhoto,
            m.director,
            m.mainPerformer,
            m.duration,
            m.area,
            m.releaseDate,
            m.price,
            m.opera,
            s.sorts,
            s.id sid
```

```xml
        FROM t_movie m
            JOIN t_sort_movie sm ON m.movieId = sm.movieid
            JOIN t_sort s ON sm.sortsid = s.id
        where m.movieName like #{name}
    </select>
    <select id="tMovieTop" resultType="com.zxl.entity.TMovie">
        SELECT *
        FROM t_movie
        ORDER BY hit desc
        LIMIT 10
    </select>

    <select id="findmoviesBySort" resultType="com.zxl.entity.TMovie">
        SELECT movieId,
               movieName,
               moviePhoto,
               director,
               mainPerformer,
               duration,
               AREA,
               releaseDate,
               price,
               opera,
               hit
        FROM t_movie
        where movieId in (select t_sort_movie.movieid from t_sort_movie where sortsid = #{id})
        ORDER BY hit DESC
    </select>

    <select id="countBysort" resultType="java.lang.Integer">
        SELECT count(movieId)
        FROM t_movie
        where movieId in (select t_sort_movie.movieid from t_sort_movie where sortsid = #{id})
        ORDER BY movieId DESC, releaseDate
    </select>
</mapper>
```

5.7.3 TSortDao 接口

TSortDao 接口对应的是 TSortDao.java 文件。在 TSortDao 接口中，主要包含了用于查询符合筛选条件的电影类型的方法。有关 TSortDao 接口中各个方法的名称及其说明，读者可参照 TSortDao 接口中的代码，代码如下：

```java
public interface TSortDao {

    /**
     * 通过 ID 查询单条数据
     *
     * @param id 主键
     * @return 实例对象
     */
    TSort queryById(Integer id);

    /**
     * 查询指定行数据
     *
     * @param offset 查询起始位置
     * @param limit 查询条数
     * @return 对象列表
```

```java
     */
    List<TSort> queryAllByLimit(@Param("offset") int offset, @Param("limit") int limit);

    /**
     * 通过实体作为筛选条件查询
     *
     * @param tSort 实例对象
     * @return 对象列表
     */
    List<TSort> queryAll(TSort tSort);

    /**
     * 根据电影 id 排除查询数据
     */
    List<TSort> findNotByMid(Integer mid);

    Integer[] findByMid(Integer movieid);

    List<TSort> findMS(Integer id);
}
```

在本项目中，TSortDao.xml 文件含有与 TSortDao 接口中各个方法绑定的 SQL 语句。TSortDao.xml 文件中的代码如下：

```xml
<?xml version="1.0" encoding="UTF-8"?>
<!DOCTYPE mapper PUBLIC "-//mybatis.org//DTD Mapper 3.0//EN" "http://mybatis.org/dtd/mybatis-3-mapper.dtd">
<mapper namespace="com.zxl.dao.TSortDao">

    <resultMap type="com.zxl.entity.TSort" id="TSortMap">
        <result property="id" column="id" jdbcType="INTEGER"/>
        <result property="sorts" column="sorts" jdbcType="VARCHAR"/>
    </resultMap>

    <!--通过电影 Id 来查询分类信息-->
    <select id="findMS" resultType="TSort">
        SELECT ts.id,ts.sorts FROM t_sort ts JOIN t_sort_movie tsm ON ts.id=tsm.sortsid WHERE tsm.movieid=#{id}
    </select>

    <!--查询单个-->
    <select id="queryById" resultMap="TSortMap">
        select id,
               sorts
        from movie.t_sort
        where id = #{id}
    </select>

    <!--查询指定行数据-->
    <select id="queryAllByLimit" resultMap="TSortMap">
        select id,
               sorts
        from movie.t_sort
        limit #{offset}, #{limit}
    </select>

    <!--通过实体作为筛选条件查询-->
    <select id="queryAll" resultMap="TSortMap">
        select
        id, sorts
        from movie.t_sort
        <where>
            <if test="id != null">
                and id = #{id}
```

```xml
            </if>
            <if test="sorts != null and sorts != ''">
                and sorts = #{sorts}
            </if>
        </where>
    </select>

    <select id="findNotByMid" resultType="com.zxl.entity.TSort">
        select *
        from t_sort
        where id not in (select sortsid from t_sort_movie where movieid = #{mid})
    </select>

    <select id="findByMid">
        select sortsid
        from t_sort_movie
        where movieid = #{movieid}
    </select>
</mapper>
```

5.7.4 TUserinfoDao 接口

TUserinfoDao 接口对应的是 TUserinfoDao.java 文件。在 TUserinfoDao 接口中，有 4 个方法，它们分别是 findbyUsername()、update()、insert()和 registered()方法。有关上述 4 个方法的说明，读者可参照 TUserinfoDao 接口中的代码，代码如下：

```java
public interface TUserinfoDao {
    //根据账户名查询用户信息
    TUserinfo findbyUsername(String name);
    //修改用户信息
    void update(TUserinfo tUserinfo);
    //新增用户信息
    void insert(TUserinfo tUserinfo);
    //注册用户信息
    void registered(TUserinfo tUserinfo);
}
```

在本项目中，与 TUserinfoDao 接口中的方法绑定的 SQL 语句被写在了 TUserinfoDao.xml 文件中。TUserinfoDao.xml 文件中的代码如下：

```xml
<?xml version="1.0" encoding="UTF-8"?>
<!DOCTYPE mapper PUBLIC "-//mybatis.org//DTD Mapper 3.0//EN" "http://mybatis.org/dtd/mybatis-3-mapper.dtd">
<mapper namespace="com.zxl.dao.TUserinfoDao">

    <resultMap type="com.zxl.entity.TUserinfo" id="TUserinfoMap">
        <result property="userName" column="user_name" jdbcType="VARCHAR"/>
        <result property="password" column="password" jdbcType="VARCHAR"/>
        <result property="userphoto" column="userPhoto" jdbcType="VARCHAR"/>
        <result property="usertype" column="userType" jdbcType="VARCHAR"/>
        <result property="regtime" column="regTime" jdbcType="VARCHAR"/>
    </resultMap>
    <update id="update" parameterType="TUserinfo">
        update t_userinfo
        <trim prefix="SET" suffixOverrides=",">
            <if test="password != null">password=#{password},</if>
            <if test="userphoto != null">userPhoto=#{userphoto},</if>
            <if test="usertype != null">userType=#{usertype}</if>
        </trim>
        <where>
```

```xml
            user_name=#{userName}
        </where>
    </update>

    <select id="findbyUsername" resultMap="TUserinfoMap">
        select user_name,password,userPhoto,userType,regTime   from t_userinfo where user_name=#{name}
    </select>

    <insert id="insert" parameterType="TUserinfo">
        insert into t_userinfo(user_name, password, userPhoto, userType, regTime)
        values (#{userName},#{password},#{userphoto},#{usertype},#{regtime})
    </insert>

    <insert id="registered" parameterType="TUserinfo">
        insert into t_userinfo
        <trim prefix="(" suffix=")" suffixOverrides=",">
            <if test="userName != null">
                user_name,
            </if>
            <if test="password != null">
                password,
            </if>
            <if test="userphoto != null">
                userPhoto,
            </if>
            <if test="usertype != null">
                userType,
            </if>
            <if test="regtime != null">
                regTime,
            </if>
        </trim>
        <trim prefix="values (" suffix=")" suffixOverrides=",">
            <if test="userName != null">
                #{userName},
            </if>
            <if test="password != null">
                #{password},
            </if>
            <if test="userphoto != null">
                #{userphoto},
            </if>
            <if test="usertype != null">
                #{usertype},
            </if>
            <if test="regtime != null">
                #{regtime},
            </if>
        </trim>
    </insert>
</mapper>
```

5.8 首页模块设计

如图 5.2 所示，嗨乐影评平台的首页包含查询输入框、隐藏按钮"我"、5 部电影的封面图片和名称、"查看更多"超链接等要素。下面将上述要素分为头部和主体两个部分，对嗨乐影评平台首页模块的设计过程进行讲解。

图 5.2　首页的效果图

5.8.1　首页页面设计

首页的头部是由查询输入框和隐藏按钮"我"组成的。其中，查询输入框用于根据电影的名称（如"迷雾"）或电影名称中的关键字（如"雾"）查询一部或者多部电影；隐藏按钮"我"用于执行登录操作，此外还包含"返回首页""退出"两个超链接。

本项目中的 header.jsp 实现的就是首页的头部，header.jsp 的关键代码如下：

```
<header>
    <div class="input" id="navbar">
        <div class="input-container">
            <form action="${pageContext.request.contextPath}/movies/findByname">
                <div class="search-box">
                    <input class="search-txt" type="text" placeholder="请输入..." name="name" />
                    <button type="submit" class="search-btn">
                        <i class="fas fa-search"></i>
                    </button>
                </div>
                <input type="hidden" name="page" value="1">
                <input type="hidden" name="pagesize" value="12">
            </form>
        </div>
        <security:authorize access="hasAnyAuthority('ROLE_ADMIN')">
            <a class="btn btn-default" href="${pageContext.request.contextPath}/pages/sysmain.jsp" role="button">
            管理员 </a>
        </security:authorize>
        <div class="login">
            <a href="${pageContext.request.contextPath}/user/find" class="login-btn">我</a>

            <div class="usrmess">
                <a href="${pageContext.request.contextPath}/movies/home">返回首页</a>
                <a href="${pageContext.request.contextPath}/logout">退出</a>
            </div>
        </div>
    </div>
</header>
```

首页的主体是由 5 部电影的封面图片和名称以及"查看更多"超链接组成的。5 部电影的封面图片和名称不是随机显示的，而是显示 t_movie（电影信息表）中最后 5 部电影的封面图片和名称。"查看更多"超链接用于分页显示 t_movie（电影信息表）中所有电影的封面图片和名称。

本项目中的 movies_home.jsp 实现的就是首页的主体，movies_home.jsp 的关键代码如下：

```html
<div class="movies_main">
    <div></div>
    <div class="section clear">
        <div class="movies_items movies_items2">
            <div class="items_header clear">
                <p><a href="${pageContext.request.contextPath}/movies/findBys?sid=0&page=1&pageSize=15">
                    查看更多</a></p>
            </div>
            <c:forEach items="${all}" var="movie">
                <div class="message">
                    <a href="${pageContext.request.contextPath}/movies/findById?id=${movie.movieid}">
                        <div class="message_img">
                            <img src="${pageContext.request.contextPath}${movie.moviephoto}" alt="">
                        </div>
                        <div class="name_price clear">
                            <h3>${movie.moviename}</h3>
                        </div>
                    </a>
                </div>
            </c:forEach>
        </div>
    </div>
</div>
```

5.8.2 控制器类设计

本项目的 UserMovieController 类是首页模块的控制器类，被@Controller 注解标注。在 UserMovieController 类中，定义了一个用于跳转到首页的 home()方法。该方法先把 t_movie（电影信息表）中最后 5 部电影的信息存储在一个 List 列表中，再把这个列表存储到 ModelMap 对象中。home()方法的代码如下：

```java
@RequestMapping("/home")
public String home(ModelMap modelMap) {
    List<TMovie> movieList = new ArrayList<>();
    List<TMovie> findall = tMovieService.findall();
    for (int i = 0; i < 5; i++) {
        TMovie tMovie = findall.get(i);
        movieList.add(tMovie);
    }
    modelMap.addAttribute("all", movieList);
    return "movies_home";
}
```

说明

ModelMap 对象用于把处理后的数据传递到结果页面上。

5.8.3 服务类设计

本项目的 TMovieService 接口为首页模块的服务接口。在 home()方法中，调用了一个抽象方法，即 findall() 方法，该方法是在 TMovieService 接口中予以定义的。findall()方法具有返回值，返回值是一个存储电影信息的 List 列表。findall()方法的代码如下：

```java
List<TMovie> findall();
```

本项目的 TMovieServiceImpl 类是 TMovieService 接口中的实现类，被@Service 注解标注，表示首页模块的服务类。在 TMovieServiceImpl 类中，重写了 TMovieService 接口中的 findall()方法。该方法的作用是调用 DAO 层（数据访问对象）来执行数据库操作。重写后的 findall()方法的代码如下：

```
@Override
public List<TMovie> findall() {
    return tMovieDao.findall();
}
```

5.8.4 DAO 层设计

本项目的 TMovieDao 接口为首页模块的数据访问对象，即首页模块的 DAO 层。在重写后的 findall()方法中，有一个同名的 findall()方法，后者是在 TMovieDao 接口中予以定义的。TMovieDao 接口中的 findall()方法具有返回值，返回值是一个存储电影信息的 List 列表。TMovieDao 接口中的 findall()方法的代码如下：

```
List<TMovie> findall();
```

说明

在本项目中，由于，与 TMovieDao 接口中的方法绑定的 SQL 语句被写在了 TMovieDao.xml 文件中，因此，读者可以在 TMovieDao.xml 文件中查找与 findall()方法绑定的 SQL 语句。

5.9 "更多"模块设计

在用户单击"查看更多"超链接后，页面将从首页跳转到"更多"页面，其效果如图 5.3 所示。

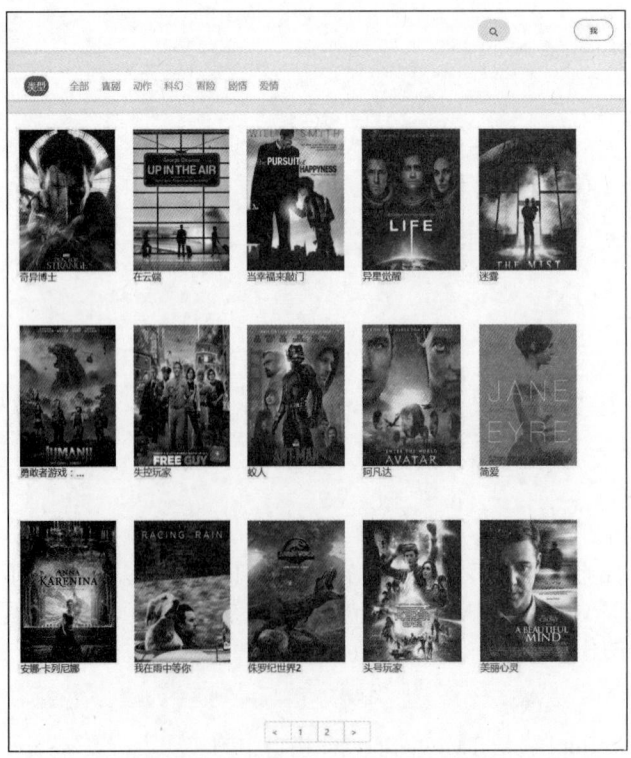

图 5.3 "更多"页面的效果图

平台上现有的全部电影被分为喜剧、动作、科幻、冒险、剧情和爱情这 6 种类型。在"更多"页面中，将分页显示全部电影或者某一种类型的电影的封面图片和名称，每一页显示 15 部电影的封面图片和名称。例如，在用户单击"喜剧"超链接后，"更多"页面就会显示"喜剧"类型的电影的封面图片和名称，其效果如图 5.4 所示。下面将介绍"更多"模块的设计过程。

图 5.4　选择"喜剧"类型后"更多"页面的效果图

5.9.1　"更多"页面设计

与首页的头部相同，"更多"页面的头部也是由查询输入框和隐藏按钮"我"组成的。因此，"更多"页面须包含 header.jsp。"更多"页面的主体是由电影类型导航栏和 JSP 分页组成的。其中，电影类型导航栏用于显示当前平台上全部电影的所有类型；JSP 分页用于分页显示全部电影或者某一种类型的电影的封面图片和名称，每一页显示 15 部电影的封面图片和名称。

本项目中的 movies_more.jsp 实现的就是"更多"页面，movies_more.jsp 的关键代码如下：

```
<body>
<jsp:c="header.jsp"/>

<div class="mod_hd" id="mod">
    <div class="mod_hd_list">
        <span class="mod">类型</span>
        <a href="${pageContext.request.contextPath}/movies/findBys?sid=0&page=1&pageSize=15" class="mod_items">
            全部</a>
        <c:forEach items="${sorts}" var="sort">
            <a href="${pageContext.request.contextPath}/movies/findBys?sid=${sort.id}&page=1&pageSize=15"
                class="mod_items">${sort.sorts}</a>
        </c:forEach>
    </div>
</div>

<div class="more-movies">
    <div class="movies_items">
        <c:forEach items="${movies.list}" var="movie">
            <div class="message">
                <a href="${pageContext.request.contextPath}/movies/findById?id=${movie.movieid}">
                    <div class="message_img">
```

```html
                    <img src="${pageContext.request.contextPath}${movie.moviephoto}" alt="">
                </div>
                <div class="name_price clear">
                    <h3>${movie.moviename}</h3>
                </div>
            </a>
        </div>
    </c:forEach>

    <!-- 分页 -->
    <c:if test="${pages !=null}">
        <div class="page">
            <ul class="pagination">
                <li class="page-item"><a class="page-link"
                    href="${pageContext.request.contextPath}/movies/findBys?sid=${sortid}&
                    page=${pages.page-1}&pageSize=${pages.pageSize}"><</a>
                </li>
                <c:forEach begin="1" end="${pages.total}" var="p">
                    <li class="page-item"><a class="page-link"
                        href="${pageContext.request.contextPath}/movies/findBys?sid=${sortid}&
                        page=${p}&pageSize=${pages.pageSize}">${p}</a>
                    </li>
                </c:forEach>
                <li class="page-item">
                    <a class="page-link"
                        href="${pageContext.request.contextPath}/movies/findBys?sid=${sortid}&
                        page=${pages.page+1}&pageSize=${pages.pageSize}">></a>
                </li>
            </ul>
        </div>
    </c:if>
</div>
</body>
```

5.9.2 控制器类设计

本项目的 UserMovieController 类既是首页模块的控制器类，也是"更多"模块的控制器类。在 UserMovieController 类中，定义了一个用于分页查询的 findByS()方法。该方法具有两个作用：一个作用是分页显示全部电影的封面图片和名称，另一个作用是分页显示某一种类型的电影的封面图片和名称。findByS() 方法的代码如下：

```java
@RequestMapping("/findBys")
public String findByS(Integer sid,Integer page,Integer pageSize,ModelMap modelMap){
    List<TSort> all = tSortService.findAll(null);
    if (sid==0){
        List<TMovie> findall = tMovieService.findall(page, pageSize);
        PageInfo<TMovie> pageInfo=new PageInfo<>(findall);
        Integer count = tMovieService.findCount();
        Page page1=new Page(pageSize,page,count);
        modelMap.addAttribute("movies",pageInfo);
        modelMap.addAttribute("pages",page1);
    }else {
        List<TMovie> bysort = tMovieService.findBysort(sid, page, pageSize);
        Integer integer = tMovieService.countSort(sid);
        PageInfo<TMovie> pageInfo=new PageInfo<>(bysort);
        Page page1=new Page(pageSize,page,integer);
        modelMap.addAttribute("movies",pageInfo);
        modelMap.addAttribute("pages",page1);
```

```
        }
        modelMap.addAttribute("sortid",sid);
        modelMap.addAttribute("sorts",all);
        return "movies_more";
}
```

在首页和"更多"页面的头部，都有查询输入框。查询输入框用于根据电影的名称（如"迷雾"）或者电影名称中的关键字（如"雾"）查询一部或者多部电影。为此，在 UserMovieController 类中，还定义了一个用于模糊查询电影信息的 findByname() 方法。该方法先把符合筛选条件的电影信息存储在一个 List 列表中，再把这个 List 列表转为页面信息，而后把页面信息存储到 ModelMap 对象中。findByname() 方法的代码如下：

```
@RequestMapping("/findByname")
public String findByname(String name, ModelMap modelMap,
                @RequestParam(value = "page", required = true) Integer page,
                @RequestParam(value = "pagesize", required = true) Integer pagesize) {
    List<TMovie> movies = tMovieService.findBYname(name, page, pagesize);
    PageInfo pageInfo = new PageInfo(movies);
    modelMap.addAttribute("movies", pageInfo);
    return "movies_more";
}
```

5.9.3 服务类设计

本项目的 TMovieService 接口既是首页模块的服务接口，又是"更多"模块的服务接口。在 findByS() 方法中，如果分页显示的是全部电影的封面图片和名称，那么需要依次调用 TMovieService 接口中的 findall(int page, int pageSize) 方法和 findCount() 方法。findall(int page, int pageSize) 方法用于获取一个存储所有电影信息的 List 列表，该方法具有两个参数：page 表示第几页，pageSize 表示每页显示的数据数量。findCount() 方法用于获取电影信息的总数。findall(int page, int pageSize) 方法和 findCount() 方法的代码分别如下：

```
List<TMovie> findall(int page, int pageSize);
Integer findCount();
```

本项目的 TMovieServiceImpl 类是 TMovieService 接口中的实现类，既是首页模块的服务类，也是"更多"模块的服务类。在 TMovieServiceImpl 类中，重写了 TMovieService 接口中的 findall(int page, int pageSize) 方法和 findCount() 方法，用于调用 DAO 层（数据访问对象）来执行数据库操作。重写后的 findall(int page, int pageSize) 方法和 findCount() 方法的代码分别如下：

```
@Override
public List<TMovie> findall(int page, int pageSize) {
    PageHelper.startPage(page, pageSize);
    List<TMovie> movies = tMovieDao.findall();
    return movies;
}
@Override
public Integer findCount() {
    return tMovieDao.findCount();
}
```

在 findByS() 方法中，如果分页显示的是某一种类型的电影的封面图片和名称，那么需要依次调用 TMovieService 接口中的 findBysort(Integer id,Integer page,Integer pageSize) 方法和 countSort(Integer sid) 方法。findBysort(Integer id,Integer page,Integer pageSize) 方法用于获取一个存储某一种类型的电影信息的 List 列表，该方法具有 3 个参数：id 表示电影类型的编号，page 表示第几页，pageSize 表示每页显示的数据数量。countSort(Integer sid) 方法用于获取某一种类型的电影信息的总数，其中 sid 表示电影类型的编号。findBysort(Integer id,Integer page,Integer pageSize) 方法和 countSort(Integer sid) 方法的代码分别如下：

```
List<TMovie> findBysort(Integer id,Integer page,Integer pageSize);
Integer countSort(Integer sid);
```

在 TMovieServiceImpl 类("更多"模块的服务类)中,重写了 TMovieService 接口中的 findBysort(Integer id,Integer page,Integer pageSize)方法和 countSort(Integer sid)方法,用于调用 DAO 层(数据访问对象)来执行数据库操作。重写后的 findBysort(Integer id,Integer page,Integer pageSize)方法和 countSort(Integer sid)方法的代码分别如下:

```
@Override
public List<TMovie> findBysort(Integer id, Integer page, Integer pageSize) {
    PageHelper.startPage(page,pageSize);
    return tMovieDao.findmoviesBySort(id);
}
@Override
public Integer countSort(Integer sid) {
    return tMovieDao.countBysort(sid);
}
```

在 UserMovieController 类的 findByname()方法中,调用了 TMovieService 接口中的 findBYname(String name,Integer page,Integer pagesize)方法。findBYname(String name,Integer page,Integer pagesize)方法用于获取一个存储符合筛选条件的电影信息的 List 列表,该方法具有 3 个参数:name 表示电影的名称(如"迷雾")或者电影名称中的关键字(如"雾"),page 表示第几页,pageSize 表示每页显示的数据数量。findBYname(String name,Integer page,Integer pagesize)方法的代码如下:

```
List<TMovie> findBYname(String name,Integer page,Integer pagesize);
```

在 TMovieServiceImpl 类("更多"模块的服务类)中,重写了 TMovieService 接口中的 findBYname(String name,Integer page,Integer pagesize)方法,用于调用 DAO 层(数据访问对象)来执行数据库操作。重写后的 findBYname(String name,Integer page,Integer pagesize)方法的代码如下:

```
@Override
public List<TMovie> findBYname(String name,Integer page ,Integer pagesize) {
    name="%"+name+"%";
    PageHelper.startPage(page,pagesize);
    return tMovieDao.findByname(name);
}
```

5.9.4　DAO 层设计

本项目的 TMovieDao 接口既是首页模块的 DAO 层,也是"更多"模块的 DAO 层。在第 5.9.3 节重写的 5 个方法中,依次调用了 TMovieDao 接口中的 5 个方法,即 findall()、findCount()、findmoviesBySort(Integer id)、countBysort(Integer id)和 findByname(String name)方法。TMovieDao 接口中的 5 个方法的代码分别如下:

```
List<TMovie> findall();
Integer findCount();
List<TMovie> findmoviesBySort(Integer id);
Integer countBysort(Integer id);
List<TMovie> findByname(String name);
```

说明

读者可以在 TMovieDao.xml 文件中查找与 findall()、findCount()、findmoviesBySort()、countBysort()和 findByname()方法绑定的 SQL 语句。

5.10 用户登录模块设计

在首页和"更多"页面的头部，都有隐藏按钮"我"。在用户单击隐藏按钮"我"后，页面将跳转到登录页面。在登录页面上，通过切换登录窗口与注册窗口，用户既可以完成登录操作，又可以完成注册操作。用户如果已经完成注册操作，就可以在登录窗口中输入账户名、密码和验证码，单击"登录"按钮，完成登录操作。登录窗口的效果如图 5.5 所示。下面将介绍用户登录模块的设计过程。

5.10.1 控制器类设计

本项目的 UserController 类为用户控制器类，被 @Controller 注解标注。在 UserController 类中，定义了一个 findme()方法。该方法先获取当前用户在登录窗口中输入的账户名，再根据这个账户名获取当前用户的信息，而后把当前用户的信息存储到 ModelMap 对象中。findme()方法的代码如下：

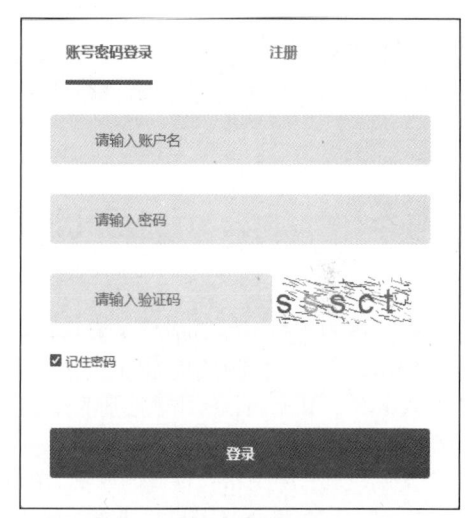

图 5.5 登录窗口的效果图

```
@RequestMapping("/find")
public String findme(ModelMap modelMap){
    String name = SecurityContextHolder.getContext().getAuthentication().getName();
    TUserinfo userByname = tUserinfoService.findUserByname(name);
    modelMap.addAttribute("me",userByname);
    return "user";
}
```

5.10.2 服务类设计

本项目的 TUserinfoService 接口为用户服务接口。在 findme()方法中，调用了 TUserinfoService 接口中的一个抽象方法，即 findUserByname(String name)方法。该方法有一个参数：name 表示当前用户在登录窗口中输入的账户名。该方法具有返回值，是一个 TUserinfo 类型的对象，表示当前用户的信息。findUserByname(String name)方法的代码如下：

```
TUserinfo findUserByname(String name);
```

本项目的 TUserinfoServiceImpl 类是 TUserinfoService 接口中的实现类，表示用户服务类，被@Service 注解标注。在 TUserinfoServiceImpl 类中，重写了 TUserinfoService 接口中的 findUserByname(String name)方法，用于调用 DAO 层（数据访问对象）来执行数据库操作。重写后的 findUserByname(String name)方法的代码如下：

```
@Override
public TUserinfo findUserByname(String name) {
    return tUserinfoDao.findbyUsername(name);
}
```

5.10.3 DAO 层设计

本项目的 TUserinfoDao 接口为用户数据访问对象，即用户 DAO 层。在重写后的 findUserByname(String name)方法中，调用了 TUserinfoDao 接口中的一个方法，即 findbyUsername(String name)。该方法有一个参数：name 表示当前用户在登录窗口中输入的账户名。该方法具有返回值，是一个 TUserinfo 类型的对象，表示当前用户的信息。TUserinfoDao 接口中的 findbyUsername(String name)方法的代码如下：

```
TUserinfo findbyUsername(String name);
```

说明 读者可以在 TUserinfoDao.xml 文件中查找与 findbyUsername()方法绑定的 SQL 语句。

5.11 用户注册模块设计

未注册的用户不能完成登录操作，须在注册窗口中完成注册操作，注册窗口的效果如图 5.6 所示。在注册窗口中，用户输入账户名和密码，单击"注册"按钮，完成注册操作，而后窗口会切换到登录窗口；在登录窗口中，完成注册操作的用户输入账户名、密码和验证码，单击"登录"按钮，即可完成登录操作。下面将介绍用户注册模块的设计过程。

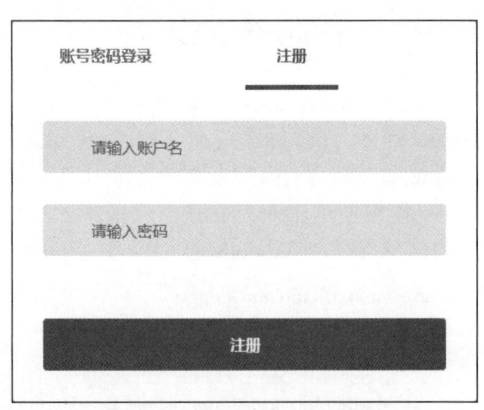

图 5.6 注册窗口的效果图

5.11.1 控制器类设计

在 UserController 类（用户控制器类）中，定义了一个 registered()方法。该方法具有一个参数，即 TUserinfo 类型的对象，用于存储用户信息。registered()方法的代码如下：

```
@RequestMapping("/registered")
public String registered(TUserinfo tUserinfo){
    System.out.println(tUserinfo);
    tUserinfoService.registered(tUserinfo);
    return "redirect:/login";
}
```

5.11.2 服务类设计

在 registered()方法中，调用了 TUserinfoService 接口（用户服务接口）的一个抽象方法，即 registered(TUserinfo tUserinfo)方法。不难发现，TUserinfoService 接口中的 registered(TUserinfo tUserinfo)方法和 UserController 类的 registered()方法不仅同名，而且同参。TUserinfoService 接口中的 registered(TUserinfo tUserinfo)方法的代码如下：

```
void registered(TUserinfo tUserinfo);
```

在 TUserinfoServiceImpl 类（用户服务类）中，重写了 TUserinfoService 接口中的 registered(TUserinfo tUserinfo)方法，用于调用 DAO 层（数据访问对象）来执行数据库操作。重写后的 registered(TUserinfo tUserinfo)方法的代码如下：

```
@Override
public void registered(TUserinfo tUserinfo) {
    tUserinfo.setPassword(BCPassward.setPassword(tUserinfo.getPassword()));
    tUserinfo.setRegtime(DateUtils.getDate(new Date()));
```

```
        tUserinfoDao.registered(tUserinfo);
    }
```

5.11.3 DAO 层设计

在重写后的 registered(TUserinfo tUserinfo)方法中，调用了 TUserinfoDao 接口（用户 DAO 层）中的一个方法，即 registered(TUserinfo tUserinfo)方法。TUserinfoDao 接口中的 registered(TUserinfo tUserinfo)方法和 TUserinfoServiceImpl 类的 registered(TUserinfo tUserinfo) 同名又同参。TUserinfoDao 接口中的 registered(TUserinfo tUserinfo)方法的代码如下：

```
void registered(TUserinfo tUserinfo);
```

> 读者可以在 TUserinfoDao.xml 文件中查找与 registered()方法绑定的 SQL 语句。

5.12 详情模块设计

不论在首页上还是在"更多"页面上，用户单击任意一部电影的封面图片或者名称，页面都可以跳转到当前电影的"详情"页面。"详情"页面不仅显示着当前电影的名称、类型、时长、上映日期、剧情介绍、演职人员等信息和评论内容，而且为用户提供了"写评论"的功能。详情模块的效果如图 5.7 所示。下面将介绍详情模块的设计过程。

图 5.7　详情模块的效果图

5.12.1 "详情"页面设计

因为"详情"页面的头部和首页的头部都是由查询输入框和隐藏按钮"我"组成的,所以"详情"页面须包含 header.jsp。"详情"页面的主体用于显示某一部电影的名称、类型、时长、上映日期、剧情介绍、演职人员等信息和评论内容。

本项目中的 movie_details.jsp 实现的就是"详情"页面,movie_details.jsp 的关键代码如下:

```jsp
<jsp:include page="header.jsp"/>

<div class="bar">
    <div class="bar-container">
        <div class="bar-img-a">
            <div class="bar-img">
                <img src="${pageContext.request.contextPath}${movie.moviephoto}" alt="">
            </div>
        </div>
        <div class="bar-message">
            <div class="bar-messae-text">
                <h3>${movie.moviename}</h3>
                <c:forEach items="${movie.sorts}" var="sort">
                    <p>${sort.sorts}</p>
                </c:forEach>
                <p>时长:${movie.duration}分钟</p>
                <p>上映日期:${movie.releasedate}</p>
            </div>
        </div>
    </div>
</div>

<div class="white"></div>

<div class="message">
    <div class="banner">
        <div class="module">
            <div class="module-title">
                <h2>剧情简介</h2>
            </div>
            <div class="module-conte">
                <span>${movie.opera}</span>
            </div>
        </div>

        <div class="module">
            <div class="module-title">
                <h2>演职人员</h2>
            </div>
            <div class="module-conte">
                <span>导演: ${movie.director}</span><br>
                <span>演员: ${movie.mainperformer}</span>
            </div>
        </div>

        <div class="module">
            <div class="module-title">
                <h2>热门短评</h2>
            </div>
            <c:forEach var="coment" items="${comments}">
```

```html
                <div class="module-conte">
                    <div class="common-list">
                        <ul>
                            <li>
                                <div class="usr-img">
                                    <div class="img">
                                        <img src="${pageContext.request.contextPath}/static/image/user.jpg" alt="">
                                    </div>
                                </div>
                                <div class="main-message">
                                    <div class="user-name">
                                        <span class="name">${coment.userobj}</span>
                                    </div>
                                    <div class="main-content">
                                        ${coment.content}
                                    </div>
                                </div>
                            </li>
                        </ul>
                    </div>
                </div>
            </c:forEach>
            <a type="button" class="commen-entry" data-toggle="modal" data-target="#myModal">写评论</a>
            <!--省略模态框的代码 -->
        </div>
    </div>
</div>
```

5.12.2 控制器类设计

本项目的 UserMovieController 类不仅是首页模块的控制器类，而且是"更多"模块的控制器类，还是详情模块的控制器类。在 UserMovieController 类中，定义了一个 findById()方法，该方法在根据某一部电影的编号查询到这部电影的信息后，会把这部电影的信息和评论内容显示在"详情"页面上。findById()方法的代码如下：

```java
@RequestMapping("/findById")
public String findById(@RequestParam(value = "id", required = true) Integer id, ModelMap modelMap) {
    TMovie tMovie = tMovieService.queryById(id);
    tMovie.setHit(tMovie.getHit() + 1);
    tMovieService.update(tMovie);
    TComment t = new TComment();
    t.setMovieobj(id);
    List<TComment> findall = tCommentService.findall(t);
    modelMap.addAttribute("movie", tMovie);
    modelMap.addAttribute("comments", findall);
    return "movie_detils";
}
```

5.12.3 服务类设计

本项目的 TMovieService 接口不仅是首页模块的服务接口，而且是"更多"模块的服务接口，还是"详情"模块的服务接口。在 findById()方法中，依次调用了 TMovieService 接口中的 queryById(Integer movieid)方法和 update(TMovie tMovie)方法。queryById(Integer movieid)方法用于根据某一部电影的编号查询这部电影的信息。update(TMovie tMovie)用于更新显示在"详情"页面上的电影信息。queryById(Integer movieid)方法和 update(TMovie tMovie)方法的代码分别如下：

```
TMovie queryById(Integer movieid);
void update(TMovie tMovie);
```

本项目的 TMovieServiceImpl 类是 TMovieService 接口中的实现类，不仅是首页模块的服务类，而且是"更多"模块的服务类，还是"详情"模块的服务类。在 TMovieServiceImpl 类中，重写了 TMovieService 接口中的 queryById(Integer movieid)方法和 update(TMovie tMovie)方法，用于调用 DAO 层（数据访问对象）来执行数据库操作。重写后的 queryById(Integer movieid)方法和 update(TMovie tMovie)方法的代码分别如下：

```
@Override
public TMovie queryById(Integer movieid) {
    return tMovieDao.queryById(movieid);
}

@Override
public void update(TMovie tMovie) {
    this.tMovieDao.update(tMovie);
}
```

5.12.4 DAO 层设计

本项目的 TMovieDao 接口不仅是首页模块的 DAO 层，而且是"更多"模块和"详情"模块的 DAO 层。在第 5.12.3 节重写的两个方法中，依次调用了 TMovieDao 接口中的两个方法，即 queryById(Integer movieid)和 update(TMovie tMovie)方法。其中，queryById(Integer movieid)方法的返回值是一个 TMovie 类型的对象，表示电影信息；update(TMovie tMovie)方法的返回值是一个 int 类型的变量，表示更新电影信息的条数。TMovieDao 接口中的这两个方法的代码分别如下：

```
TMovie queryById(Integer movieid);
int update(TMovie tMovie);
```

说明

读者可以在 TMovieDao.xml 文件中查找与 queryById()和 update()方法绑定的 SQL 语句。

5.13 写评论模块设计

在"详情"页面上，用户单击"写评论"超链接，程序将弹出评论模态框，其效果如图 5.8 所示。在评论模态框中，用户只有完成登录操作，才能提交已经写好的评论。对于本项目而言，写评论模块设计由两个部分组成：一个是"写评论"超链接，另一个是评论模态框。下面将介绍写评论模块的设计过程。

图 5.8 评论模态框的效果图

5.13.1 评论模态框设计

模态框是一种用于在网页上展示重要信息或者功能的交互式窗口，它通常在页面顶部或者页面中部弹出，覆盖在页面之上，使页面部分内容不可见，直到模态框被关闭。模态框可以包含文本、图像、表单、按钮等元素，用于向用户展示信息、获取用户输入或者执行其他操作。

评论模态框可以分为头部、主体和底部 3 个部分。其中，评论模态框的头部用于设置标题；评论模态框的主体是一个文本域；评论模态框的底部除了包含一个默认的"提交"按钮，还包含一个"取消"按钮。

在第 5.12.1 节的 movie_details.jsp（"详情"页面）中，为了实现写评论模块，除了已经添加的"写评论"超链接，还需要添加评论模态框。评论模态框的代码如下：

```html
<!-- 模态框 -->
<form action="${pageContext.request.contextPath}/comment/save">
    <div class="modal fade" id="myModal">
        <div class="modal-dialog modal-lg">
            <div class="modal-content">

                <!-- 模态框头部 -->
                <div class="modal-header">
                    <h4 class="modal-title">评论</h4>
                    <button type="button" class="close" data-dismiss="modal">&times;</button>
                </div>

                <!-- 模态框主体 -->
                <div class="modal-body">
                    <textarea cols="90" rows="3" name="content"></textarea>
                </div>
                <input type="hidden" name="movieobj" value="${movie.movieid}">

                <!-- 模态框底部 -->
                <div class="modal-footer">
                    <input type="submit" class="btn btn-primary">
                    <button    data-dismiss="modal" class="btn btn-secondary">取消</button>
                </div>
            </div>
        </div>
    </div>
</form>
```

5.13.2 控制器类设计

本项目的 CommentController 类是写评论模块的控制器类，被@Controller 注解标注。在 CommentController 类中，定义了一个用于保存电影评论的 save()方法。电影评论包含评论的编号、电影的编号、评论的内容、账户名、评论的时间等信息。save()方法的代码如下：

```java
@RequestMapping("/save")
public String save(TComment tComment){
    String name = SecurityContextHolder.getContext().getAuthentication().getName();
    tComment.setUserobj(name);                              //账户名
    Date date=new Date();
    String date1 = DateUtils.getDate(date);
    tComment.setCommenttime(date1);                         //评论的时间
    tCommentService.insert(tComment);                       //电影评论对象（含评论的编号和内容）
    Integer movieobj = tComment.getMovieobj();
```

```
            System.out.println(movieobj);
            String id=String.valueOf(movieobj);                        //电影的编号
            return "redirect:/movies/findById?id="+id;
}
```

5.13.3　服务类设计

本项目的 TCommentService 接口为写评论模块的服务接口。在 save()方法中，调用了一个抽象方法，即 insert(TComment tComment)方法，该方法是在 TCommentService 接口中予以定义的。insert(TComment tComment)方法具有返回值，是一个 TComment 类型的对象。insert(TComment tComment)方法的代码如下：

```
TComment insert(TComment tComment);
```

本项目的 TCommentServiceImpl 类是 TCommentService 接口的实现类，被@Service 注解标注，表示写评论模块的服务类。在 TCommentServiceImpl 类中，重写了 TCommentService 接口中的 insert(TComment tComment)方法。该方法的作用是调用 DAO 层（数据访问对象）来执行数据库操作。重写后的 insert(TComment tComment)方法的代码如下：

```
@Override
public TComment insert(TComment tComment) {
    this.tCommentDao.insert(tComment);
    return tComment;
}
```

5.13.4　DAO 层设计

本项目的 TCommentDao 接口为写评论模块的数据访问对象，即写评论模块的 DAO 层。在重写后的 insert(TComment tComment)方法中，有一个同名且同参的 insert(TComment tComment)方法，后者是在 TCommentDao 接口中予以定义的。TMovieDao 接口中的 insert(TComment tComment)方法具有返回值，返回值是一个 int 类型的变量，表示新增电影评论的条数。TMovieDao 接口中的 insert(TComment tComment)方法的代码如下：

```
int insert(TComment tComment);
```

说明

读者可以在 TCommentDao.xml 文件中查找与 insert()方法绑定的 SQL 语句。

5.14　电影管理模块设计

用户在登录嗨乐影评平台时，程序会验证当前用户的身份是普通用户，还是管理员。如果当前用户的身份是管理员（账户名为 admin，密码为 123456），则当前用户就可以进入嗨乐影评平台的后台。通过后台，管理员可以对平台上的电影进行管理。电影管理模块由两个功能模块组成：一个是分页显示平台上所有电影的信息（支持模糊查询），如图 5.9 所示；一个是添加电影信息，如图 5.10 所示。下面将分别介绍电影管理模块中的两个功能模块。

添加电影

电影 导演 演员 查询

电影名字	导演	主演	时长	上映时间	地区	点击量
奇异博士	斯科特·德瑞克森	本尼迪克特·康伯巴奇	115	2016-11-04 00:00:00	美国	2
在云端	贾森·雷特曼	乔治·克鲁尼	109	2009-12-23 00:00:00	美国	0
当幸福来敲门	加布里尔·穆奇诺	威尔·史密斯 / 贾登·史密斯	117	2006-12-15 00:00:00	美国	10
异星觉醒	丹尼尔·伊斯皮诺萨	杰克·吉伦哈尔	104	2017-03-24 00:00:00	美国	6
迷雾	弗兰克·德拉邦特	托马斯·简	126	2007-11-21 00:00:00	美国	4
勇敢者游戏：决战丛林	杰克·卡斯丹	道恩·强森 / 凯文·哈特	119	2017-12-20 00:00:00	美国	0
失控玩家	肖恩·利维	瑞安·雷诺兹	115	2021-08-13 00:00:00	美国、加拿大	1
蚁人	佩顿·里德	保罗·路德 / 迈克尔·道格拉斯	117	2015-07-17 00:00:00	美国	0
阿凡达	詹姆斯·卡梅隆	萨姆·沃辛顿 / 佐伊·索尔达娜	162	2009-12-18 00:00:00	美国	1
简爱	凯瑞·福永	迈克尔·法斯宾德	120	2011-03-11 00:00:00	英国、美国	1

上一页 1 2 下一页

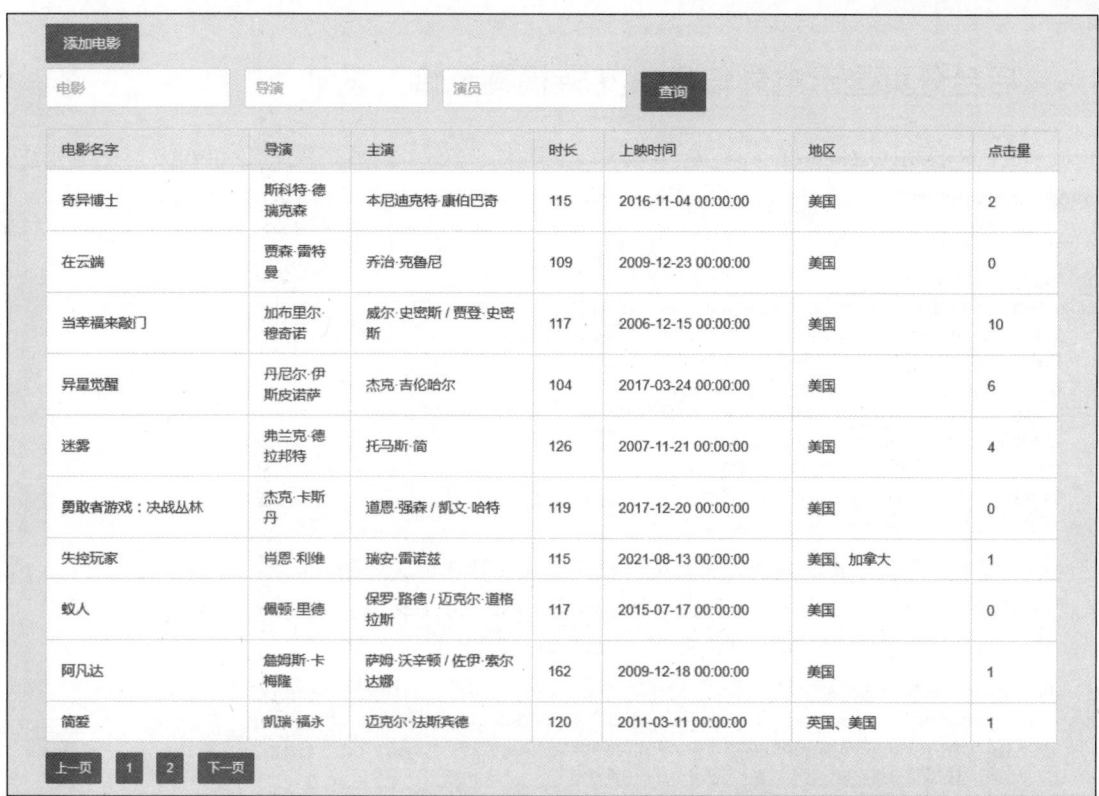

图 5.9　分页显示平台上所有电影的信息（支持模糊查询）的效果图

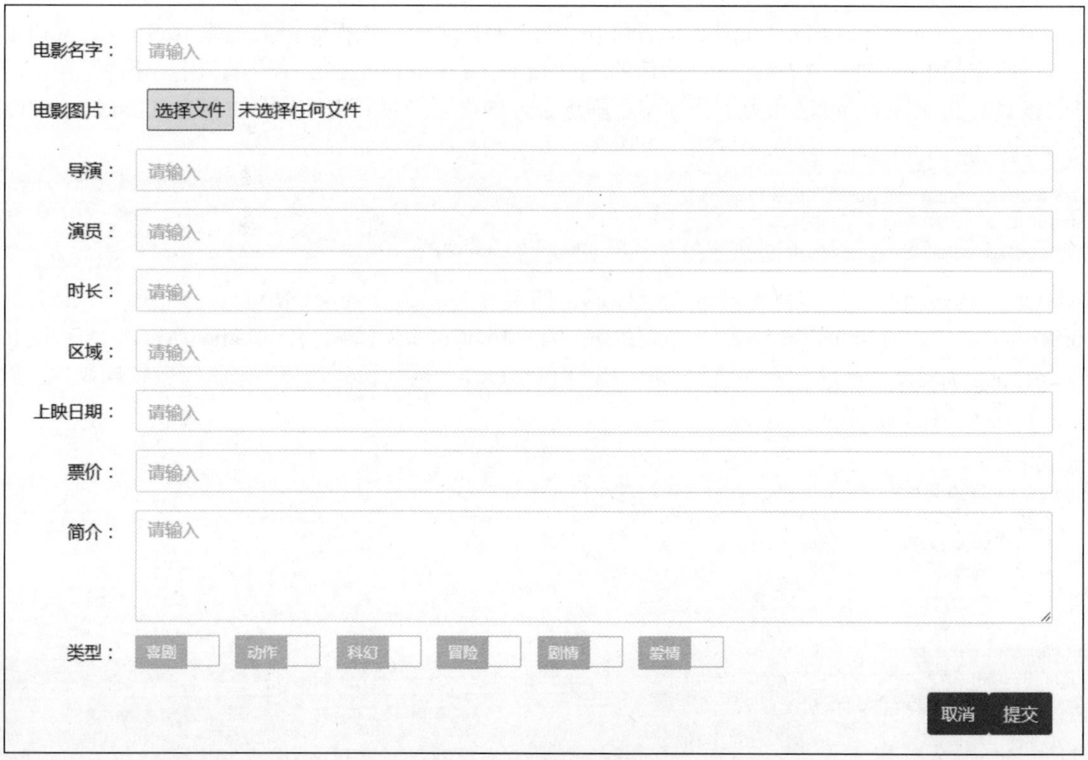

图 5.10　添加电影的效果图

5.14.1 后台分页显示电影信息（支持模糊查询）设计

本项目的 SysMoviesController 类是电影管理模块的控制器类，被 @Controller 注解标注。在 SysMoviesController 类中，定义了一个用于查找并分页显示电影信息的 findAll()方法。如图 5.9 所示，这里的电影信息包含电影名字、导演、主演、时长、上映时间、地区、点击量等信息。findAll()方法的代码如下：

```java
@RequestMapping("/findAll")
public String findAll(Integer page,Integer pageSize,ModelMap modelMap){
    List<TMovie> findall = tMovieService.findall(page, pageSize);
    PageInfo<TMovie> pageInfo=new PageInfo<>(findall);
    Integer count = tMovieService.findCount();
    Page page1=new Page(pageSize,page,count);
    modelMap.addAttribute("movies",pageInfo);
    modelMap.addAttribute("pages",page1);
    return "sysmovie";
}
```

后台不仅能够分页显示电影信息，而且支持模糊查询（管理员通过输入电影、导演、演员的名字或者其中的关键字可以在平台上查询电影）。为此，在 SysMoviesController 类（电影管理模块的控制器类）中，定义了一个用于实现模糊查询的 findSome()方法。findSome()方法的代码如下：

```java
@RequestMapping("/findSome")
public String findSome(TMovie tMovie,ModelMap modelMap,Integer page,Integer pageSize){
    List<TMovie> movies=tMovieService.findSome(tMovie,page,pageSize);
    PageInfo<TMovie> pageInfo=new PageInfo<>(movies);
    modelMap.addAttribute("movies",pageInfo);
    return "sysmovie";
}
```

在 SysMoviesController 类的 findAll()方法和 findSome()方法中，依次调用了 TMovieService 接口中用于查询平台上所有的电影信息并初始化分页插件的 findall(int page, int pageSize)方法、用于查询平台上所有电影信息的条数的 findCount()方法和用于查询符合筛选条件的电影信息并初始化分页插件的 findSome(TMovie tMovie, Integer page, Integer pageSize)方法。上述 3 个方法的代码分别如下：

```java
List<TMovie> findall(int page, int pageSize);
Integer findCount();
List<TMovie> findSome(TMovie tMovie, Integer page, Integer pageSize);
```

TMovieServiceImpl 类是 TMovieService 接口中的实现类。在 TMovieServiceImpl 类中，依次重写了 TMovieService 接口中的 findall(int page, int pageSize)方法、findCount()方法和 findSome(TMovie tMovie, Integer page, Integer pageSize)方法。重写后的方法都用于调用 DAO 层（数据访问对象）来执行数据库操作。重写后的上述 3 个方法的代码分别如下：

```java
@Override
public List<TMovie> findall(int page, int pageSize) {
    PageHelper.startPage(page, pageSize);
    List<TMovie> movies = tMovieDao.findall();
    return movies;
}

@Override
public Integer findCount() {
    return tMovieDao.findCount();
}

@Override
```

```
public List<TMovie> findSome(TMovie tMovie, Integer page, Integer pageSize) {
    TSchedule tSchedule=new TSchedule();
    tSchedule.setScheduleid(111);
    tMovie.setMoviename("%"+tMovie.getMoviename()+"%");
    tMovie.setDirector("%"+tMovie.getDirector()+"%");
    tMovie.setMainperformer("%"+tMovie.getMainperformer()+"%");
    PageHelper.startPage(page,pageSize);
    List<TMovie> some = tMovieDao.findSome(tMovie);
    for (TMovie movie : some) {
        movie.settSchedule(tSchedule);
    }
    return some;
}
```

在 TMovieServiceImpl 类被重写的 findall(int page, int pageSize)方法、findCount()方法和 findSome(TMovie tMovie, Integer page, Integer pageSize)方法中，依次调用了 TMovieDao 接口中用于查询平台上所有的电影信息的 findall()方法、用于查询平台上所有电影信息的条数的 findCount()方法和用于查询符合筛选条件的电影信息的 findSome(TMovie tMovie)方法。上述 3 个方法的代码分别如下：

```
List<TMovie> findall();
Integer findCount();
List<TMovie> findSome(TMovie tMovie);
```

说明

读者可以在 TMovieDao.xml 文件中查找与 findall()方法、findCount()方法和 findSome(TMovie tMovie)方法绑定的 SQL 语句。

5.14.2 添加电影信息设计

在 SysMoviesController 类（电影管理模块的控制器类）中，定义了一个用于添加电影信息的 add()方法。如图 5.10 所示，在添加电影信息时，需要为其选择封面图片，因此在 add()方法中务必正确设置封面图片的路径。add()方法的代码如下：

```
@RequestMapping("/add")
public String add(HttpServletRequest request,MultipartFile photo,TMovie tMovie) throws IOException {
    String path = request.getSession().getServletContext().getRealPath("/");
    path=path+"/static/image";
    String filename=photo.getOriginalFilename();
    String movieP="/static/image/"+filename;
    tMovie.setMoviephoto(movieP);
    System.out.println(tMovie);
    tMovieService.insert(tMovie);
    photo.transferTo(new File(path,filename));
    return "redirect:/sysMovies/findAll?page=1&pageSize=10";
}
```

如图 5.10 所示，在新增电影信息时，除了选择封面图片和输入必要的信息，还要为新增的电影勾选类型。在勾选类型之前，需要获取平台上所有电影的类型。为此，在 SysMoviesController 类（电影管理模块的控制器类）中，定义了一个用于获取平台上所有电影的类型的 toAdd()方法。toAdd()方法的代码如下：

```
@RequestMapping("/toAdd")
public String toAdd(ModelMap modelMap){
    List<TSort> all = tSortService.findAll(null);
    modelMap.addAttribute("sorts",all);
    return "movie_add";
}
```

在 SysMoviesController 类的 add()方法中，调用了 TMovieService 接口中用于新增电影信息的 insert(TMovie tMovie)方法。insert(TMovie tMovie)方法的代码如下：

TMovie insert(TMovie tMovie);

在 SysMoviesController 类的 toAdd()方法中，调用了 TSortService 接口中用于查询平台上所有电影的类型的 findAll(TSort tSort)方法。findAll(TSort tSort)方法的代码如下：

List<TSort> findAll(TSort tSort);

在 TMovieServiceImpl 类（TMovieService 接口中的实现类）中，重写了 TMovieService 接口中的 insert(TMovie tMovie)方法。重写后的 insert(TMovie tMovie)方法用于调用 DAO 层（数据访问对象）来执行数据库操作。重写后的 insert(TMovie tMovie)方法的代码如下：

```
@Override
public TMovie insert(TMovie tMovie) {
    this.tMovieDao.insert(tMovie);
    System.out.println(tMovie);
    Integer[] sortid = tMovie.getSortid();
    for (Integer integer : sortid) {
        tSortDao.insertm_s(tMovie.getMovieid(),integer);
    }
    return tMovie;
}
```

在 TSortServiceImpl 类（TSortService 接口中的实现类）中，重写了 TSortService 接口中的 findAll(TSort tSort)方法。重写后的 findAll(TSort tSort)方法用于调用 DAO 层（数据访问对象）来执行数据库操作。重写后的 findAll(TSort tSort)方法的代码如下：

```
@Override
public List<TSort> findAll(TSort tSort) {
    return tSortDao.queryAll(tSort);
}
```

在 TMovieServiceImpl 类被重写的 insert(TMovie tMovie)方法中，调用了 TMovieDao 接口中用于新增电影信息的 insert(TMovie tMovie)方法，该方法的返回值是一个 int 类型的变量，表示新增电影信息的条数。insert(TMovie tMovie)方法的代码如下：

int insert(TMovie tMovie);

在 TMovieServiceImpl 类被重写的 insert(TMovie tMovie)方法中，还调用了 TSortDao 接口中的 insertm_s(@Param("mid") Integer mid,@Param("sid") Integer sid)方法，该方法用于新增电影和类型的关联信息。insertm_s(@Param("mid") Integer mid,@Param("sid") Integer sid)方法的代码如下：

void insertm_s(@Param("mid") Integer mid,@Param("sid") Integer sid);

在 TSortServiceImpl 类被重写的 findAll(TSort tSort)方法中，还调用了 TSortDao 接口中的 queryAll(TSort tSort)方法，该方法用于查询平台上所有电影的类型。queryAll(TSort tSort)方法的代码如下：

List<TSort> queryAll(TSort tSort);

说明

读者可以在 TMovieDao.xml 文件中查找与 insert()方法绑定的 SQL 语句，在 TSortDao.xml 文件中查找与 insertm_s()和 queryAll()方法绑定的 SQL 语句。

5.15 项目运行

通过前述步骤,设计并完成了嗨乐影评平台项目的开发。下面运行本项目,以检验我们的开发成果。如图 5.11 所示,在 IntelliJ IDEA 中,单击▶快捷图标,即可运行本项目。

图 5.11 IntelliJ IDEA 的快捷图标

启动项目后,嗨乐影评平台的首页将被自动打开,如图 5.12 所示。首页会显示 5 部电影的封面图片和名称。用户单击"查看更多"超链接,页面将跳转到"更多"页面。在"更多"页面上,将分页显示全部电影或者某一种类型的电影的封面图片和名称,每一页显示 15 部电影的封面图片和名称。不论在首页上还是在"更多"页面上,用户单击任意一部电影的封面图片或者名称,页面都可以跳转到当前电影的"详情"页面。"详情"页面不仅显示着当前电影的名称、类型、剧情介绍、演职人员等信息和评论内容,而且为用户提供"写评论"的功能。已经登录的用户可以在任意一部电影的"详情"页面上通过"写评论"的功能发表自己对当前电影的观后感。

图 5.12 嗨乐影评平台的首页

SSM 框架是比较标准的 MVC 模式。使用 Spring 实现业务对象的管理,使用 SpringMVC 负责请求的转发和视图的管理,使用 MyBatis 作为数据对象的持久化引擎。标准的 SSM 框架有 4 层,它们分别为 Mapper 层、Service 层、Controller 层和 Entity 层。其中,Mapper 层主要负责对数据库进行数据持久化操作,它的方法都是用于数据库操作的;Service 层是针对使用者的,Service 层的.impl 文件是把 Service 层和 Mapper 层进行整合的文件;Controller 层就是控制器,Controller 通过接收前端传过来的参数进行业务操作,再把处理结果返回前端界面中;Entity 层用来存放实体类,与数据库中的属性值保持一致。

5.16 源码下载

虽然本章详细地讲解了如何编码实现"嗨乐影评平台"项目的各个功能,但给出的代码都是代码片段,而非源码。为了方便读者学习,本书提供了完整的项目源码,扫描右侧二维码即可下载。

第3篇

Spring Boot 应用项目

Spring Boot 是一种基于 Spring 框架的新型开源框架，旨在简化 Spring 应用的创建、运行、调试和部署过程。具体而言，Spring Boot 可以根据项目中的依赖关系自动配置 Spring 框架，减少了大量冗余的 XML 或 Java 配置；它内嵌了 Tomcat、Jetty 或 Undertow 等 Web 服务器，避免了单独部署 Web 服务器的复杂步骤；它支持以 jar 包的形式独立运行 Spring 项目，便于构建和部署；它提供并推荐使用 POM 文件以简化 Maven 配置，使得项目构建更加简单和标准化。Spring Boot 是当下 Java Web 开发中一个非常流行的框架，是企业级应用项目的首选框架，这是因为 Spring Boot 具有快速启动和简单部署的特点。此外，Spring Boot 还特别适合微服务架构和云原生应用项目的开发。

本篇主要使用 Spring Boot 框架、Vue 和 MySQL 等技术开发 3 个 Java Web 应用项目，具体如下：

- ☑ 电瓶车品牌信息管理系统
- ☑ 寻物启事网站
- ☑ 明日之星物业管理系统

第 6 章
电瓶车品牌信息管理系统

——Spring Boot + Vue + MySQL

电瓶车又被称作电动自行车或电动助力车,是一种结合了传统自行车和电动车技术的交通工具。电瓶车作为一种环保、节能、便捷的交通工具,不仅适合短途出行,而且可以作为公共交通的接驳工具,可以减少人们对汽车的依赖,缓解城市交通拥堵和空气污染问题。本章将使用 Java Web 开发中的 Spring Boot 框架、Vue.js 和 MySQL 数据库等关键技术开发一个简单的电瓶车品牌信息管理系统,用于管理电瓶车的品牌信息。

项目微视频

本项目的核心功能及实现技术如下:

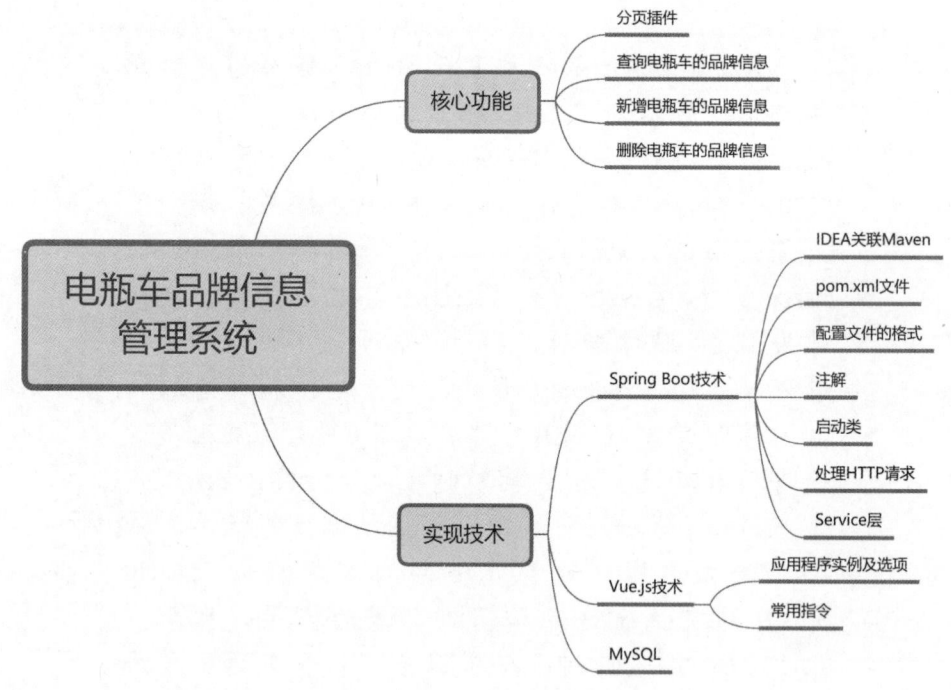

6.1 开发背景

随着科学技术的进步,人们的环保意识不断提高,电瓶车市场不断扩大,不断有新的制造商加入。同时,为了更好地满足消费者的需求,扩大市场占有率,各品牌纷纷推出新款电瓶车车型,市场竞争日趋激烈。为了在市场竞争中脱颖而出,电瓶车制造商需要不断加大研发投入,不断推出更多符合市场需求的产品,不断提升产品质量和服务水平。例如,雅迪、爱玛等传统品牌通过提升产品性能和智能化水平来巩固市场地位;小牛电动、九号等新兴品牌通过推出符合年轻人品位的产品来聚焦高端市场。电瓶车品牌信息管理系统是基于对电瓶车的品牌信息进行管理的网络平台,主要实现了对电瓶车品牌信息的查看、新增、删除等功能。

电瓶车品牌信息管理系统将实现以下目标：
- ☑ 页面简洁、功能明确、操作方便；
- ☑ 用户可以查看各个电瓶车的品牌信息（如品牌名称、品牌评分、好评率和品牌介绍等）；
- ☑ 用户可以新增电瓶车的品牌信息；
- ☑ 用户可以删除某个电瓶车的品牌信息。

6.2 系统设计

6.2.1 开发环境

本项目的开发及运行环境如下：
- ☑ 操作系统：推荐 Windows 10、11 及以上，兼容 Windows 7（SP1）。
- ☑ 开发工具：IntelliJ IDEA。
- ☑ 开发语言：Java EE。
- ☑ 数据库：MySQL 8.0。
- ☑ Web 服务器：Tomcat 9.0 及以上版本。

6.2.2 业务流程

启动项目后，打开浏览器，访问 http://localhost:8080/pages/bikes.html，即可看到电瓶车品牌信息管理系统的主页面。

在主页面上，程序将分页显示 10 条数据，共分两页，每一页最多显示 7 条数据，用户通过分页导航可以随意切换并访问这两个分页的数据。

用户单击"新增电瓶车品牌信息"按钮后，程序将弹出一个窗口，用户在这个窗口中依次输入电瓶车的品牌名称、品牌评分、好评率和品牌介绍等信息，单击"确定"按钮，完成新增电瓶车品牌信息的操作。

在主页面上显示的每一条数据的后面，都有一个"删除"按钮，用户先单击某一个"删除"按钮，再单击删除提示窗口中的"确定"按钮，程序将删除对应的电瓶车品牌信息。

电瓶车品牌信息管理系统的业务流程如图 6.1 所示。

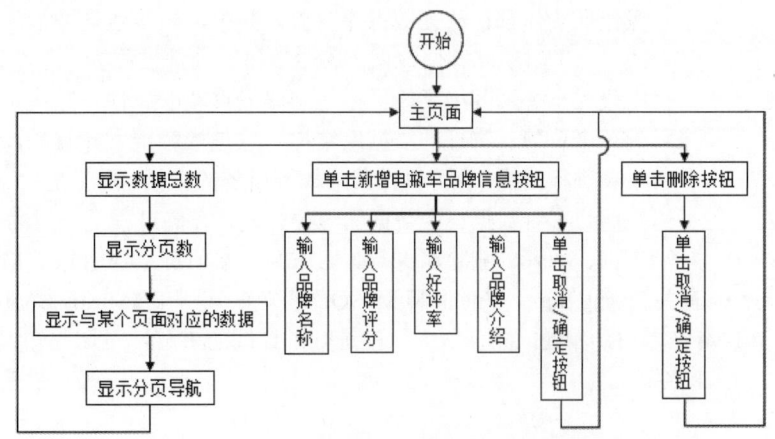

图 6.1　电瓶车品牌信息管理系统的业务流程图

6.2.3 功能结构

本项目的功能结构已经在章首页中给出，作为基于对电瓶车的品牌信息进行管理的网络平台，本项目实现的具体功能如下：

- ☑ 分页插件：用于显示数据总数、分页数、与某个页面对应的数据和分页导航等信息。
- ☑ 查询电瓶车的品牌信息：查询数据库中所有电瓶车的品牌名称、品牌评分、好评率和品牌介绍等信息。
- ☑ 新增电瓶车的品牌信息：向数据库添加新的电瓶车品牌信息。
- ☑ 删除电瓶车的品牌信息：从数据库删除某一条电瓶车品牌信息。

6.3 技术概览

- ☑ Spring Boot：Spring Boot 是在 Spring 的基础上发展而来、全新、开源的框架。开发 Spring Boot 的主要动机是简化部署 Web 项目的配置过程。那么，Spring Boot 是如何以更简单、更灵活的方式开发 Web 项目的呢？它可以自动创建各种工厂类，程序开发人员直接通过依赖注入就可以获取各个类的对象。程序开发人员只需在 Spring Boot 的配置文件中编写一些配置项就可以影响整个项目，即使不编写任何配置，项目也可以采用一套默认配置予以启动。Spring Boot 自带 Tomcat 服务器，在项目启动的过程中可以自动完成所有资源的部署操作。Spring Boot 项目启动的速度很快，即使包含庞大的依赖库，也能够在几秒内完成部署和启动。
- ☑ Vue.js：Vue.js 是一套用于构建用户界面的渐进式框架。与其他重量级框架不同的是，它只关注视图层，采用自底向上增量开发的设计方式。Vue.js 的核心目标之一，是通过尽可能简单的 API 实现响应的数据绑定和可组合的视图组件。它不仅容易上手，还非常容易与其他库或已有项目进行整合。Vue.js 实际上是一个用于开发 Web 前端界面的库，其本身具有响应式编程和组件化的特点。所谓响应式编程，就是保持状态和视图的同步。响应式编程允许将相关模型的变化自动反映到视图上，反之亦然。和其他前端框架一样，Vue.js 同样采用"一切都是组件"的理念，即将一个网页分割成多个可复用的组件。
- ☑ MySQL：在 MySQL 数据库中，表、字段、索引、视图、存储过程等用于存储数据或对数据进行操作的实体都被称为数据库对象。其中，表是包含数据库中所有数据的数据库对象，由行和列组成，用于组织和存储数据；表中的每一列都被称为一个字段，字段具有自己的属性，如字段类型、字段大小等，其中字段类型是字段最重要的属性，它决定了字段能够存储哪种类型的数据；索引是一个单独、物理的数据库结构，它是依赖于表建立的，在数据库中索引使数据库程序无须对整个表进行扫描，就可以在其中找到所需的数据；视图是从一张或多张表中导出的表（也称虚拟表），是用户查看数据表中数据的一种方式；存储过程是一组为了完成特定功能的 SQL 语句集合（包含查询、插入、删除和更新等操作）。本项目主要用到了 MySQL 数据库中的表和字段。

有关 Spring Boot、Vue.js 和 MySQL 的知识分别在《Spring Boot 从入门到精通》《Vue.js 从入门到精通》《Java 从入门到精通（第 7 版）》中进行了详细的讲解，对这些知识不太熟悉的读者可以参考这 3 本书中的相关内容。

6.4 Spring Boot 技术基础

在第 6.3 节中，只是对 Spring Boot 进行了简明扼要的概述。下面将对本项目用到的 Spring Boot 中的重点知识进行必要介绍，以确保读者可以顺利完成本项目。

6.4.1 IDEA 关联 Maven

因为在使用 Spring Boot 开发 Web 项目的过程中会依赖一些第三方 jar 包，所以本书选用了 Maven 管理这些第三方 jar 包。程序开发人员只需要在 XML 文件中填写 Web 项目所需 jar 包的名称和版本号等信息，Maven 就可以自动从服务器下载并向 Web 项目导入这些 jar 包。在 IntelliJ IDEA 中关联 Maven 的具体步骤如下：

（1）在成功安装 IntelliJ IDEA 后，创建一个名为 Demo 的 Java 项目（只创建项目即可，不需要在其中创建类）。

（2）在 IntelliJ IDEA 的菜单栏中，依次单击 File→Settings，打开 Build, Execution, Deployment→Build Tools，再单击 Maven，如图 6.2 所示。

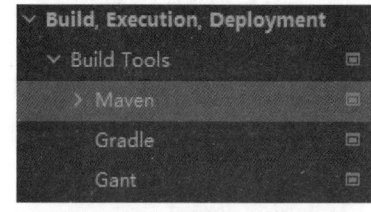

图 6.2　在 IntelliJ IDEA 的菜单栏中找到并单击 Maven

（3）本书使用的 Maven 版本是 3.6.3，笔者把 Maven 安装在了 D:\maven 目录下。读者需要根据 Maven 的安装目录（不要被笔者的安装目录误导），对如图 6.3 所示的 3 个目录进行设置。其中，Maven home path 表示 Maven 的安装目录，User settings file 表示 settings.xml 文件所在的目录，Local repository 表示本地仓库的目录。设置完毕后，单击 OK 按钮。

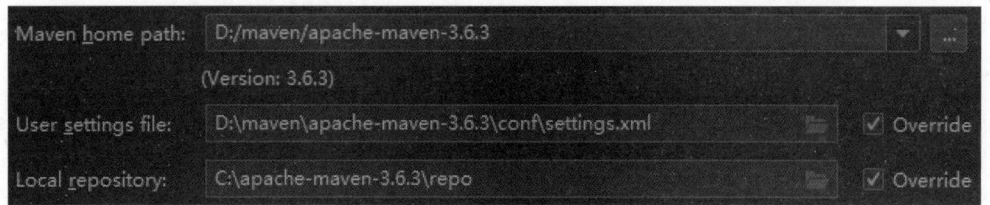

图 6.3　设置 3 个目录

6.4.2 pom.xml 文件

pom.xml 是 Maven 构建项目的核心配置文件，程序开发人员可以在此文件中为项目添加新的依赖，将新的依赖添加在<dependencies>标签内部，作为其子标签，格式如下：

```
<dependency>
    <groupId>所属团队</groupId>
    <artifactId>项目 ID</artifactId>
    <version>版本号</version>
    <scope>使用范围（可选）</scope>
</dependency>
```

注意

<dependency>是<dependencies>的子标签，dependencies 是 dependency 的复数形式。

例如，Spring Boot 项目自带的 Web 依赖和 Junit 单元测试依赖，其在 pom.xml 中的代码如下：

```
<dependency>
    <groupId>org.springframework.boot</groupId>
    <artifactId>spring-boot-starter-web</artifactId>
</dependency>
```

```xml
<dependency>
    <groupId>org.springframework.boot</groupId>
    <artifactId>spring-boot-starter-test</artifactId>
    <scope>test</scope>
</dependency>
```

程序开发人员只需要仿照这种格式在<dependencies>标签内部添加其他依赖，而后保存 pom.xml 文件，Maven 就会自动下载依赖中的 jar 文件并自动将其引入项目中。

6.4.3 配置文件的格式

程序开发人员在配置 Spring Boot 项目的过程中，会在配置文件中配置该项目所需的数据信息。这些数据信息被称作"配置信息"。那么，配置信息都包含哪些内容呢？在实际开发中，配置信息的内容非常丰富，这里仅举例予以说明。

- ☑ Tomcat 服务器。
- ☑ 数据库的连接信息，即用于连接数据库的用户名和密码。
- ☑ Spring Boot 项目的启动端口。
- ☑ 第三方系统或者接口的调用密钥信息。
- ☑ 打印用于发现和定位问题的日志。

Spring Boot 支持多种格式的配置文件，最常用的是 properties 格式（默认格式）和比较新颖的 yml 格式。下面将分别介绍这两种格式的特点。

1. properties 格式

properties 格式是经典的键值对文本格式。也就是说，如果某一个配置文件的格式是 properties 格式，那么这个配置文件的文本格式为键值对。键值对的语法非常简单，具体如下：

```
key=value
```

"="左侧为键（key），"="右侧为值（value）。在配置文件中，每个键独占一行。如果多个键之间存在层级关系，就需要使用"父键.子健"的格式予以表示。例如，在配置文件中，为一个有三层关系的键赋值的语法如下：

```
key1.key2.key3=value
```

例如，启动 Spring Boot 项目的 Tomcat 端口号为 8080，那么在这个项目的 application.properties 文件中就能够找到如下内容：

```
server.port=8080
```

启动这个项目后，即可在控制台看到如下一行日志：

```
Tomcat started on port(s): 8080 (http) with context path ''
```

这行日志表明 Tomcat 根据配置开启的是 8080 端口。

在 application.properties 文件中，"#"被称作注释符号，用于向其中添加注释信息。例如：

```
# Tomcat 端口
server.port=8080
```

application.properties 文件不支持中文。如果程序开发人员在 application.properties 文件中编写中文，IntelliJ IDEA 会自动将其转化为 Unicode 码，将鼠标悬停在 Unicode 码处可以看到对应的中文。

2. yml 格式

yml 是 YAML 的缩写，它是一种可读性高、用于表达数据序列化的文本格式。对于 yml 格式的配置文件，其文本格式也是键值对。只不过，键值对的语法与 Python 语言中的键值对的语法非常相似，具体如下：

key: value

英文格式的"："与值之间必须有至少一个空格。

英文格式的"："左侧为键（key），英文格式的"："右侧为值（value）。需要注意的是，英文格式的"："与值之间只能用空格缩进，不能用 tab 缩进；空格数量表示各层的层级关系。例如，在配置文件中，为一个有三层关系的键赋值的语法如下：

```
key1:
 key2:
  key3: value
```

在 properties 格式的配置文件中，即使父键相同，在为每一个的子健赋值时也要单独占一行，还要把父键写完整，例如：

```
com.mr.strudent.name=tom
com.mr.strudent.age=21
```

但是在 yml 格式的配置文件中，只需要编写一次父键，并保证两个子健缩进关系相同即可。例如，把上述 properties 格式的键值对修改为 yml 格式的键值对的语句如下：

```
com:
 mr:
  student:
   name: Tom
   age: 21
```

对于 Spring Boot 项目的配置文件，不论是采用 properties 格式，还是采用 yml 格式，都由程序开发人员自行决定。但是，在同一个 Spring Boot 项目中，尽量只使用一种格式的配置文件。否则，这个 Spring Boot 项目中的 yml 格式的配置文件将被忽略。

6.4.4 注解

在给出注解的概念之前，须明确什么是元数据。所谓元数据，指的是用于描述数据的数据。下面结合某个配置文件里的一行信息，举例说明什么是元数据。

```
<string name="app_name">AnnotionProject</string>
```

上述信息中的数据 app_name 是用于描述数据 AnnotionProject 的。也就是说，数据 app_name 就是元数据。那么，什么是注解呢？注解又被称作标注，是一种被加入源码、具有特殊语法的元数据。需要特别说明的是：

- ☑ 注解仅是元数据，和业务逻辑无关。
- ☑ 虽然注解不属于程序本身，但是可以对程序作出解释。
- ☑ 应用程序中的类、方法、变量、参数、包等程序元素都可以被注解。

理解了"什么是注解"后，再来了解一下在应用程序中注解的应用体现在哪些方面。

- ☑ 在编译时进行格式检查。例如，如果被@Override 标记的方法不是父类的某个方法，编译器就会报错。
- ☑ 减少配置。依据代码的依赖性，使用注解替代配置文件。
- ☑ 减少重复工作。在程序开发的过程中，通过注解减少对某个方法的调用次数。

Spring Boot 是一个支持海量注解的框架，其自带的常用注解如表 6.1 所示。

表 6.1　Spring Boot 的常用注解

注　　解	标 注 位 置	功　　能
@Autowired	成员变量	自动注入依赖
@Bean	方法	用@Bean 标注方法等价于 XML 中配置的 bean，用于注册 Bean
@Component	类	用于注册组件。当不清楚注册类属于哪个模块时就用这个注解
@ComponentScan	类	开启组件扫描器
@Configuration	类	声明配置类
@ConfigurationProperties	类	用于加载额外的 properties 配置文件
@Controller	类	声明控制器类
@ControllerAdvice	类	可用于声明全局异常处理类和全局数据处理类
@EnableAutoConfiguration	类	开启项目的自动配置功能
@ExceptionHandler	方法	用于声明处理全局异常的方法
@Import	类	用于导入一个或者多个 @Configuration 注解标注的类
@ImportResource	类	用于加载 xml 配置文件
@PathVariable	方法参数	让方法参数从 URL 中的占位符中取值
@Qualifier	成员变量	与@Autowired 配合使用，当 Spring 容器中有多个类型相同的 Bean 时，可以用@Qualifier("name")来指定注入哪个名称的 Bean
@RequestMapping	方法	指定方法可以处理哪些 URL 请求
@RequestParam	方法参数	让方法参数从 URL 参数中取值
@Resource	成员变量	与@AutoWired 功能类似，但有 name 和 type 两个参数，可根据 Spring 配置的 bean 的名称进行注入
@ResponseBody	方法	表示方法的返回结果直接写入 HTTP response body 中。如果返回值是字符串，则直接在网页上显示该字符串
@RestController	类	相当于@Controller 和@ResponseBody 的合集，表示这个控制器下的所有方法都被@ResponseBody 标注
@Service	服务的实现类	用于声明服务的实现类
@SpringBootApplication	主类	用于声明项目主类
@Value	成员变量	动态注入，支持"#{ }"与"$ { }"表达式

这些注解的编码位置是非常灵活的。当注解用于标注类、成员变量和方法时，注解的编码位置可以在成员变量的上边，例如：

```
@Autowired
private String name;
```

也可以在成员变量的左边，例如：

```
@Autowired private String name;
```

在 Spring Boot 的常用注解中，需特别说明的是，使用@RequestParam 能够标注方法中的参数。例如：

```
@RequestMapping("/user")
@ResponseBody
public String getUser(@RequestParam Integer id) {
    return "success";
}
```

6.4.5 启动类

使用注解能够启动一个 Spring Boot 项目,这是因为在每一个 Spring Boot 项目中都有一个启动类,并且启动类必须被@SpringBootApplication 注解标注,进而能够调用用于启动一个 Spring Boot 项目的 SpringApplication.run()方法。

在本项目中,com.mr 包下的 RunApplication 类就是启动类。代码如下:

```
package com.mr;

import org.springframework.boot.SpringApplication;
import org.springframework.boot.autoconfigure.SpringBootApplication;

@SpringBootApplication
public class RunApplication {                                    //启动类
    public static void main(String[] args) {                     //主方法
        SpringApplication.run(RunApplication.class, args);
    }
}
```

@SpringBootApplication 注解虽然重要,但使用起来非常简单,因为这个注解是由多个功能强大的注解整合而成的。打开@SpringBootApplication 注解的源码可以看到它被很多其他注解标注,其中最核心的 3 个注解如下:

☑ @SpringBootConfiguration 注解,让项目采用基于 Java 注解的配置方式,而不是传统的 XML 文件配置。当然,如果程序开发人员写了传统的 XML 配置文件,Spring Boot 也是能够读取这些 XML 文件并识别里面的内容的。

☑ @EnableAutoConfiguration 注解,开启自动配置。这样 Spring Boot 在启动的时候就可以自动加载所有配置文件和配置类了。

☑ @ComponentScan 注解,启用组件扫描器。这样项目才能自动发现并创建各个组件的 Bean,包括 Web 控制器(@Controller)、服务(@Service)、配置类(@Configuration)和其他组件(@Component)。

注意

一个项目可以有多个启动类,但这样的代码毫无意义。一个项目应该只使用一次@SpringBootApplication 注解。

6.4.6 处理 HTTP 请求

在开发 Spring Boot 项目的过程中,Spring Boot 的典型应用是处理 HTTP 请求。所谓处理 HTTP 请求,就是 Spring Boot 把用户通过 URL 地址发送的请求交给不同的业务代码进行处理的过程。

Spring Boot 提供了用于声明控制器类的@Controller 注解。也就是说,在 Spring Boot 项目中,把被@Controller 注解标注的类称作控制器类。控制器类在 Spring Boot 项目中发挥的作用是处理用户发送的 HTTP 请求。Spring Boot 会把不同的用户请求交给不同的控制器进行处理,而控制器则会把处理后得到的结果反馈给用户。

说明

控制器(Controller)定义了应用程序的行为,它负责对用户发送的请求进行解释,并把这些请求映射成相应的行为。

因为@Controller 注解本身被@Component 注解标注,所以控制器类属于组件。这说明在启动 Spring Boot

项目时，控制器类会被扫描器自动扫描。这样，程序开发人员就可以在控制器类中注入 Bean。例如，在控制器中注入 Environment 环境组件，代码如下：

```
@Controller
public class TestController {
    @Autowired
    Environment env;
}
```

Spring Boot 提供了用于映射 URL 地址的@RequestMapping 注解。@RequestMapping 注解可以标注类和方法。如果一个类或者方法被@RequestMapping 注解标注，那么这个类或者方法就能够处理用户通过@RequestMapping 注解映射的 URL 地址发送的请求。

> **注意**
> @Controller 注解要结合@RequestMapping 注解一起使用。

@RequestMapping 有几个常用属性，下面主要对 value 属性进行介绍。

value 属性是@RequestMapping 注解的默认属性，用于指定映射的 URL 地址。在单独使用 value 属性时，value 属性可以被隐式调用。调用 value 属性的语法如下：

```
@RequestMapping("test")
@RequestMapping("/test")
@RequestMapping(value= "/test")
@RequestMapping(value={"/test"})
```

上面这 4 种语法所映射的 URL 地址均为"域名/test"。其中，域名指的是当前 Spring Boot 项目所在的域。如果在 IntelliJ IDEA 中启动一个 Spring Boot 项目，那么域名就是 127.0.0.1:8080。

@RequestMapping 注解映射的 URL 地址可以是多层的。例如：

```
@RequestMapping("/shop/books/computer")
```

上述代码映射的完整的 URL 地址是 http://127.0.0.1:8080/shop/books/computer。需要特别注意的是，这个 URL 地址中的任何一层都是不可或缺的，否则将引发 404 错误。

@RequestMapping 注解允许一个方法同时映射多个 URL 地址。其语法如下：

```
@RequestMapping(value = { "/address1", "/address2", "/address3", ....... })
```

6.4.7 Service 层

Spring Boot 中的 Service 层是业务逻辑层，其作用是处理业务需求，封装业务方法，执行 Dao 层中用于访问、处理数据的操作。Service 层通常由一个接口和这个接口的实现类组成。其中，Service 层的接口可以在 Controller 层中被调用，用于实现数据的传递和处理；Service 层的实现类须使用@Service 注解予以标注。

在 Spring Boot 中，把被@Service 注解标注的类称作服务类。@Service 注解属于 Component 组件，可以被 Spring Boot 的组件扫描器扫描到。当启动 Spring Boot 项目时，服务类的对象会被自动创建，并被注册成 Bean。

Service 层的实现过程如图 6.4 所示。

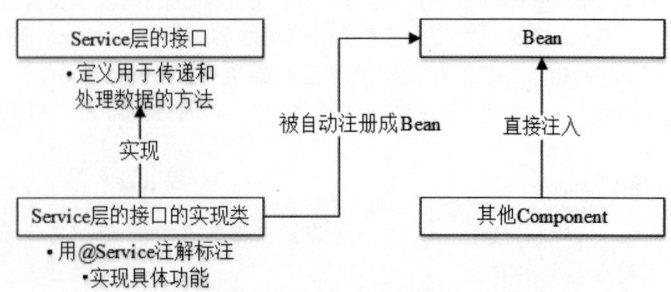

图 6.4 Service 层的实现过程

（1）定义一个 Service 层的接口，在这个接口中定义用于传递和处理数据的方法。例如，定义一个 Service 层的接口 ProductService，代码如下：

```
public interface ProductService {
    …                                       //省略用于传递和处理数据的方法
}
```

（2）定义一个 Service 层的接口的实现类，使用@Service 注解予以标注。这个实现类的作用有两个：一个是实现 Service 层的接口中的业务方法；另一个是执行 Dao 层中用于访问、处理数据的操作。例如，使用@Service 注解标注实现 ProductService 接口的 ProductServiceImpl 类，代码如下：

```
@Service
public class ProductServiceImpl implements ProductService {
    …                    //省略用于实现接口的业务方法和用于执行访问处理数据的操作的代码
}
```

（3）在服务类的对象被自动创建并被注册成 Bean 之后，其他 Component 组件即可直接注入这个 Bean。

6.5　Vue.js 技术基础

通过第 6.3 节，读者已经明确了什么是 Vue.js，本节将对本项目中用到的 Vue.js 中的重点知识进行必要介绍，以确保读者可以顺利完成本项目。

6.5.1　应用程序实例及选项

每个 Vue.js 的应用都需要创建一个应用程序的实例对象并挂载到指定 DOM 上。在 Vue.js 3.0 中，创建一个应用程序实例的语法格式如下：

```
Vue.createApp(App)
```

createApp()是一个全局 API，它接收一个根组件选项对象作为参数。选项对象中包括数据、方法、生命周期钩子函数等。创建应用程序实例后，可以调用实例的 mount()方法，将应用程序实例的根组件挂载到指定的 DOM 元素上。这样，该 DOM 元素中的所有数据变化都会被 Vue.js 所监控，从而实现数据的双向绑定。例如，要绑定的 DOM 元素的 id 属性值为 app，创建一个应用程序实例并绑定到该 DOM 元素的代码如下：

```
Vue.createApp(App).mount('#app')
```

下面分别对组件选项对象中的几个常用选项进行介绍。

1. 数据

在组件选项对象中有一个 data 选项，该选项是一个函数，Vue.js 在创建组件实例时会调用该函数。data()函数可以返回一个数据对象，应用程序实例本身会代理数据对象中的所有数据。例如，创建一个根组件实例 vm，在实例的 data 选项中定义一个数据。代码如下：

```
<div id="app">
    <h2>{{text}}</h2>
</div>
<script src="https://unpkg.com/vue@next"></script>
<script type="text/javascript">
    //创建应用程序实例
    const vm = Vue.createApp({
        //返回数据对象
```

```
        data(){
            return {
                text:'千里之行，始于足下。'              //定义数据
            }
        }
    //挂载应用程序实例的根组件
    }).mount('#app');
</script>
```

上述代码中，将创建的根组件实例赋值给变量 vm，在实际开发中并不要求一定要将根组件实例赋值给某个变量。

2. 方法

在创建的应用程序实例中，通过 methods 选项可以定义方法。应用程序实例本身也会代理 methods 选项中的所有方法，因此也可以像访问 data 数据那样来调用方法。例如，在根组件实例的 methods 选项中定义一个 showInfo()方法，代码如下：

```
<div id="app">
    <p>{{showInfo()}}</p>
</div>
<script src="https://unpkg.com/vue@next"></script>
<script type="text/javascript">
    //创建应用程序实例
    const vm = Vue.createApp({
        //返回数据对象
        data(){
            return {
                text : '静以修身，俭以养德。',
                author : ' —— 诸葛亮'
            }
        },
        methods : {
            showInfo : function(){
                return this.text + this.author;       //连接字符串
            }
        }
    //挂载应用程序实例的根组件
    }).mount('#app');
</script>
```

3. 生命周期钩子函数

每个应用程序实例在创建时都有一系列的初始化步骤。例如，创建数据绑定、编译模板、将实例挂载到 DOM 并在数据变化时触发 DOM 更新、销毁实例等。在这个过程中会运行一些叫作生命周期钩子的函数，通过这些钩子函数可以定义业务逻辑。应用程序实例中几个主要的生命周期钩子函数说明如下：

- ☑ beforeCreate：在实例初始化之后，数据观测和事件/监听器配置之前调用。
- ☑ created：在实例创建之后进行调用，此时尚未开始 DOM 编译。在需要初始化处理一些数据时会比较有用。
- ☑ beforeMount：在挂载开始之前进行调用，此时 DOM 还无法操作。
- ☑ mounted：在 DOM 文档渲染完毕之后进行调用。相当于 JavaScript 中的 window.onload()方法。
- ☑ beforeUpdate：在数据更新时进行调用，适合在更新之前访问现有的 DOM，如手动移除已添加的事件监听器。
- ☑ updated：在数据更改导致的虚拟 DOM 被重新渲染时进行调用。
- ☑ beforeDestroy：在销毁实例前进行调用，此时实例仍然有效。此时可以解绑一些使用 addEventListener 监听的事件等。

☑ destroyed：在实例被销毁之后进行调用。

下面通过一个示例来了解 Vue.js 内部的运行机制。代码如下：

```
<div id="app">
    <p>{{text}}</p>
</div>
<script src="https://unpkg.com/vue@next"></script>
<script type="text/javascript">
    //创建应用程序实例
    const vm = Vue.createApp({
        //返回数据对象
        data(){
            return {
                text : '山不在高，有仙则名。'
            }
        },
        beforeCreate : function(){
            console.log('beforeCreate');
        },
        created : function(){
            console.log('created');
        },
        beforeMount : function(){
            console.log('beforeMount');
        },
        mounted : function(){
            console.log('mounted');
        },
        beforeUpdate : function(){
            console.log('beforeUpdate');
        },
        updated : function(){
            console.log('updated');
        }
    //挂载应用程序实例的根组件
    }).mount('#app');
    setTimeout(function(){
        vm.text = "水不在深，有龙则灵。";
    },2000);
</script>
```

在浏览器控制台中运行上述代码，页面渲染完成后，结果如图 6.5 所示。

经过 2 秒后调用 setTimeout()方法，修改 text 的内容，触发 beforeUpdate 和 updated 钩子函数，结果如图 6.6 所示。

6.5.2 常用指令

在 Vue.js 中，为了实现渲染视图的功能，指令是必不可少的。例如，在视图中经常需要通过条件判断控制 DOM 的显示状态，这时就需要使用 v-if、v-else、v-else-if 等指令。下面将对 Vue.js 中的常用指令进行介绍。

1. v-if 指令

v-if 指令可以根据表达式的值来判断是否输出 DOM 元素及其包含的子元素。如果表达式的值为 true，就输出 DOM 元素及其包含的子元素；否则，就将 DOM 元素及其包含的子元素移除。

v-if 是一个指令，必须将它添加到一个元素上，根据表达式的结果判断是否输出该元素。如果需要对一

图 6.5 页面渲染后的效果

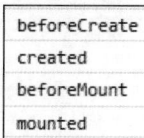

图 6.6 页面最终效果

组元素进行判断，则需要使用<template>元素作为包装元素，并在该元素上使用 v-if 指令，最后的渲染结果里不会包含<template>元素。

例如，根据表达式的结果判断是否输出一组单选按钮。代码如下：

```
<div id="app">
    <template v-if="show">
        <input type="radio" value="手机">手机
        <input type="radio" value="电脑">电脑
        <input type="radio" value="家电">家电
        <input type="radio" value="家具">家具
    </template>
</div>
<script src="https://unpkg.com/vue@next"></script>
<script type="text/javascript">
    //创建应用程序实例
    const vm = Vue.createApp({
        //返回数据对象
        data(){
            return {
                show : true
            }
        }
    //挂载应用程序实例的根组件
    }).mount('#app');
</script>
```

2. v-else 指令

v-else 指令的作用相当于 JavaScript 中的 else 语句部分。可以将 v-else 指令配合 v-if 指令一起使用。

例如，输出用户的年龄，并判断该年龄是否小于 18，如果是则输出用户未成年，否则输出用户已成年。代码如下：

```
<div id="app">
    <p>Tom 的年龄是{{age}}</p>
    <p v-if="age<18">Tom 未成年</p>
    <p v-else>Tom 已成年</p>
</div>
<script src="https://unpkg.com/vue@next"></script>
<script type="text/javascript">
    //创建应用程序实例
    const vm = Vue.createApp({
        //返回数据对象
        data(){
            return {
                age: 20
            }
        }
    //挂载应用程序实例的根组件
    }).mount('#app');
</script>
```

3. v-else-if 指令

v-else-if 指令的作用相当于 JavaScript 中的 else if 语句部分，应用该指令可以进行更多的条件判断，不同的条件对应不同的输出结果。

例如，输出数据对象中的属性 m 和 n 的值，并根据比较两个属性的值，输出比较的结果。代码如下：

```
<div id="app">
    <p>m 的值是{{m}}</p>
```

```
        <p>n 的值是{{n}}</p>
        <p v-if="m<n">m 小于 n</p>
        <p v-else-if="m===n">m 等于 n</p>
        <p v-else>m 大于 n</p>
    </div>
    <script src="https://unpkg.com/vue@next"></script>
    <script type="text/javascript">
        //创建应用程序实例
        const vm = Vue.createApp({
            //返回数据对象
            data(){
                return {
                    m: 16,
                    n: 16
                }
            }
        //挂载应用程序实例的根组件
        }).mount('#app');
    </script>
```

4. v-show 指令

v-show 指令是根据表达式的值来判断是否显示或隐藏 DOM 元素。当表达式的值为 true 时，元素将被显示；当表达式的值为 false 时，元素将被隐藏，此时为元素添加了一个内联样式 style="display:none"。与 v-if 指令不同，使用 v-show 指令的元素，无论表达式的值为 true 还是 false，该元素始终都会被渲染并保留在 DOM 中。绑定值的改变只是简单地切换元素的 CSS 属性 display。

> **注意**
>
> v-show 指令不支持<template>元素，也不支持 v-else 指令。

5. v-for 指令

Vue.js 提供了列表渲染的功能，可将数组或对象中的数据循环渲染到 DOM 中。在 Vue.js 中，列表渲染使用的是 v-for 指令，其效果类似于 JavaScript 中的遍历。

v-for 指令将根据接收的数组中的数据重复渲染 DOM 元素。该指令需要使用 item in items 形式的语法。其中，items 为数据对象中的数组名称，item 为数组元素的别名，通过别名可以获取当前数组遍历的每个元素。

例如，应用 v-for 指令将<li>标签循环渲染，输出数组中存储的职位名称。代码如下：

```
<div id="app">
    <ul>
        <li v-for="item in items">{{item.position}}</li>
    </ul>
</div>
<script src="https://unpkg.com/vue@next"></script>
<script type="text/javascript">
    const vm = Vue.createApp({
        data(){
            return {
                items : [                                           //定义职位数组
                    { position : '前端工程师'},
                    { position : '一二线运维'},
                    { position : '项目经理'}
                ]
            }
```

```
        }).mount('#app');
    </script>
```

在应用 v-for 指令遍历数组时，还可以指定一个参数作为当前数组元素的索引，语法格式为(item,index) in items。其中，items 为数组名称，item 为数组元素的别名，index 为数组元素的索引。

例如，应用 v-for 指令将<li>标签循环渲染，输出数组中存储的职位名称和相应的索引。代码如下：

```
<div id="app">
    <ul>
        <li v-for="(item,index) in items">{{index}} - {{item.position}}</li>
    </ul>
</div>
<script src="https://unpkg.com/vue@next"></script>
<script type="text/javascript">
    const vm = Vue.createApp({
        data(){
            return {
                items : [                                          //定义职位数组
                    { position : '前端工程师'},
                    { position : '一二线运维'},
                    { position : '项目经理'}
                ]
            }
        }
    }).mount('#app');
</script>
```

6.6 数据库设计

本项目使用的是 MySQL 数据库。在 MySQL 数据库中创建一个名为 db_e-bike 的库，在这个库中创建与电瓶车品牌信息对应的表。

电瓶车品牌信息表的名称为 bike，主要用于存储电瓶车的品牌编号、品牌名称、品牌评分、好评率、品牌介绍等，其结构如表 6.2 所示。

表 6.2 bike 表结构

字段名称	数据类型	长 度	是否主键	说 明
id	INT		主键	电瓶车的品牌编号
brand_name	VARCHAR	20		电瓶车的品牌名称
brand_rating	VARCHAR	10		电瓶车的品牌评分
favorable_rate	VARCHAR	10		电瓶车的好评率
brand_intro	VARCHAR	255		电瓶车的品牌介绍

6.7 添加依赖和配置信息

在开发 Spring Boot 项目的过程中，程序开发人员不仅需要为当前项目手动添加依赖，而且需要为当前项目手动添加配置信息。下面将分别介绍如何为本项目添加依赖和配置信息。

6.7.1 在 pom.xml 文件中添加依赖

因为本项目把 Maven 作为项目构建工具，而 pom.xml 是 Maven 构建项目的核心配置文件，所以需要在 pom.xml 文件中为本项目添加依赖，这些依赖会被添加到 pom.xml 文件中的<dependencies>标签内部。代码如下：

```xml
<?xml version="1.0" encoding="UTF-8"?>
<project xmlns="http://maven.apache.org/POM/4.0.0" xmlns:xsi="http://www.w3.org/2001/XMLSchema-instance"
         xsi:schemaLocation="http://maven.apache.org/POM/4.0.0 https://maven.apache.org/xsd/maven-4.0.0.xsd">
    <modelVersion>4.0.0</modelVersion>
    <parent>
        <groupId>org.springframework.boot</groupId>
        <artifactId>spring-boot-starter-parent</artifactId>
        <version>2.6.3</version>
    </parent>

    <groupId>com.gs</groupId>
    <artifactId>springboot_crud</artifactId>
    <version>0.0.1-SNAPSHOT</version>

    <properties>
        <java.version>1.8</java.version>
    </properties>

    <dependencies>

        <dependency>
            <groupId>com.baomidou</groupId>
            <artifactId>mybatis-plus-boot-starter</artifactId>
            <version>3.4.3</version>
        </dependency>

        <dependency>
            <groupId>com.alibaba</groupId>
            <artifactId>druid-spring-boot-starter</artifactId>
            <version>1.2.6</version>
        </dependency>

        <dependency>
            <groupId>org.springframework.boot</groupId>
            <artifactId>spring-boot-starter-web</artifactId>
        </dependency>

        <dependency>
            <groupId>mysql</groupId>
            <artifactId>mysql-connector-java</artifactId>
        </dependency>

        <dependency>
            <groupId>org.springframework.boot</groupId>
            <artifactId>spring-boot-starter-test</artifactId>
            <scope>test</scope>
        </dependency>

        <!--lombok-->
        <dependency>
            <groupId>org.projectlombok</groupId>
            <artifactId>lombok</artifactId>
```

```xml
            </dependency>
        </dependencies>

        <build>
            <plugins>
                <plugin>
                    <groupId>org.springframework.boot</groupId>
                    <artifactId>spring-boot-maven-plugin</artifactId>
                </plugin>
            </plugins>
        </build>
</project>
```

6.7.2 在 application.yml 文件中添加配置信息

本项目采用的是 yml 格式的配置文件，并在 application.yml 文件中添加如下的配置信息：

```yml
server:
  port: 8080
spring:
  datasource:
    druid:
      driver-class-name: com.mysql.cj.jdbc.Driver
      url: jdbc:mysql://localhost:3306/db_e-bike?useUnicode=true&characterEncoding=UTF-8&serverTimezone=GMT%2b8
      username: root
      password: root

mybatis-plus:
  global-config:
    db-config:
      id-type: auto
  configuration:
    log-impl: org.apache.ibatis.logging.stdout.StdOutImpl
```

6.8 工具类设计

将一些反复调用的代码封装成工具类，不仅可以提高开发效率，还可以提高代码的可读性。本项目具有两个工具类，分别是全局异常处理类和通用返回类。下面将分别介绍这两个类。

6.8.1 全局异常处理类

当一个 Spring Boot 项目没有对用户触发的异常进行拦截时，用户触发的异常就会触发最底层异常。在实际开发中，程序开发人员必须对最底层异常进行拦截。

拦截全局最底层异常的方式非常简单，只需在全局异常处理类中单独写一个"兜底"的、用于处理异常的方法，并使用@ExceptionHandler(Exception.class)注解予以标注。

本项目的全局异常处理类是 com.mr.controller.util 工具包下的 ProjectExceptionAdvice 类，该类的代码如下：

```java
package com.mr.controller.util;

import org.springframework.web.bind.annotation.ExceptionHandler;
```

```java
public class ProjectExceptionAdvice {
    //拦截所有的异常信息
    @ExceptionHandler
    public R doException(Exception ex){
        ex.printStackTrace();
        return new R("服务器故障，请稍后再试！");
    }
}
```

6.8.2　通用返回类

在实际开发过程中，需要编写很多个控制器。虽然这些控制器中的方法各不相同，但是这些控制器的作用都是先让后端处理由前端发送的请求，再把由后端返回的结果传递给前端。程序开发人员习惯把由后端返回的所有结果都统一封装成一个类，并把这个类称作"通用返回类"，同时定义这个类为 R 类，这样由后端传递给前端的结果的类型就都是 R 类型了。也就是说，在控制器中，R 类不仅接收了由后端处理的结果，而且被传递给前端，进而统一了返回的类型。

在本项目的 R 类中，包含了 3 个私有的属性，它们分别是表示 Boolean 型对象的 bool、表示实体类对象的 obj 和表示字符串信息的 str。为了方便外部类访问这 3 个私有的属性，需要为它们添加 Getter/Setter 方法。

此外，在 R 类中还包含了 1 个无参构造方法和 4 个有参构造方法，这 4 个有参构造方法分别为只含有 Boolean 型对象的 bool 的构造方法、含有 Boolean 型对象的 bool 和实体类对象的 obj 的构造方法、含有 Boolean 型对象的 bool 和字符串信息的 str 的构造方法、只含有字符串信息的 str 的构造方法。com.mr.controller.util 工具包下的 R 类的代码如下：

```java
package com.mr.controller.util;
public class R {                                    //通用返回值类
    private Boolean bool;                           //Boolean 型对象
    private Object obj;                             //实体类对象
    private String str;                             //字符串信息

    //为通用返回值类添加无参构造方法和有参构造方法
    public R() {
    }

    public R(Boolean flag) {
        this.bool = flag;
    }

    public R(Boolean flag, Object data) {
        this.bool = flag;
        this.obj = data;
    }

    public R(Boolean flag, String msg) {
        this.bool = flag;
        this.str = msg;
    }

    public R(String msg) {
        this.str = msg;
    }

    //分别为上述 3 个属性添加 Getter/Setter 方法
```

```java
    public Boolean getFlag() {
        return bool;
    }

    public void setFlag(Boolean flag) {
        this.bool = flag;
    }

    public Object getData() {
        return obj;
    }

    public void setData(Object data) {
        this.obj = data;
    }

    public String getMsg() {
        return str;
    }

    public void setMsg(String msg) {
        this.str = msg;
    }
}
```

6.9 实体类设计

实体类又称数据模型类。顾名思义,实体类是一种专门用于保存数据模型的类。每一个实体类都对应着一种数据模型,通常会将类的属性与数据表的字段相对应。虽然实体类的属性都是私有的,但是通过每一个属性的 Getter/Setter 方法,外部类就能够获取或修改实体类的某一个属性值。实体类通常都会提供无参构造方法,并根据具体情况确定是否提供有参构造方法。

本项目只有一个实体类,这个实体类对应的是 com.mr.pojo 包下的 Bike.java 文件,表示电瓶车类。电瓶车类中的品牌编号、品牌名称、品牌评分、好评率、品牌介绍这 5 个属性,与 db_e-bike 库的 bike 表中的 5 个字段相对应。为了方便外部类访问这 5 个私有的属性,需要为它们添加 Getter/Setter 方法。com.mr.pojo 包下的 Bike 类代码如下:

```java
package com.mr.pojo;

public class Bike {
    private Integer id;                    //编号
    private String brandName;              //品牌名称
    private String brandRating;            //品牌评分
    private String favorableRate;          //好评率
    private String brandIntro;             //品牌介绍
    //分别为上述 5 个属性添加 Getter/Setter 方法
    public Integer getId() {
        return id;
    }

    public void setId(Integer id) {
        this.id = id;
    }

    public String getBrandName() {
```

```java
        return brandName;
    }

    public void setBrandName(String brandName) {
        this.brandName = brandName;
    }

    public String getBrandRating() {
        return brandRating;
    }

    public void setBrandRating(String brandRating) {
        this.brandRating = brandRating;
    }

    public String getFavorableRate() {
        return favorableRate;
    }

    public void setFavorableRate(String favorableRate) {
        this.favorableRate = favorableRate;
    }

    public String getBrandIntro() {
        return brandIntro;
    }

    public void setBrandIntro(String brandIntro) {
        this.brandIntro = brandIntro;
    }
}
```

6.10 DAO 层设计

本项目的 DAO 层与前几章 SSM 框架中的 DAO 层在设计上有所不同，不同之处在于本项目用到了 MyBatisPlus。MyBatisPlus，简称 MP，是一个 MyBatis 的增强工具。MyBatisPlus 在 MyBatis 的基础上只做增强不做改变，专为简化开发、提高开发效率而生。

何以体现 MyBatisPlus 能够简化开发、提高开发效率呢？当使用 MyBatis 时，在编写 Mapper 接口后，不仅需要手动编写对数据执行增、删、改、查等操作的方法，还需要手动编写与每个方法对应的 SQL 语句。例如，使用 MyBatis 读取 t_people 表中的数据，并把读取的数据封装在实体对象中。为此，创建 PeopleMapper 接口作为映射器，在映射器中实现以下 3 个业务：

- ☑ 向 t_people 表添加一个新人员，该人员的数据如下：小丽，女性，20 岁。
- ☑ 将小丽的年龄修改为 19 岁。
- ☑ 删除小丽的所有数据。

PeopleMapper 接口的代码如下：

```java
import org.apache.ibatis.annotations.Delete;
import org.apache.ibatis.annotations.Insert;
import org.apache.ibatis.annotations.Update;
public interface PeopleMapper {
    @Insert("insert into t_people(name,gender,age) values('小丽','女',20)")
    boolean addXiaoLi();
```

```
    @Update("update t_people set age = 19 where name = '小丽'")
    boolean updateXiaoLi();

    @Delete("delete from t_people where name = '小丽'")
    boolean delXiaoLi();
}
```

当使用 MyBatisPlus 时，只需要创建 Mapper 接口并继承 BaseMapper 接口，此时当前的 Mapper 接口就会获得由 BaseMapper 接口提供的对数据执行增、删、改、查等操作的方法。也就是说，在创建 Mapper 接口后，既不需要手动编写对数据执行增、删、改、查等操作的方法，也不需要手动编写与每个方法对应的 SQL 语句，从而实现简化开发、提高开发效率的目的。在使用 MyBatisPlus 的情况下，可以把上述代码作如下修改：

```
import com.baomidou.mybatisplus.core.mapper.BaseMapper;
import com.mr.po.People;

@Mapper
@Repository
public interface PeopleMapper extends BaseMapper<People> {

}
```

简而言之，当使用 MyBatisPlus 时，创建的 Mapper 接口是一个空接口。

本项目的 BikeDao 接口就是 DAO 层。因为 BikeDao 接口继承了 BaseMapper 接口，所以 BikeDao 接口是一个空接口。BikeDao 接口的代码如下：

```
package com.mr.dao;

import com.baomidou.mybatisplus.core.mapper.BaseMapper;
import com.mr.pojo.Bike;
import org.apache.ibatis.annotations.Mapper;
import org.springframework.stereotype.Repository;

@Mapper
@Repository
public interface BikeDao extends BaseMapper<Bike> {                                    // Dao 层

}
```

为了让读者能够更深入地理解 BaseMapper 接口，这里给出 BaseMapper 接口的相关代码：

```
package com.baomidou.mybatisplus.core.mapper;

import com.baomidou.mybatisplus.core.conditions.Wrapper;
import com.baomidou.mybatisplus.core.metadata.IPage;
import java.io.Serializable;
import java.util.Collection;
import java.util.List;
import java.util.Map;
import org.apache.ibatis.annotations.Param;

public interface BaseMapper<T> extends Mapper<T> {
    int insert(T entity);
    int deleteById(Serializable id);
    int deleteByMap(@Param("cm") Map<String, Object> columnMap);
    int delete(@Param("ew") Wrapper<T> queryWrapper);
    int deleteBatchIds(@Param("coll") Collection<? extends Serializable> idList);
    int updateById(@Param("et") T entity);
    int update(@Param("et") T entity, @Param("ew") Wrapper<T> updateWrapper);
```

```
    T selectById(Serializable id);
    List<T> selectBatchIds(@Param("coll") Collection<? extends Serializable> idList);
    List<T> selectByMap(@Param("cm") Map<String, Object> columnMap);
    T selectOne(@Param("ew") Wrapper<T> queryWrapper);
    Integer selectCount(@Param("ew") Wrapper<T> queryWrapper);
    List<T> selectList(@Param("ew") Wrapper<T> queryWrapper);
    List<Map<String, Object>> selectMaps(@Param("ew") Wrapper<T> queryWrapper);
    List<Object> selectObjs(@Param("ew") Wrapper<T> queryWrapper);
    <E extends IPage<T>> E selectPage(E page, @Param("ew") Wrapper<T> queryWrapper);
    <E extends IPage<Map<String, Object>>> E selectMapsPage(E page, @Param("ew") Wrapper<T> queryWrapper);
}
```

6.11 分页插件模块设计

在主页面上，本项目通过分页插件显示了数据总数、分页数、与某个页面对应的数据和分页导航等信息。程序分页显示了 10 条数据，共分两页，每一页最多显示 7 条数据，用户通过分页导航可以随意切换并访问这两个分页的数据，如图 6.7 所示。下面将介绍分页插件模块的实现过程。

编号	品牌名称	品牌评分	好评率	品牌介绍	操作
1	九号	98.5分	97%	质量好、耐用、驾驶体验与安全性能尤为出色	删除
2	雅迪	95.0分	97%	注重品质和外观设计、超长续航	删除
3	小牛电动	92.6分	96%	强调智能化体验和创新科技，具有独特的改装文化	删除
4	绿源	82.0分	98%	注重安全和科技研发	删除
5	台铃	78.7分	99%	主推节能和长续航	删除
6	新日	76.1分	100%	曾经的行业龙头	删除
7	五羊-本田	69.4分	97%	亲民耐用、主打轻便	删除

共 10 条 < 1 2 > 前往 1 页

图 6.7 分页插件的效果图

6.11.1 分页插件的页面设计

本项目中的 bikes.html 即为主页面。在初始化主页面时，主页面还没有来得及从数据库中获取数据，此时数据总数为 0，并且电瓶车的品牌名称、品牌评分、好评率和品牌介绍的值均为空的字符串。因此，在 bikes.html 中初始化分页插件相关数据的代码如下：

```
pagination: {                           //分页插件的相关数据
    currentPage: 1,                     //当前页码
    pageSize: 7,                        //每页显示的记录数
    total: 0,                           //总记录数
    brandName: "",
    brandRating: "",
    favorableRate: "",
    brandIntro: ""
}
```

6.11.2 分页插件配置类设计

com.mr.config 包下的 PageConfig 类为分页插件配置类，这个类可以为本项目配置分页插件，进而分页显示与每个页面对应的数据。下面将介绍分页插件的出处及其配置过程。

在介绍分页插件的出处之前，先了解一下 MyBatis 插件机制。所谓 MyBatis 插件机制，指的是 MyBatis 插件会拦截 Executor、StatementHandler、ParameterHandler 和 ResultSetHandler 这 4 个接口的方法，为了执行自定义的拦截逻辑，需要先利用 JDK 动态代理机制为这些接口的实现类创建代理对象，再执行代理对象的方法。

- Executor：MyBatis 的内部执行器，它负责调用 StatementHandler 操作数据库，并把结果集通过 ResultSetHandler 予以自动映射。
- StatementHandler：MyBatis 直接让数据库执行 SQL 脚本的对象。
- ParameterHandler：MyBatis 为了实现 SQL 入参而设置的对象。
- ResultSetHandler：MyBatis 把 ResultSet 集合映射成 POJO 的接口对象。

MyBatisPlus 依据 MyBatis 插件机制，为程序开发人员提供了 PaginationInnerInterceptor、BlockAttackInnerInterceptor、OptimisticLockerInnerInterceptor 等常用的插件，以便在实际开发中使用。不难发现，这些插件都实现了 InnerInterceptor 接口。

- PaginationInnerInterceptor：用于实现自动分页的插件。
- BlockAttackInnerInterceptor：用于防止全表更新与删除的插件。
- OptimisticLockerInnerInterceptor：用于实现乐观锁的插件。

在明确分页插件的出处后，下面将介绍分页插件的配置过程。因为 PageConfig 类是分页插件配置类，所以须使用@Configuration 注解标注 PageConfig 类。在 PageConfig 类中，有一个用于返回 MybatisPlus 插件对象的 mybatisPlusInterceptor()方法。在这个方法中，首先创建一个 MybatisPlus 插件对象，然后让这个 MybatisPlus 插件对象实现自动分页的功能。com.mr.config 包下的 PageConfig 类的代码如下：

```
package com.mr.config;

import com.baomidou.mybatisplus.extension.plugins.MybatisPlusInterceptor;
import com.baomidou.mybatisplus.extension.plugins.inner.PaginationInnerInterceptor;
import org.springframework.context.annotation.Bean;
import org.springframework.context.annotation.Configuration;

@Configuration
public class PageConfig {                                               //分页插件配置类
    @Bean
    public MybatisPlusInterceptor mybatisPlusInterceptor() {
        MybatisPlusInterceptor interceptor = new MybatisPlusInterceptor();   //MybatisPlus 插件对象
        interceptor.addInnerInterceptor(new PaginationInnerInterceptor());   //配置分页插件
        return interceptor;
    }
}
```

6.12　查询电瓶车品牌信息模块设计

在 db_e-bike 库的 bike 表中一共有 10 条电瓶车的品牌信息，这些信息将被分页显示在主页面上。其中，主页面上的第 1 个分页显示了 7 条数据，如图 6.8 所示；主页面上的第 2 个分页显示了 3 条数据，如图 6.9

所示。下面将介绍查询电瓶车品牌信息模块的实现过程。

编号	品牌名称	品牌评分	好评率	品牌介绍	操作
1	九号	98.5分	97%	质量好、耐用，驾驶体验与安全性能尤为出色	删除
2	雅迪	95.0分	97%	注重品质和外观设计、超长续航	删除
3	小牛电动	92.6分	96%	强调智能化体验和创新科技，具有独特的改装文化	删除
4	绿源	82.0分	98%	注重安全和科技研发	删除
5	台铃	78.7分	99%	主推节能和长续航	删除
6	新日	76.1分	100%	曾经的行业龙头	删除
7	五羊-本田	69.4分	97%	亲民耐用、主打轻便	删除

图 6.8　主页面第 1 个分页的效果图

编号	品牌名称	品牌评分	好评率	品牌介绍	操作
1	小刀	65.6分	92%	爬坡很猛	删除
2	爱玛	63.1分	99%	智能化和舒适性方面表现突出	删除
3	凤凰	51.7分	91.2%	品质过硬、价格亲民	删除

图 6.9　主页面第 2 个分页的效果图

6.12.1　查询模块的页面设计

如图 6.8 和图 6.9 所示，电瓶车的品牌信息都显示在了一个表格模型中，这个表格模型的表头分别为编号、品牌名称、品牌评分、好评率、品牌介绍和操作。在分页插件的作用下，这些信息会被分页显示在主页面上。在第 6.11.1 节中，分页插件的相关数据（如当前页码、每页显示的记录数、总记录数等）虽然已经被初始化，但是需要以电瓶车的品牌信息为依据予以重置。代码如下：

```
<el-table size="small" current-row-key="id" :data="dataList" stripe highlight-current-row>
    <el-table-column type="index" align="center" label="编号"></el-table-column>
    <el-table-column prop="brandName" label="品牌名称" align="center"></el-table-column>
    <el-table-column prop="brandRating" label="品牌评分" align="center"></el-table-column>
    <el-table-column prop="favorableRate" label="好评率" align="center"></el-table-column>
    <el-table-column prop="brandIntro" label="品牌介绍" align="center"></el-table-column>
    <el-table-column label="操作" align="center">
        <template slot-scope="scope">
            <el-button type="danger" size="mini" @click="handleDelete(scope.row)">删除</el-button>
        </template>
    </el-table-column>
</el-table>

<!--分页插件-->
<div class="pagination-container">
    <el-pagination
            class="pagiantion"
            @current-change="handleCurrentChange"
            :current-page="pagination.currentPage"
            :page-size="pagination.pageSize"
            layout="total, prev, pager, next, jumper"
            :total="pagination.total">
```

```
        </el-pagination>
    </div>
```

Vue.js 在创建组件实例时会调用 getAll()方法和 handleCurrentChange()方法。getAll()方法用于分页查询，通过对 db_e-bike 库的 bike 表执行查询操作，获取其中所有的电瓶车品牌信息（10 条），并以此为依据确定分页插件的"当前页码""每页显示的记录数""总记录数"，即"当前页码"为 1、"每页显示的记录数"为 7、"总记录数"为 10。因此，10 条电瓶车品牌信息将被分页插件分为 2 页，第 1 个分页有 7 条数据，第 2 个分页有 3 条数据。handleCurrentChange()方法用于切换页面，即可以从第 1 个分页切换至第 2 个分页，也可以从第 2 个分页切换至第 1 个分页。代码如下：

```javascript
//分页查询
getAll() {
    param = "?"
    param += "brandName="+this.pagination.brandName;
    param += "&brandRating="+this.pagination.brandRating;
    param += "&favorableRate="+this.pagination.favorableRate;
    param += "&brandIntro="+this.pagination.brandIntro;
    //发送异步请求
    axios.get("/bikes/"+this.pagination.currentPage+"/"+this.pagination.pageSize+param).then((res) => {
        this.pagination.pagesize = res.data.data.size;
        this.pagination.currentPage = res.data.data.current;
        this.pagination.total = res.data.data.total;
        this.dataList = res.data.data.records;
    });
},
//切换页码
handleCurrentChange(currentPage) {
    //修改页面值为当前选中页码值
    this.pagination.currentPage = currentPage;
    //执行查询
    this.getAll();
},
```

6.12.2 查询模块控制器类设计

本项目的 BikeController 类为控制器类，被@RestController 注解标注。在 BikeController 类中，定义了两个方法，它们分别是 getAll()方法和 getPage()方法。其中，getAll()方法用于查询所有的电瓶车品牌信息；getPage()方法用于获取显示数据的某个分页。getAll()方法和 getPage()方法的代码分别如下：

```java
@GetMapping
public R getAll() {
    return new R(true, bikeService.list());                        //查询所有电瓶车的品牌信息
}

@GetMapping("{currentPage}/{pageSize}")
public R getPage(@PathVariable int currentPage, @PathVariable int pageSize, Bike bike) {
    IPage<Bike> page = bikeService.getPage(currentPage, pageSize, bike);    //获取显示数据的某个分页
    //如果当前页码值大于总页码值，那么重新执行操作，使最大页码值为当前页码
    if (currentPage > page.getPages()) {
        page = bikeService.getPage((int) page.getPages(), pageSize, bike);
    }
    return new R(true, page);
}
```

@GetMapping 用于处理 get 请求，通常在查询数据时使用，@GetMapping 的语法如下：

```
@GetMapping("path")
```

@GetMapping 等价于处理 get 请求的@RequestMapping，@RequestMapping 的语法如下：
@RequestMapping(value = "path" , method = RequestMethod.GET)

6.12.3 查询模块服务类设计

本项目的 BikeService 接口为服务接口。在 BikeService 接口中，定义了一个用于获取显示数据的某个分页的 getPage()方法。在 getPage()方法中有 3 个参数，它们分别为 int 类型的表示"当前页码"的 currentPage、int 类型的表示"每页显示的记录数"的 pageSize、Bike 类型的表示"电瓶车对象（内含电瓶车品牌信息）"的 bike。此外，该方法具有返回值，返回值是显示"电瓶车对象（内含电瓶车品牌信息）"的某个分页。getPage()方法的代码如下：

```
IPage<Bike> getPage(int currentPage, int pageSize, Bike bike);     //获取用于显示数据的某个分页
```

本项目的 BikeServiceImpl 类是 BikeService 接口的实现类，被@Service 注解标注，即服务类。在 BikeServiceImpl 类中，重写了 BikeService 接口中的 getPage()方法，该方法的主要作用是调用 DAO 层（数据访问对象）来执行数据库操作。重写后的 getPage()方法的代码如下：

```
@Override
public IPage<Bike> getPage(int currentPage, int pageSize, Bike bike) {     //获取用于显示数据的某个分页
    LambdaQueryWrapper<Bike> lqw = new LambdaQueryWrapper<Bike>();
    lqw.like(Strings.isNotEmpty(bike.getBrandName()), Bike::getBrandName, bike.getBrandName());
    lqw.like(Strings.isNotEmpty(bike.getBrandRating()), Bike::getBrandRating, bike.getBrandRating());
    lqw.like(Strings.isNotEmpty(bike.getFavorableRate()), Bike::getFavorableRate, bike.getFavorableRate());
    lqw.like(Strings.isNotEmpty(bike.getBrandIntro()), Bike::getBrandIntro, bike.getBrandIntro());
    IPage page = new Page(currentPage,pageSize);
    bikeDao.selectPage(page,lqw);
    return page;
}
```

上述代码中，LambdaQueryWrapper 是 MyBatisPlus 中的一个功能类，用于构建 lambda 表达式风格的查询条件，它提供了类型安全的条件构造器，可以减少编写字段名称的错误。

6.13 新增电瓶车品牌信息模块设计

如图 6.10 所示，在主页面的头部有一个"新增电瓶车品牌信息"按钮。用户单击这个按钮，程序将弹出"新增电瓶车品牌信息"窗口，如图 6.11 所示。用户在这个窗口中依次输入电瓶车的品牌名称、品牌评分、好评率和品牌介绍等信息，单击"确定"按钮，完成新增电瓶车品牌信息的操作。下面将介绍新增电瓶车品牌信息模块的实现过程。

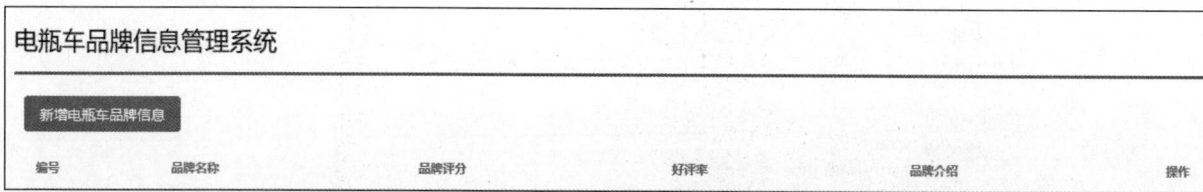

图 6.10　主页面头部的效果图

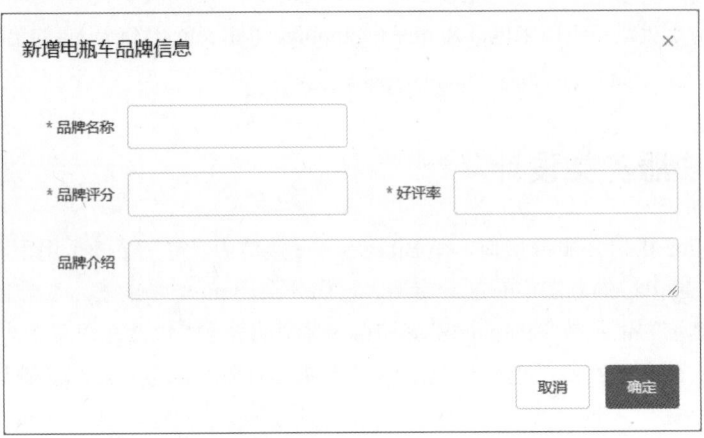

图 6.11　"新增电瓶车品牌信息"窗口的效果图

6.13.1　新增模块的页面设计

bikes.html 是本项目的主页面，因此在 bikes.html 的头部添加"新增电瓶车品牌信息"按钮，代码如下：

```
<div class="filter-container">
    <el-button type="primary" class="butT" @click="handleCreate()">新增电瓶车品牌信息</el-button>
</div>
```

用户单击"新增电瓶车品牌信息"按钮，程序将弹出"新增电瓶车品牌信息"窗口。在这个窗口中，包含 4 个标签、3 个文本框、1 个文本域、1 个"取消"按钮和 1 个"确定"按钮。代码如下：

```
<div class="add-form">
    <el-dialog title="新增电瓶车品牌信息" :visible.sync="dialogFormVisible">
        <el-form ref="dataAddForm" :model="formData" :rules="rules" label-position="right"
                label-width="100px">
            <el-row>
                <el-col :span="12">
                    <el-form-item label="品牌名称" prop="brandName">
                        <el-input v-model="formData.brandName"/>
                    </el-form-item>
                </el-col>
            </el-row>

            <el-row>
                <el-col :span="12">
                    <el-form-item label="品牌评分" prop="brandRating">
                        <el-input v-model="formData.brandRating"/>
                    </el-form-item>
                </el-col>
                <el-col :span="12">
                    <el-form-item label="好评率" prop="favorableRate">
                        <el-input v-model="formData.favorableRate"/>
                    </el-form-item>
                </el-col>
            </el-row>

            <el-row>
                <el-col :span="24">
                    <el-form-item label="品牌介绍">
                        <el-input v-model="formData.brandIntro" type="textarea"></el-input>
                    </el-form-item>
                </el-col>
```

```
            </el-row>
        </el-form>

        <div slot="footer" class="dialog-footer">
            <el-button @click="cancel()">取消</el-button>
            <el-button type="primary" @click="handleAdd()">确定</el-button>
        </div>
    </el-dialog>
</div>
```

Vue.js 在创建组件实例时会分别调用 handleCreate()方法、resetForm()方法、handleAdd()方法和 cancel() 方法。其中，handleCreate()方法用于显示"新增电瓶车品牌信息"窗口；resetForm()方法用于重置表格模型中的数据；handleAdd()方法的作用是用户在"新增电瓶车品牌信息"窗口中依次输入电瓶车的品牌名称、品牌评分、好评率和品牌介绍等信息，单击"确定"按钮，如果操作成功，那么表格模型中的数据将被重置，进而新增的电瓶车品牌信息会显示在第 2 个分页上；cancel()方法的作用是如果用户单击窗口中的"取消"按钮，那么窗口将被关闭，并弹出"当前操作取消"的信息。代码如下：

```
//弹出添加窗口
handleCreate() {
    this.dialogFormVisible = true;
    this.resetForm();
},
//重置表单
resetForm() {
    this.formData = {};
},
//添加
handleAdd() {
    axios.post("/bikes", this.formData).then((res) => {
        //判断添加是否成功
        if (res.data.flag) {
            //关闭弹层
            this.dialogFormVisible = false;
            this.$message.success(res.data.msg);
        } else {
            this.$message.error(res.data.msg);
        }
    }).finally(() => {
        //重新加载数据
        this.getAll();
    });
},
//取消
cancel() {
    this.dialogFormVisible = false;
    this.dialogFormVisible4Edit = false;
    this.$message.info("当前操作取消");
},
```

6.13.2 新增模块控制器类设计

在 BikeController 类（控制器类）中，定义了一个 insert()方法，该方法用于判断是否成功地执行了新增电瓶车品牌信息的操作。如果操作成功，就返回"添加成功"；否则，就返回"添加失败"。insert()方法的代码如下：

```
@PostMapping
public R insert(@RequestBody Bike bike) {
```

```
        boolean flag = bikeService.insertBike(bike);            //是否成功执行新增电瓶车品牌信息的操作
        return new R(flag, flag ? "添加成功" : "添加失败");
}
```

@PostMapping：处理 post 请求，通常在新增数据时使用，@PostMapping 的语法如下：

@PostMapping("path")

@PostMapping 等价于处理 post 请求的@RequestMapping，@RequestMapping 的语法如下：

@RequestMapping(value = "path", method = RequestMethod.POST)

6.13.3 新增模块服务类设计

在 BikeService 接口（服务接口）中，定义了一个 insertBike()方法，该方法用于判断是否成功地执行了新增电瓶车品牌信息的操作。在 insertBike()方法中，有 1 个参数，即 Bike 类型的表示"电瓶车对象（内含电瓶车品牌信息）"的 bike。此外，该方法具有返回值，返回值是一个布尔值。insertBike()方法的代码如下：

```
boolean insertBike(Bike bike);                                  //是否执行新增电瓶车品牌信息的操作
```

在 BikeServiceImpl 类（服务类）中，重写了 BikeService 接口中的 insertBike()方法。该方法的主要作用是调用 DAO 层（数据访问对象）来执行数据库操作。重写后的 insertBike()方法的代码如下：

```
@Override
public boolean insertBike(Bike bike) {                          //是否成功执行新增电瓶车品牌信息的操作
    return bikeDao.insert(bike)>0;
}
```

6.14 删除电瓶车品牌信息模块设计

如图 6.8 和图 6.9 所示，在主页面上显示的每一条数据的后面，都有一个"删除"按钮，用户先单击某一个"删除"按钮，再单击删除提示窗口中的"确定"按钮（如图 6.12 所示），程序将删除与这个"删除"按钮对应的电瓶车品牌信息，以完成删除电瓶车品牌信息的操作。下面将介绍删除电瓶车品牌信息模块的实现过程。

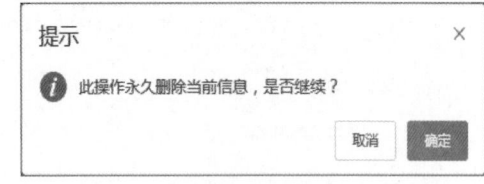

图 6.12　删除提示窗口的效果图

6.14.1 删除模块的页面设计

在第 6.12.1 节中，已经设计完成了表格模型中的表头。在表头"操作"中，已经添加了"删除"按钮。为了让"删除"按钮发挥作用，Vue.js 在创建组件实例时会调用 handleDelete()方法：用户在单击"删除"按钮后，如果继续单击删除提示窗口中的"确定"按钮，程序将删除与这个"删除"按钮对应的电瓶车品牌信息，并且表格模型中的数据会被重置；如果单击删除提示窗口中的"取消"按钮，那么窗口将被关闭，并弹出"取消操作"的信息。代码如下：

```
//删除
handleDelete(row) {
    this.$confirm("此操作永久删除当前信息，是否继续？", "提示", {type: "info"}).then((res) => {
        axios.delete("/bikes/" + row.id).then((res) => {
```

```
            if (res.data.flag) {
                this.$message.success("删除成功");
            } else {
                this.$message.error("数据同步失败，自动刷新");
            }
        }).finally(() => {
            //重新加载数据
            this.getAll();
        });
    }).catch(() => {
        this.$message.info("取消操作");
    });
},
```

6.14.2 删除模块控制器类设计

在 BikeController 类（控制器类）中，定义了一个用于根据电瓶车的品牌编号删除电瓶车品牌信息的 delete()方法。delete()方法的代码如下：

```
@DeleteMapping("{id}")
public R delete(@PathVariable Integer id) {
    return new R(bikeService.deleteBike(id));          //删除电瓶车的品牌信息
}
```

@DeleteMapping：处理 delete 请求，通常在删除数据时使用，@DeleteMapping 的语法如下：

`@DeleteMapping("path")`

@GetMapping 等价于处理 delete 请求的@RequestMapping，@RequestMapping 的语法如下：

`@RequestMapping(value = "path",method = RequestMethod.DELETE)`

6.14.3 删除模块服务类设计

在 BikeService 接口（服务接口）中，定义了一个 deleteBike()方法，该方法用于判断是否成功地执行了删除电瓶车品牌信息的操作。在 deleteBike()方法中，有 1 个参数，即 Integer 类型的表示"电瓶车的品牌编号"的 id。此外，该方法具有返回值，返回值是一个布尔值。deleteBike()方法的代码如下：

`boolean deleteBike(Integer id);                //是否执行删除电瓶车品牌信息的操作`

在 BikeServiceImpl 类（服务类）中，重写了 BikeService 接口中的 deleteBike()方法。该方法的主要作用是调用 DAO 层（数据访问对象）来执行数据库操作。重写后的 deleteBike()方法的代码如下：

```
@Override
public boolean deleteBike(Integer id) {            //是否成功执行删除电瓶车品牌信息的操作
    return bikeDao.deleteById(id)>0;
}
```

6.15 项 目 运 行

通过前述步骤，设计并完成了"电瓶车品牌信息管理系统"项目的开发。下面运行本项目，以检验我们的开发成果。如图 6.13 所示，在 IntelliJ IDEA 中，单击▶快捷图标，即可运行本项目。

图 6.13　IntelliJ IDEA 的快捷图标

成功运行本项目，打开浏览器，访问 http://localhost:8080/pages/bikes.html，即可看到如图 6.14 所示的电瓶车品牌信息管理系统的主页面。

主页面一共显示 10 条数据。因为一个分页只能显示 7 条数据，所以需要两个分页，即第 1 个分页显示 7 条数据，第 2 个分页显示 3 条数据。用户通过分页导航可以随意切换并访问这两个分页的数据。

用户在单击"新增电瓶车品牌信息"按钮后，需要先在弹出的窗口中依次输入电瓶车的品牌名称、品牌评分、好评率和品牌介绍等信息，再单击"确定"按钮，进而完成新增电瓶车品牌信息的操作。

在主页面上显示的每一条数据的后面，都有一个"删除"按钮，用户先单击某一个"删除"按钮，再单击删除提示窗口中的"确定"按钮，进而完成删除电瓶车品牌信息的操作。

这样，我们就成功地检验了本项目的运行。

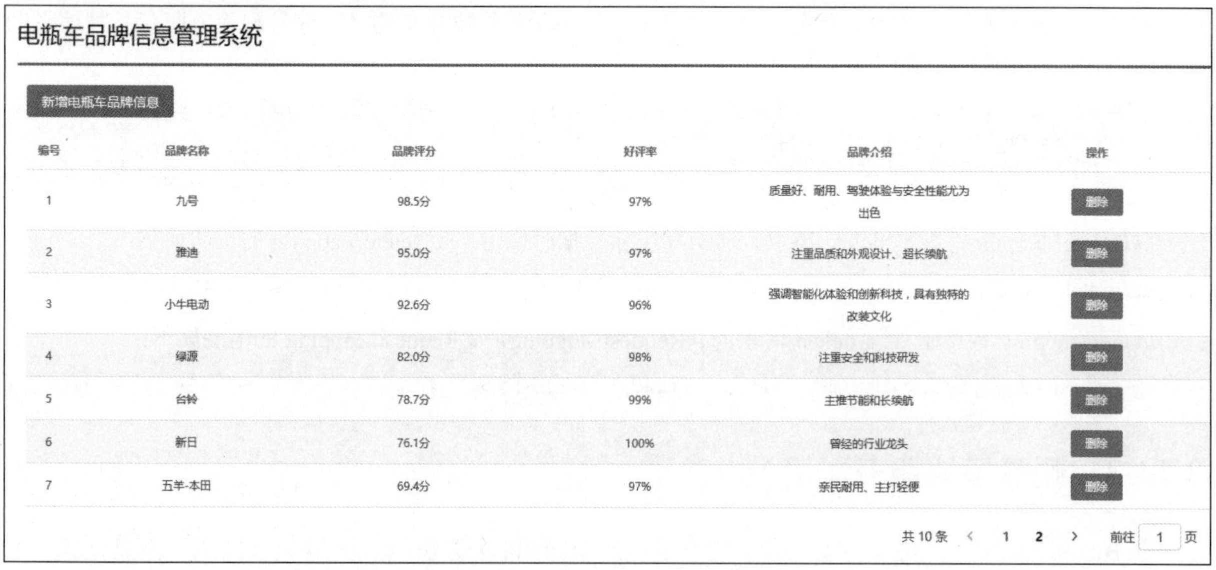

图 6.14　电瓶车品牌信息管理系统的主页面

本项目是一个简单的应用单表查询的 Spring Boot 项目，但是麻雀虽小，五脏俱全。本项目在设计页面时使用了 Vue.js，Vue.js 在创建组件实例时会调用特定的方法，进而达到渲染页面的目的。此外，本项目在处理业务时层次分明、结构清晰，Controller 层主要负责具体的业务模块流程的控制，Service 层主要负责业务模块的逻辑应用设计，DAO 层主要负责与数据库进行交互。需要说明的是，本项目使用了 MyBatisPlus，在创建 Mapper 接口时，让其继承 BaseMapper 接口，这样 Mapper 接口就是一个空接口，既不需要手动编写对数据执行增、删、改、查等操作的方法，也不需要手动编写与每个方法对应的 SQL 语句。

6.16　源码下载

虽然本章详细地讲解了如何编码实现"电瓶车品牌信息管理系统"项目的各个功能，但给出的代码都是代码片段，而非源码。为了方便读者学习，本书提供了完整的项目源码，扫描右侧二维码即可下载。

第 7 章
寻物启事网站

——Spring Boot + Vue + MySQL

在现实生活中，人们丢失物品后，往往难于寻找失物。与此同时，如果拾得人希望归还物品，也是苦于寻找失主。于是，搭建一个良好的交流平台，让失主可以在上面发布丢失物品的信息，而拾得人在看到信息后能够及时地联系到平台的管理者或者失主并归还物品，无疑是十分有意义的。为此，本章将使用 Java Web 开发中的 Spring Boot 框架、Vue 和 MySQL 数据库等关键技术开发一个寻物启事网站项目。

项目微视频

本项目的核心功能及实现技术如下：

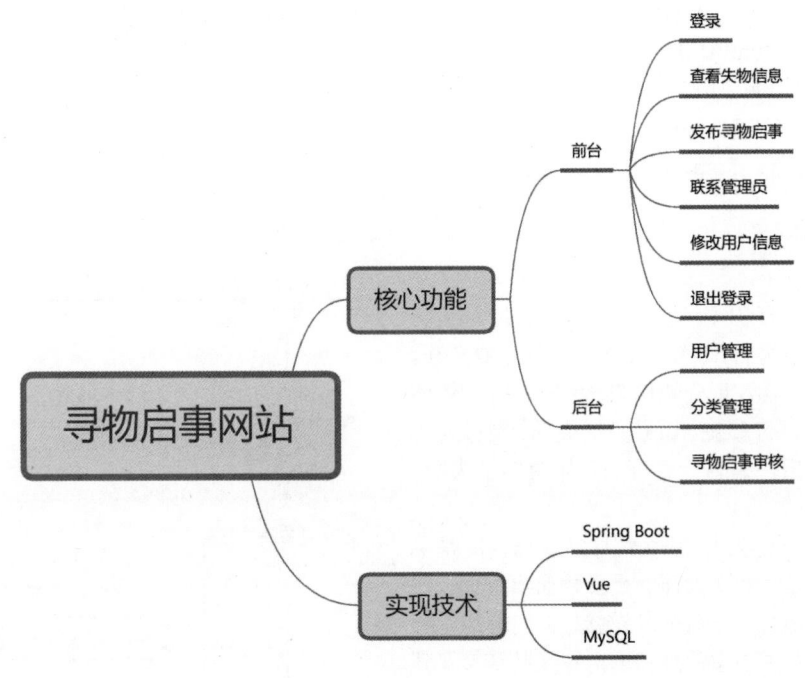

7.1 开发背景

在日常生活中，常常有人丢三落四，等他发现了以后回过头去找，自己的物品早已不见踪影。由于信息沟通不畅，失主很难联系到拾得人，往往不能及时地找回失物，导致经济损失。同时，即使拾得人有心将物品归还失主，也往往苦于不知道如何联系到失主，手足无措，无法归还。基于这一现实需求，本章将开发一个寻物启事网站，在失主和拾得人之间搭建一个信息交流的平台。为了降低项目开发难度，暂定该平台的用户群体是学校的学生。失主在网站上可以发布自己丢失的物品信息；拾得人看到并确认失物信息后，先联系网站管理员，再把失物交给网站管理员，而后由网站管理员把失物归还给失主。由网站管理员充当中间人的好处在于可以确保失主和拾得人都保持诚信，进而避免失主和拾得人之间产生不必要的纠纷。

寻物启事网站将实现以下目标：
- ☑ 让失物信息清晰化、透明化，并且易于管理；
- ☑ 失主在线发布失物信息后，能够让更多人看到信息，并帮忙寻找；
- ☑ 拾得人看到消息后，能够及时联系网站管理员，归还失物；
- ☑ 不让失主和拾得人直接接触，避免不必要的纠纷；
- ☑ 网站界面友好，功能实用，用户体验良好。

7.2 系统设计

7.2.1 开发环境

本项目的开发及运行环境如下：
- ☑ 操作系统：推荐 Windows 10、11 及以上，兼容 Windows 7（SP1）。
- ☑ 开发工具：IntelliJ IDEA。
- ☑ 开发语言：Java EE。
- ☑ 数据库：MySQL 8.0。
- ☑ Web 服务器：Tomcat 9.0 及以上版本。

7.2.2 业务流程

启动项目后，打开浏览器，访问地址 http://localhost:8080/#/，寻物启事网站的登录页面将被打开。在登录页面上，用户输入正确的用户名和密码后，程序需要对用户的身份进行验证。如果验证成功，则进入下一步验证，即判断用户是否为管理员；如果验证失败，则提示错误。

如果用户的身份是学生，那么程序默认访问的是寻物启事超链接，即程序默认打开的是寻物启事页面。在寻物启事页面的头部有 3 个导航超链接，它们分别是寻物启事超链接、个人中心超链接和退出登录超链接。其中，在寻物启事页面上，学生既可以查看失物信息，也可以发布寻物启事，还可以联系管理员归还失物；在个人中心页面上，学生可以修改用户信息，如联系方式、登录密码等；通过退出登录超链接，则可以退出登录。

如果用户的身份是管理员，那么程序默认访问的是管理中心超链接，即程序默认打开的是管理中心页面。在管理中心页面的头部有 4 个导航超链接，它们分别是寻物启事超链接、个人中心超链接、管理中心超链接和退出登录超链接。其中，在管理中心页面上，包含了 3 个功能，分别是用户管理、分类管理和寻物启事审核。

寻物启事网站的业务流程如图 7.1 所示。

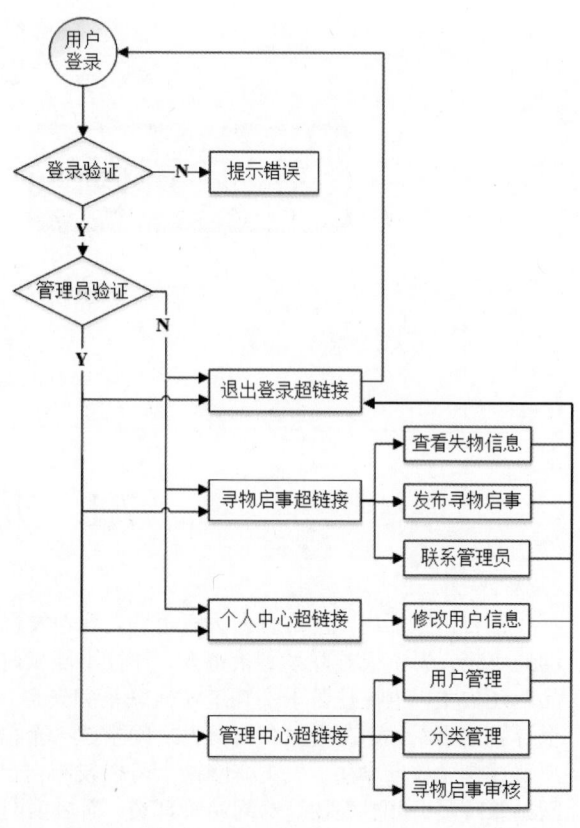

图 7.1 寻物启事网站的业务流程图

7.2.3 功能结构

本项目的功能结构已经在章首页中给出，作为一个让失主和拾得人相互交流信息的平台，本项目实现的具体功能如下：
- ☑ 用户登录：用户输入正确的用户名和密码后，即可登录到寻物启事网站。
- ☑ 查看失物信息：已登录的用户可以查看经由管理员审核并且通过审核的失物信息。
- ☑ 发布寻物启事：已登录的、丢失物品的用户可以发布失物信息，并等待管理员审核。
- ☑ 联系管理员：已登录的、拾得失物的用户看到并确认失物信息后，可以联系管理员归还失物。
- ☑ 修改用户信息：已登录的用户可以修改用户名、微信号、手机号码、登录密码等信息。
- ☑ 用户管理：管理员既可以查看用户信息，也可以把用户设置为管理员。
- ☑ 分类管理：管理员对失物进行分类。
- ☑ 寻物启事审核：管理员对已发布并等待审核的失物信息进行审核，通过审核的失物信息将显示在寻物启事页面上。

7.3 技术准备

- ☑ Spring Boot：Spring Boot 是一个轻量级的框架，它是 Spring 生态圈的一部分，旨在简化 Spring 应用项目的初始搭建和开发过程。Spring Boot 支持多种流行的数据访问框架，并且可以与 Spring 的事务管理一起使用，为数据访问提供灵活的抽象。此外，Spring Boot 还解决了现有 Web 框架在呈现层和请求处理层之间的分离不足的问题，通过创建 Spring MVC 改进了这一点。在本项目中，Spring MVC 的表现层、控制层、业务逻辑层和数据访问层都发挥着重要作用，共同实现项目中各个模块的功能。
- ☑ Vue：Vue 是一个框架，也是一个生态，其功能覆盖了大部分前端开发常见的需求。但 Web 世界是十分多样化的，不同的程序开发人员在 Web 上构建的东西可能在形式和规模上会有很大的不同。考虑到这一点，Vue 的设计非常注重灵活性和"可以被逐步集成"这个特点。本项目的页面都是由 Vue 设计的。例如，登录页面对应的是 Login.vue 文件，寻物启事页面对应的是 SearchGoods.vue 文件。为了便于管理项目，本项目采用了前后端分离的方式，将所有.vue 文件单独置于一个包中。
- ☑ MySQL：MySQL 主要用于在 Web 应用方面，它能够提供高效、稳定和易扩展的数据存储和管理服务。作为一种开源软件，用户可以免费获取和使用 MySQL，这使其成为许多开发者和企业的首选数据库系统。MySQL 采用结构化查询语言（SQL），这是一种标准化的数据库访问语言，使得数据的存取更加方便和统一。MySQL 的多种存储引擎支持不同的功能需求，如 InnoDB 支持事务处理，而 MyISAM 则适用于读多写少的场景。MySQL 不仅支持多线程操作，还优化了查询算法，提高了数据处理速度。综上所述，MySQL 因其高效、灵活且免费的特性，在全球范围内被广泛应用于各种规模的项目中，从小网站到大型企业系统。这使得它成为目前最流行的关系型数据库之一。本项目将继续在 MySQL 中创建数据库，并依次在数据库中创建 5 张数据表，关于这 5 张数据表的内容将在后面的内容中予以介绍。

有关 Spring Boot、Vue 和 MySQL 的知识，分别在《Spring Boot 从入门到精通》《Vue.js 从入门到精通》《Java 从入门到精通（第 7 版）》中进行了详细的讲解，对这些知识不太熟悉的读者可以参考这 3 本书中的相关内容。

7.4 数据库设计

7.4.1 数据库概述

本项目采用的数据库主要包含 5 张数据表，如表 7.1 所示。

表 7.1 寻物启事网站的数据库结构

表 名	表 说 明	表 名	表 说 明
t_user	用户信息表	sys_role	用户身份表
t_lost_info	失物信息表	sys_role_user	用户身份明细表
t_kind	失物类别表		

7.4.2 数据表设计

下面将详细介绍本项目使用的 5 张表的结构设计。

- ☑ t_user（用户信息表）：主要用于存储用户编号、用户名、真实姓名、学号、登录密码、手机号码、微信号、身份证号码、用户信息的创建时间、用户信息的更新时间等信息。该数据表的结构如表 7.2 所示。

表 7.2 用户信息表

字 段 名 称	数 据 类 型	长 度	是 否 主 键	说 明
user_id	VARCHAR	50	主键	用户编号
user_nick	VARCHAR	50		用户名
user_name	VARCHAR	50		真实姓名
stu_num	VARCHAR	50		学号
user_pwd	VARCHAR	100		登录密码
user_phone	VARCHAR	50		手机号码
wechat_id	VARCHAR	50		微信号
num_id	VARCHAR	50		身份证号码
create_time	DATETIME			用户信息的创建时间
update_time	DATETIME			用户信息的更新时间

- ☑ t_lost_info（失物信息表）：主要用于存储失物编号、失物分类编号、失物名称、丢失地点、丢失时间、失物描述、失物信息发布人编号、失物信息的发布时间、失物信息的审核状态等信息。该数据表的结构如表 7.3 所示。

表 7.3 失物信息表

字 段 名 称	数 据 类 型	长 度	是 否 主 键	说 明
lost_id	VARCHAR	50	主键	失物编号
kind_id	VARCHAR	50		失物分类编号
lost_name	VARCHAR	50		失物名称

续表

字 段 名 称	数 据 类 型	长 度	是 否 主 键	说 明
lost_place	VARCHAR	50		丢失地点
lost_time	DATETIME			丢失时间
lost_decp	VARCHAR	100		失物描述
user_id	VARCHAR	50		失物信息发布人编号
lost_release_time	DATETIME			失物信息的发布时间
check_status	INT			失物信息的审核状态

- ☑ t_kind（失物类别表）：主要用于存储失物分类编号、失物分类名称、失物分类的创建时间等信息。该数据表的结构如表 7.4 所示。

表 7.4 失物类别表

字 段 名 称	数 据 类 型	长 度	是 否 主 键	说 明
kind_id	VARCHAR	50	主键	失物分类编号
kind_name	VARCHAR	50		失物分类名称
create_time	DATETIME			失物分类的创建时间

- ☑ sys_role（用户身份表）：主要用于存储用户身份编号、用户身份名称等信息。该数据表的结构如表 7.5 所示。

图 7.5 用户身份表

字 段 名 称	数 据 类 型	长 度	是 否 主 键	说 明
role_id	VARCHAR	50	主键	用户身份编号
role_name	VARCHAR	50		用户身份名称

- ☑ sys_role_user（用户身份明细表）：主要用于存储用户身份明细编号、用户编号、用户身份编号等信息。该数据表的结构如表 7.6 所示。

图 7.6 用户身份明细表

字 段 名 称	数 据 类 型	长 度	是 否 主 键	说 明
role_user_id	VARCHAR	50	主键	用户身份明细编号
user_id	VARCHAR	50		用户编号
role_id	VARCHAR	50		用户身份编号

7.5 添加依赖和配置信息

在开发 Spring Boot 项目的过程中，需要为项目添加依赖和配置信息。下面将分别介绍如何为本项目添加依赖和配置信息。

7.5.1 在 pom.xml 文件中添加依赖

pom.xml 是 Maven 构建项目的核心配置文件，其中存储的是本项目所需的依赖，这些依赖会被添加到

<dependencies>标签的内部。代码如下：

```xml
<?xml version="1.0" encoding="UTF-8"?>
<project xmlns="http://maven.apache.org/POM/4.0.0" xmlns:xsi="http://www.w3.org/2001/XMLSchema-instance"
  xsi:schemaLocation="http://maven.apache.org/POM/4.0.0 http://maven.apache.org/xsd/maven-4.0.0.xsd">
    <modelVersion>4.0.0</modelVersion>

    <groupId>org.example</groupId>
    <artifactId>LostAndFound</artifactId>
    <packaging>pom</packaging>
    <version>1.0-SNAPSHOT</version>

    <parent>
        <groupId>org.springframework.boot</groupId>
        <artifactId>spring-boot-starter-parent</artifactId>
        <version>2.0.2.RELEASE</version>
    </parent>

    <modules>
        <module>core</module>
        <module>edu-business</module>
    </modules>

    <name>LostAndFound</name>

    <properties>
        <spring-cloud.version>Finchley.RELEASE</spring-cloud.version>
        <mysql.version>8.0.21</mysql.version>
        <mybatis.version>2.1.3</mybatis.version>
    </properties>

    <dependencies>
        <dependency>
            <groupId>org.projectlombok</groupId>
            <artifactId>lombok</artifactId>
            <version>1.18.10</version>
            <scope>provided</scope>
        </dependency>
    </dependencies>

    <dependencyManagement>
        <dependencies>
            <!--springcloud-->
            <dependency>
                <groupId>org.springframework.cloud</groupId>
                <artifactId>spring-cloud-dependencies</artifactId>
                <version>${spring-cloud.version}</version>
                <type>pom</type>
                <scope>import</scope>
            </dependency>
            <!--mysql-->
            <dependency>
                <groupId>mysql</groupId>
                <artifactId>mysql-connector-java</artifactId>
                <version>${mysql.version}</version>
            </dependency>
            <!--mybatis-->
            <dependency>
                <groupId>org.mybatis.spring.boot</groupId>
                <artifactId>mybatis-spring-boot-starter</artifactId>
                <version>${mybatis.version}</version>
```

```
        </dependency>
    </dependencies>
</dependencyManagement>
</project>
```

7.5.2 在 application.yml 文件中添加配置信息

本项目的配置文件是 application.yml，该配置文件采用的是 yml 格式，其中包含如下配置信息：

```yml
server:
  port: 8661

spring:
  application:
    name: lost
  datasource:
    url:
      jdbc:mysql://localhost:3306/lost_and_found?useUnicode=true&\n
      characterEncoding=UTF-8&serverTimezone=GMT%2b8
    username: root
    password: root
    driver-class-name: com.mysql.cj.jdbc.Driver

mybatis:
  mapper-locations: classpath:/mapper/*
  configuration:
    map-underscore-to-camel-case: true
    log-impl: org.apache.ibatis.logging.stdout.StdOutImpl
```

7.6 实体类设计

在 Java 语言中，一个实体类对应着一种数据模型。在第 7.4.2 节中，已经介绍了本项目使用的 5 张数据表。下面将着重介绍与 t_user（用户信息表）和 t_lost_info（失物信息表）这两张数据表对应的实体类。

7.6.1 用户信息类

本项目与 t_user（用户信息表）相对应的实体类是 UserInfo 类，表示用户信息类。除用户身份（roleName）外，UserInfo 类中的每一个属性都与 t_user（用户信息表）中的各个字段是一一对应的。UserInfo 类代码如下：

```java
@Data
public class UserInfo {
    @ApiModelProperty("用户编号")
    private String userId;

    @ApiModelProperty("用户名")
    private String userNick;

    @ApiModelProperty("真实姓名")
    private String userName;

    @ApiModelProperty("学号")
    private String stuNum;
```

```
    @ApiModelProperty("密码")
    private String userPwd;

    @ApiModelProperty("手机号码")
    private String userPhone;

    @ApiModelProperty("微信号")
    private String wechatId;

    @ApiModelProperty("身份证号码")
    private String numId;

    @ApiModelProperty("创建时间")
    @JsonFormat(pattern = "yyyy-MM-dd")
    private Date createTime;

    @ApiModelProperty("更新时间")
    @JsonFormat(pattern = "yyyy-MM-dd")
    private Date updateTime;

    @ApiModelProperty("用户身份")
    private String roleName;
}
```

@Data 是一个由 Lombok 提供的注解，在 UserInfo 类上添加@Data 注解后，Lombok 会自动为其生成所有字段的 Getter 和 Setter 方法、equals()方法、hashCode()方法、toString()方法等。

@ApiModelProperty 注解用于对 Java 类中的属性进行标注，表示这个属性是一个 Swagger 模型的属性。@ApiModelProperty 注解用于描述属性的名称、说明、数据类型等信息。

说明

Swagger 模型主要指的是在 API 文档中定义和展示的数据模型。通过 Swagger，程序开发人员可以定义 API 的数据结构，包括请求和响应的格式，这些定义可以帮助其他程序开发人员或客户端应用程序理解如何与 API 进行交互。

7.6.2 失物信息类

本项目与 t_lost_info（失物信息表）相对应的实体类是 LostInfo 类，表示失物信息类。LostInfo 类中的每一个属性都与 t_lost_info（失物信息表）中的各个字段是一一对应的。LostInfo 类代码如下：

```
@Data
public class LostInfo {
    @ApiModelProperty("失物编号")
    private String lostId;

    @ApiModelProperty("失物分类编号")
    @NotBlank(message = "失物分类编号不能为空")
    private String kindId;

    @ApiModelProperty("失物名称")
    @NotBlank(message = "失物名称不能为空")
    private String lostName;

    @ApiModelProperty("丢失地点")
    private String lostPlace;
```

```
    @ApiModelProperty("丢失时间")
    @JsonFormat(pattern = "yyyy-MM-dd")
    private Date lostTime;

    @ApiModelProperty("失物描述")
    private String lostDecp;

    @ApiModelProperty("失物信息发布人编号")
    private String userId;

    @ApiModelProperty("失物信息的发布时间")
    @JsonFormat(pattern = "yyyy-MM-dd")
    private Date lostReleaseTime;

    @ApiModelProperty("审核状态")
    private Integer checkStatus;
}
```

7.7　登录模块设计

如图 7.2 所示，寻物启事网站的登录页面被打开后，用户需要先在页面上依次输入用户名和密码，再单击"登录"按钮，以完成登录操作。"登录"按钮被单击后，程序将验证当前用户输入的用户名和密码是否正确。如果用户输入的用户名和密码是正确的，则程序将以此为依据判断当前用户的身份是"管理员"还是"学生"。下面将介绍登录模块的设计过程。

图 7.2　登录页面的效果图

7.7.1　展示层对象设计

展示层对象（view object，简称 VO）主要用于展示层，其作用是把某个指定前端页面的所有数据封装起来。这样，既可以减少传输的数据量，又可以保护数据库中的隐私数据（如用户名、密码等信息），还可以避免数据库结构的外泄。

在本项目中，UserLoginVO 类的对象是与登录模块相对应的展示层对象，因为登录页面仅需要使用用户名和密码，所以 UserLoginVO 类只包含用户名和密码两个属性。UserLoginVO 类的代码如下：

```
@Data
public class UserLoginVO {
```

```
    @ApiModelProperty("用户名")
    private String username;

    @ApiModelProperty("密码")
    private String password;
}
```

7.7.2 登录页面设计

寻物启事网站的登录页面由标题、输入框、密码框和"登录"按钮组成。其中，输入框供用户输入用户名，不能为空；密码框供用户输入密码，也不能为空；"登录"按钮则先判断用户输入的用户名和密码是否正确，再判断用户的身份。

寻物启事网站的登录页面对应的是 Login.vue 文件，代码如下：

```
<template>
  <div class="login-content">
     <h1 style="text-align: center;margin-top: 7vh;margin-bottom: 4vh;color: white">寻物启事网站</h1>

     <el-form class="form-content" ref="form" :model="form">
        <el-form-item>
           <el-input placeholder="用户名" v-model="form.username"></el-input>
        </el-form-item>

        <el-form-item>
           <el-input type="password" placeholder="密码" v-model="form.password"></el-input>
        </el-form-item>

        <el-form-item>
           <el-button type="primary" @click="login" style="width: 100%">登录</el-button>
        </el-form-item>
     </el-form>
  </div>
</template>

<script>
  export default {
    name: "Login",
    data(){
      return{
        form: {
          username: '',
          password: '',
        },
      }
    },
    methods:{
      login(){
        let flag = false
        if(this.form.username===''){
          this.$message.error("用户名不能为空！")
          flag = true
        }
        else if(this.form.password===''){
          this.$message.error("密码不能为空！")
          flag = true
        }
        if(flag){
```

```
            this.$router.push({
              path: '/'
            })
          }else{
            let that = this
            this.axios.post("http://localhost:8661/core/user/login",{
              username: this.form.username,
              password: this.form.password,
            })
            .then(function (res) {
              if(res.data.code === 200){
                console.log(res.data);
                sessionStorage.setItem('userId',res.data.data.userId);
                sessionStorage.setItem('userNick',res.data.data.userNick);
                sessionStorage.setItem("token",res.data.data.token);
                sessionStorage.setItem("userPhone",res.data.data.userPhone);
                sessionStorage.setItem("roleName",res.data.data.roleName);
                if(res.data.data.roleName=='student'){
                  that.$router.push({
                    path: '/searchGoods'
                  });
                }else{
                  that.$router.push({
                    path: '/userManage'
                  });
                }
                that.$message.success(res.data.msg);
              }
              else{
                that.$message.error(res.data.msg);
              }
            }).catch((err)=>{
              that.$message.error("登录失败,信息："+err);
            });
          }
        }
      }
    }
</script>

<style scoped>
  .login-content{
    border: 1px solid #ee8d35;
    background-color: #ee8d35;
    width: 700px;
    height: 50vh;
    padding-right: 10px;
    margin: 200px auto;
  }
  .form-content{
    margin: 0 auto;
    padding-top: 20px;
    width: 450px;
    text-align: center;
  }
</style>
```

7.7.3 控制器类设计

UserController 类是登录模块的控制器类，被@RestController 注解标注。在 UserController 类中，定义了

一个用于登录的 login()方法，其中包含了一个参数，即 UserLoginVO 类的对象。login()方法的代码如下：

```
@PostMapping("/user/login")
@ApiOperation("登录")
public Result<UserInfo> login(@RequestBody UserLoginVO userLoginVO){
    UserInfo userInfo = userService.login(userLoginVO);
    if(userInfo == null){
        return Result.fail("用户名或者密码错误");
    }
    try {
        String token = JwtTokenUtil.createToken(userInfo.getUserId());
        userInfo.setToken(token);
    } catch (Exception e) {
        e.printStackTrace();
    }
    return Result.success(userInfo);
}
```

7.7.4 服务类设计

UserService 接口为登录模块的服务接口。在 UserController 类的 login()方法中，调用了 UserService 接口的 login()方法。UserService 接口的 login()方法不仅有一个参数（即 UserLoginVO 类的对象），而且具有返回值，返回值是一个 UserInfo 类型的只包含用户名和密码的对象。UserService 接口的 login()方法的代码如下：

```
UserInfo login(UserLoginVO userLoginVO);
```

UserServiceImpl 类是 UserService 接口的实现类，被@Service 注解标注，表示登录模块的服务类。在 UserServiceImpl 类中，重写了 UserService 接口的 login()方法。该方法的作用是调用 DAO 层（数据访问对象）来执行数据库操作。UserServiceImpl 类的 login()方法的代码如下：

```
@Override
public UserInfo login(UserLoginVO userLoginVO) {
    UserInfo userInfo = userMapper.selectByPhone(userLoginVO.getUsername());
    if(userInfo == null){
        return null;
    }
    if(userLoginVO.getPassword().equals(userInfo.getUserPwd())){
        return userInfo;
    }
    return null;
}
```

7.7.5 DAO 层设计

UserMapper 接口为登录模块的 DAO 层。在 UserServiceImpl 类的 login()方法中，调用了 UserMapper 接口的 selectByPhone()方法，该方法用于根据手机号码获取用户信息。因此，selectByPhone()方法的参数是 String 类型的表示手机号码的 userPhone；返回值是一个 UserInfo 类型的对象。UserMapper 接口的 selectByPhone()方法的代码如下：

```
UserInfo selectByPhone(String userPhone);
```

> **说明**
> 在本项目中，与 UserMapper 接口中的方法绑定的 SQL 语句被写在了 UserMapper.xml 文件中。因此，读者可以在 UserMapper.xml 文件中查找与 selectByPhone()方法绑定的 SQL 语句。

7.8　头部导航链接设计

在登录页面上，用户输入正确的用户名和密码后，程序还要对用户的身份进行判断。如图 7.3 所示，如果用户的身份是学生，那么程序默认访问的是寻物启事超链接。在寻物启事页面的头部有 3 个导航超链接，它们分别是寻物启事超链接、个人中心超链接和退出登录超链接。

如图 7.4 所示，如果用户的身份是管理员，那么程序默认访问的是管理中心超链接。在管理中心页面的头部有 4 个导航超链接，它们分别是寻物启事超链接、个人中心超链接、管理中心超链接和退出登录超链接。

图 7.3　供学生访问的导航链接 　　　　图 7.4　供管理员访问的导航连接

寻物启事网站的导航链接对应的是 HeaderIndex.vue 文件，代码如下：

```vue
<template>
  <div class="header">
    <div class="content">

      <router-link :to="{name:'SearchGoods'}"
                   style="cursor: pointer;display: inline-block;">
        <!--<i class="iconfont icon-renwu l"></i>-->
        寻物启事
      </router-link>
      <router-link :to="{name:'PersonalInfo'}"
                   style="cursor: pointer;display: inline-block;">
        <!--<i class="iconfont icon-renwu l"></i>-->
        个人中心
      </router-link>
      <router-link v-if="roleName=='admin'" :to="{name:'UserManage'}"
                   style="cursor: pointer;display: inline-block;">
        <!--<i class="iconfont icon-renwu l"></i>-->
        管理中心
      </router-link>
      <div class="user-info">
        <span >{{userNick}}</span>
        <el-button type="text" @click="loginOut">退出登录</el-button>
      </div>
    </div>
  </div>
</template>

<script>
export default {
  name: "HeaderIndex",
  data(){
    return{
```

```
        userNick: sessionStorage.getItem("userNick"),
        roleName: sessionStorage.getItem("roleName")
      }
    },
    methods:{
      loginOut(){
        this.$confirm('确认退出登录？')
         .then(_ => {
           sessionStorage.removeItem("token")
           this.$router.push({
             path: '/'
           })
         }).catch(_ => {});
      }
    }
  }
</script>

<style scoped>
  .header{
    height: 60px;
    line-height: 60px;
  }
  .header .content{
    text-align: right;
  }
  .header .content a{
    text-decoration: none;
    color: white;
    font-size: large;
    font-weight: bold;
  }
  .header .content a.router-link-active{
    color: #ffffff;
    background: #ea692b;
  }
  .parent-info{
    width: 200px;
    float: right;
    text-align: center;
  }
  .user-info{
    width: 200px;
    float: right;
    text-align: center;
  }
</style>
```

7.9 查看失物信息模块设计

在寻物启事页面上，先通过卡片组件显示每一条失物信息，再通过分页插件显示与某个页面相对应的全部失物信息、失物信息的总条数、每个分页可显示的记录条数、分页数和分页导航等信息，如图 7.5 所示。此外，通过下拉列表还能够对每页可显示的记录条数进行设置，如图 7.6 所示。下面将介绍查看失物信息模块的设计过程。

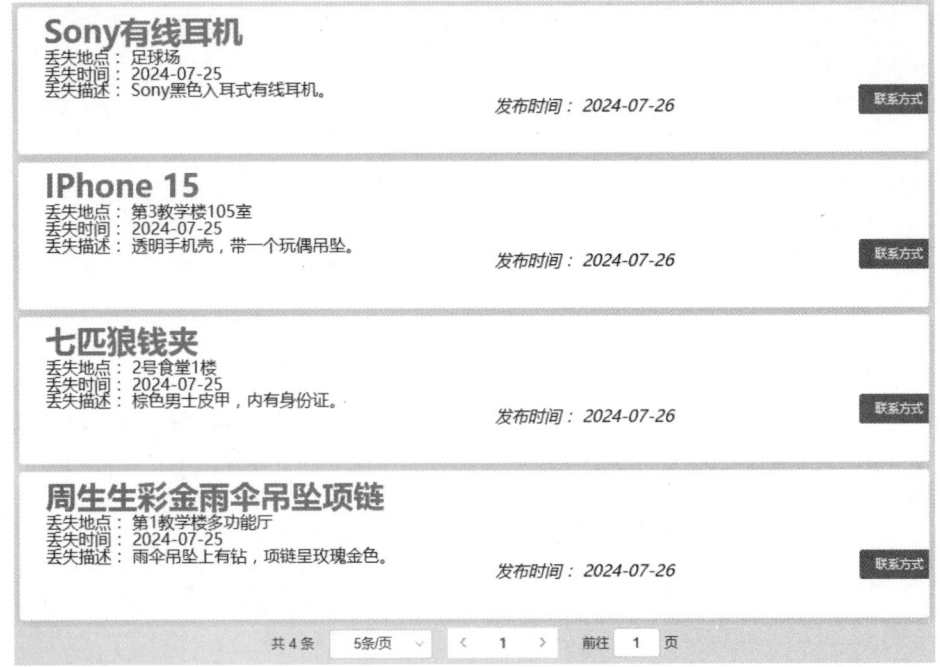
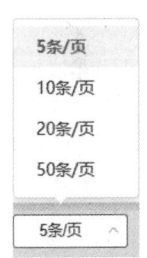

图 7.5　查看失物信息模块的效果图

图 7.6　设置每页可显示的
记录条数

7.9.1　数据传输对象设计

数据传输对象（data transfer object，DTO）用于在展示层与服务层之间进行数据传输。在本项目中，PageInfoDTO 类的对象是与查看失物信息模块相对应的数据传输对象，代码如下：

```
public class PageInfoDTO<T> extends PageVO implements Serializable {
    @ApiModelProperty("总记录数")
    private long total;

    @ApiModelProperty("总页数")
    private int pages;

    @ApiModelProperty("记录数据")
    private List<T> list;

    public void setTotal(long total) {
        this.total = total;
    }

    public void setPages(int pages) {
        this.pages = pages;
    }

    public List<T> getList() {
        return list;
    }

    public void setList(List<T> list) {
        this.list = list;
    }
```

```
    public static <T> PageInfoDTO<T> to(PageInfo<T> pageInfo) {
        PageInfoDTO<T> page = new PageInfoDTO<>();
        page.setList(pageInfo.getList());
        page.setPageSize(pageInfo.getPageSize());
        page.setPageNum(pageInfo.getPageNum());
        page.setTotal(pageInfo.getTotal());
        page.setPages(pageInfo.getPages());
        return page;
    }
}
```

7.9.2 寻物启事页面设计

在寻物启事页面上，包含了卡片组件和分页插件。卡片组件用于显示每一条失物信息，每一条失物信息的内容都是相同的，即失物名称、丢失地点、丢失时间、丢失描述和发布时间。分页插件用于把全部的失物信息按照每个分页可显示的记录条数予以显示。

寻物启事页面对应的是 SearchGoods.vue 文件，代码如下：

```
<template>
  <el-container>
    <el-header>
      <HeaderIndex/>
    </el-header>
    <el-main>
              //省略部分代码
              <div class="card_header" style="border: 0px solid red;height: 5vh;line-height: 5vh">
                <span class="title" style="float: left;font-weight: bolder;font-size: 30px;color: #f86d0f;">
                  {{ item.lostName }}
                </span>
              </div>
              <div class="clearfix" style="border: 0px solid #0452d9;height: 14vh;">
                <ul style="text-align: left">
                  <li>
                    <label>丢失地点：</label>
                    <span class="weight">{{ item.lostPlace }}</span>
                  </li>
                  <li>
                    <label>丢失时间：</label>
                    <span class="weight">{{ item.lostTime }}</span>
                  </li>
                  <li>
                    <label>丢失描述：</label>
                    <span class="weight">{{ item.lostDecp }}</span>
                  </li>

                  <li style="text-align: right">
                    <i>
                      <label>发布时间：</label>
                      <span class="weight">{{ item.lostReleaseTime }}</span>
                    </i>
                  </li>
                </ul>
              </div>
              //省略部分代码
            </el-card>
          </li>
        </ul>
```

```html
      <el-pagination
        @size-change="handleSizeChange"
        @current-change="handleCurrentChange"
        :current-page="page.pageNum"
        :page-sizes="[5, 10, 20, 50]"
        :page-size="page.pageSize"
        layout="total, sizes, prev, pager, next, jumper"
        :total="page.total">
      </el-pagination>
    </div>
  </el-main>
 </el-container>
</template>

<script>
  import HeaderIndex from "./HeaderIndex";
  export default {
    name: "SearchGoods",
    components:{HeaderIndex},
    data () {
      return {

        dataTable: [],
        checkList: 1,
        dialogVisible: false,
        dialogCommentVisible: false,
        page: {
          pageSize: 5,
          pageNum: 1,
          pages: 4,
          total: 20
        },
        dialogData: null,
        date: [],
        searchContent: '',
        kindId: '',
        kinds: {},
        form: {

        },
        userPhone: sessionStorage.getItem("userPhone"),
        code: '',
        comments: [],
        send: {
          content: '',
        },
        searchGood: {},
      }
    },
    //省略部分代码
</script>
```
//省略部分代码

7.9.3 控制器类设计

LostController 类是查看失物信息模块的控制器类，被@RestController 注解标注。在 LostController 类中，定义了一个用于列表查询寻物启事的 getList()方法。getList()方法的代码如下：

```
@PostMapping(value = "/losts")
@ApiOperation("列表查询寻物启事")
public Result<PageInfoDTO<LostInfo>> getList(@RequestBody LostInfoVO lostInfoVO){
    PageInfoDTO<LostInfo> list = lostService.getList(lostInfoVO);
    return Result.success(list);
}
```

在本项目中，LostInfoVO 类的对象是与查看失物信息模块相对应的展示层对象，读者可以在 LostInfoVO.java 文件中查看相关代码。

7.9.4 服务类设计

LostService 接口为查看失物信息模块的服务接口。在 LostController 类的 getList()方法中，调用了 LostService 接口的 getList()方法。LostService 接口的 getList()方法不仅有一个参数（即 LostInfoVO 类的对象），而且具有返回值，返回值是一个 PageInfoDTO 类的对象。LostService 接口的 getList()方法的代码如下：

```
PageInfoDTO<LostInfo> getList(LostInfoVO lostInfoVO);
```

LostServiceImpl 类是 LostService 接口的实现类，被@Service 注解标注，表示查看失物信息模块的服务类。在 LostServiceImpl 类中，重写了 LostService 接口的 getList()方法。该方法的作用是调用 DAO 层（数据访问对象）来执行数据库操作。LostServiceImpl 类的 getList()方法的代码如下：

```
@Override
public PageInfoDTO<LostInfo> getList(LostInfoVO lostInfoVO) {
    PageHelper.startPage(lostInfoVO.getPageNum(),lostInfoVO.getPageSize());
    List<LostInfo> lostInfoList = lostMapper.selectList(lostInfoVO);
    PageInfo<LostInfo> pageInfo=new PageInfo<>(lostInfoList);
    return PageInfoDTO.to(pageInfo);
}
```

7.9.5 DAO 层设计

LostMapper 接口为查看失物信息模块的 DAO 层。在 LostServiceImpl 类的 getList()方法中，调用了 LostMapper 接口的 selectList()方法，该方法用于列表查询寻物启事。selectList()方法的参数是 LostInfoVO 类的对象，返回值是一个用于存储失物信息的列表。LostMapper 接口的 selectList()方法的代码如下：

```
List<LostInfo> selectList(LostInfoVO lostInfoVO);
```

在本项目中，与 LostMapper 接口中的方法绑定的 SQL 语句被写在了 LostMapper.xml 文件中。因此，读者可以在 LostMapper.xml 文件中查找与 selectList()方法绑定的 SQL 语句。

7.10 发布寻物启事模块设计

在寻物启事页面上，有一个"发布寻物启事"超链接，如图 7.7 所示。已登录的用户在单击"发布寻物启事"超链接后，寻物启事页面将跳转到如图7.8所示的"发布寻物启事"页面。在发布寻物启事页面上，用户需要先输入失物信息，再单击"确定发布"按钮，以完成发布寻物启事的操作。下面将介绍发布寻物启事模块的设计过程。

图7.7 "发布寻物启事"超链接的效果图

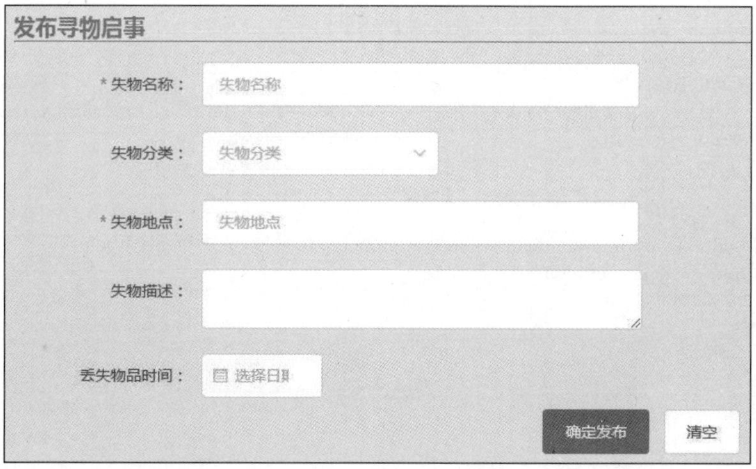

图7.8 "发布寻物启事"页面的效果图

7.10.1 发布寻物启事页面设计

在发布寻物启事页面上，包含标签、输入框、下拉列表、文本域、时间选择器、按钮等组件。其中，每一个组件都设置了提示信息，如时间选择器上的提示信息是"选择日期"。

发布寻物启事页面对应的是 AddSearchGoods.vue 文件，代码如下：

```
<template>
  <el-container>
    <el-header>
      <HeaderIndex/>
    </el-header>
    <el-main>
      <div class="center" style="border: 0px solid red;width: 80%;margin: 0 auto">
        <div class="header-info" style="font-size:20px;font-weight:bolder;
          color:#f86d0f;border:0px solid black;height: 4vh;line-height: 4vh;text-align: left;">
          发布寻物启事
        </div>
        <hr>
        <div class="form-content">
          <el-form ref="form" :model="form" label-width="120px" :rules="rules">
            <div class="basic-info" style="border: 0px solid red;width: 90%;margin:3vh auto;height: 50vh;text-align: left;">
              <el-form-item label="失物名称：" prop="lostName">
                <el-input v-model="form.lostName" style="width: 400px" placeholder="失物名称"></el-input>
```

219

```html
            </el-form-item>
            <el-form-item label="失物分类：">
              <el-select v-model="form.kindId" placeholder="失物分类">
                <el-option v-for="item in kinds" :key="item.kindId" :label="item.kindName" :value="item.kindId">
                </el-option>
              </el-select>
            </el-form-item>
            <el-form-item label="失物地点：" prop="lostPlace">
              <el-input v-model="form.lostPlace" placeholder="失物地点" style="width: 400px"></el-input>
            </el-form-item>
            <el-form-item label="失物描述：">
              <el-input type="textarea" v-model="form.lostDecp" style="width: 50%"></el-input>
            </el-form-item>
            <el-form-item label="丢失物品时间：">
              <el-col :span="11">
                <el-date-picker type="date" placeholder="选择日期" v-model="form.lostTime" style="width: 30%">
                </el-date-picker>
              </el-col>
              <!--<el-col class="line" :span="1">-</el-col>
              <el-col :span="11">
                <el-time-picker placeholder="选择时间" v-model="form.date2" style="width: 30%;"></el-time-picker>
              </el-col>-->
            </el-form-item>
          </div>
          <el-form-item>
            <el-button type="primary" @click="AddSearchGoods">确定发布</el-button>
            <el-button>清空</el-button>
          </el-form-item>
        </el-form>
      </div>
    </div>
  </el-main>
</el-container>
</template>

<script>
  import HeaderIndex from "./HeaderIndex";
  export default {
    name: "AddSearchGoods",
    components: {HeaderIndex},
    data() {
      return {
        form: {
          lostName:'',
          kindId: '',
          lostPlace: '',
          lostDecp: '',
          lostTime: '',
          lostPhoto: ''
        },
        kinds: {},
        rules:{
          lostName:[{required: true, message: '请填写物品名称', trigger: 'blur'}],
          lostPlace: [{required: true, message: '请填写丢失地点信息', trigger: 'blur'}],
        }
      }
    },
    created() {
      this.queryKind()
    },
    methods: {
      queryKind(){
```

```
            let that = this
            this.axios.get('http://localhost:8661/core/kinds')
                .then(function (res) {
                    if (res.data.code === 200) {
                        console.log(res.data)
                        that.kinds = res.data.data
                    } else {
                        that.$message.error(res.data.msg)
                    }
                })
        },
        handleAvatarSuccess(res, file) {
            console.log(res.data)
            console.log(res.data)
            this.form.lostPhoto = 'http://localhost:8661/'+res.data
            console.log(this.form.lostPhoto)
        },
        AddSearchGoods(){
            let that = this
            this.axios.post('http://localhost:8661/core/lost',{
                kindId: this.form.kindId,
                lostName: this.form.lostName,
                lostPlace: this.form.lostPlace,
                lostTime:this.form.lostTime,
                lostDecp: this.form.lostDecp,
                lostPhoto: this.form.lostPhoto,
                userId: sessionStorage.getItem("userId"),
            })
                .then(function (res) {
                    if (res.data.code === 200) {
                        console.log(res.data)
                        that.$message.success(res.data.msg)
                        that.$router.push({
                            path: '/searchGoods'
                        })
                    } else {
                        that.$message.error(res.data.msg)
                    }
                })
        }
    }
}
//省略部分代码
</style>
```

7.10.2 控制器类设计

LostController 类是发布寻物启事模块的控制器类。在 LostController 类中，定义了一个用于发布寻物启事的 addLost() 方法。addLost() 方法的代码如下：

```
@PostMapping(value = "/lost")
@ApiOperation("发布寻物启事")
public Result<String> addLost(@RequestBody @Valid LostInfo lostInfo){
    int result = lostService.addLost(lostInfo);
    if(result==1){
        return Result.success("发布寻物启事成功");
    }
    return Result.fail("发布寻物启事失败");
}
```

7.10.3 服务类设计

LostService 接口为发布寻物启事模块的服务接口。在 LostController 类的 addLost()方法中，调用了 LostService 接口的 addLost()方法。LostService 接口的 addLost()方法不仅有一个参数（即 LostInfo 类的对象），而且具有返回值，返回值是一个 int 类型、表示是否成功发布寻物启事的变量（1 表示成功，0 表示失败）。LostService 接口的 addLost()方法的代码如下：

```
int addLost(LostInfo lostInfo);
```

LostServiceImpl 类是 LostService 接口的实现类，即发布寻物启事模块的服务类。在 LostServiceImpl 类中，重写了 LostService 接口的 addLost()方法。该方法的作用是调用 DAO 层（数据访问对象）来执行数据库操作。LostServiceImpl 类的 addLost()方法的代码如下：

```java
@Override
public int addLost(LostInfo lostInfo) {
    lostInfo.setLostId(UUID.randomUUID().toString().replace("-",""));
    lostInfo.setLostReleaseTime(new Date());
    lostInfo.setCreateTime(new Date());
    lostInfo.setLostStatus(0);
    lostInfo.setCheckStatus(0);
    lostInfo.setCommentNum(0);
    int result = lostMapper.insertLostInfo(lostInfo);
    return result;
}
```

7.10.4 DAO 层设计

LostMapper 接口为发布寻物启事模块的 DAO 层。在 LostServiceImpl 类的 addLost()方法中，调用了 LostMapper 接口的 insertLostInfo()方法，该方法用于发布寻物启事。insertLostInfo()方法的参数是 LostInfo 类的对象；返回值是一个 int 类型、表示是否成功发布寻物启事的变量（1 表示成功，0 表示失败）。LostMapper 接口的 insertLostInfo()方法的代码如下：

```
int insertLostInfo(LostInfo lostInfo);
```

说明

> 在本项目中，与 LostMapper 接口中的方法绑定的 SQL 语句被写在了 LostMapper.xml 文件中。因此，读者可以在 LostMapper.xml 文件中查找与 insertLostInfo()方法绑定的 SQL 语句。

7.11 联系管理员模块设计

在寻物启事页面的卡片组件上，每一条失物信息的右下角都有一个"联系方式"按钮，如图 7.9 所示。已登录的用户在单击"联系方式"按钮后，程序将弹出"物品信息"对话框，如图 7.10 所示。在"物品信息"对话框中，将显示失物信息和管理员的联系方式。用户打开"物品信息"对话框并确认失物信息后，即可联系管理员归还失物。

因为联系管理员模块的控制器类、服务类和 DAO 层与查看失物信息模块的相同，所以下面将只介绍联系管理员模块的页面设计。

图 7.9　查看失物信息的效果图

图 7.10　"物品信息"对话框

联系管理员模块的页面设计仅涉及寻物启事页面的卡片组件，而寻物启事页面对应的是 SearchGoods.vue 文件，因此在 SearchGoods.vue 文件中不仅需要为卡片组件添加"联系方式"按钮，还需要添加"物品信息"对话框。代码如下：

```html
<div style="border: 0px solid #bc0aff;width: 10vh;float:right;height: 19vh;">
  <div style="">
    <el-button style="width: 80px" type="primary" size="mini" @click="queryOne(index)" v-on:click="dialogVisible=true">
      联系方式
    </el-button>
    <el-dialog title="物品信息" :visible.sync="dialogVisible">
      <div class="basic-info" style="border: 0px solid red;width: 90%;margin:0 auto;text-align: left;">
        <ul style="text-align: left;">
          <li>
            <label>失物名称：</label>
            <span class="weight">{{ form.lostName }}</span>
          </li>
          <li>
            <label>丢失地点：</label>
            <span class="weight">{{ form.lostPlace }}</span>
          </li>
          <li>
            <label>丢失时间：</label>
            <span class="weight">{{ form.lostTime }}</span>
          </li>
          <li>
            <label>丢失描述：</label>
            <span class="weight">{{ form.lostDecp }}</span>
          </li>
        </ul>
```

```
        </div>
        <div style="border: 0px solid red;width: 90%;height: 5vh;margin:0 auto;">
            <span>请再次确认物品信息，如果确认，请联系管理员，管理员电话13307118899。</span>
        </div>
        <el-button @click="dialogVisible = false">关闭</el-button>
      </el-dialog>
    </div>
</div>
```

7.12 修改用户信息模块设计

不论是供学生访问的导航链接，还是供管理员访问的导航链接，都有"个人中心"超链接。以供学生访问的导航链接为例，程序默认访问的是"寻物启事"超链接，当用户单击"个人中心"超链接时，寻物启事页面将跳转到如图 7.11 所示的个人中心页面。在个人中心页面上，用户可以先单击图标按钮，对微信号、手机号码或者密码进行修改，再单击"确定修改个人信息"按钮，以完成修改用户信息的操作。下面将介绍修改用户信息模块的设计过程。

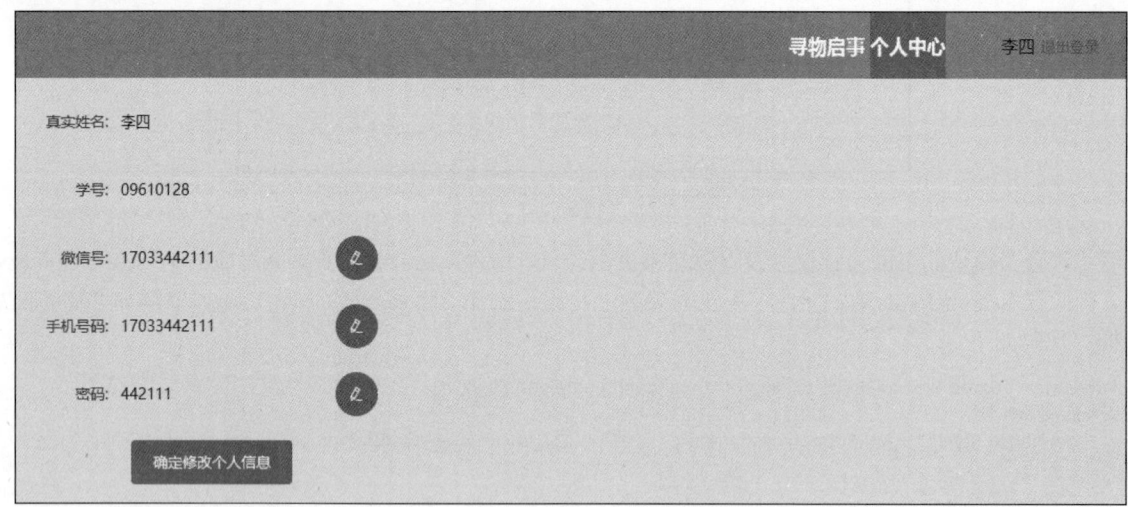

图 7.11　个人中心页面的效果图

7.12.1 个人中心页面设计

在个人中心页面上，包含了标签、输入框、（图标）按钮等组件。个人中心页面为已登录的用户提供了修改微信号、手机号码或者密码的功能。

个人中心页面对应的是 PersonalInfo.vue 文件，代码如下：

```
<template>
  <el-container>
    <el-header>
      <HeaderIndex/>
    </el-header>
    <el-container>
      <el-main>
        <div>
          <el-form :label-position='left' label-width="80px" :model="tableData">
```

```html
            <el-form-item label="真实姓名:">
              <span v-show="!inputLock.realname" class="content">{{ tableData.userName }}</span>
              <el-input v-model="tableData.userName" v-if="inputLock.realname" ></el-input>
            </el-form-item>
            <el-form-item label="学号:">
              <span v-show="!inputLock.num" class="content">{{ tableData.stuNum }}</span>
              <el-input v-model="tableData.stuNum" v-if="inputLock.num"></el-input>
            </el-form-item>
            <el-form-item label="微信号:">
              <span v-show="!inputLock.wx" class="content">{{ tableData.wechatId }}</span>
              <el-input v-model="tableData.wechatId" v-if="inputLock.wx"></el-input>
              <el-button class="button" type="primary" icon="el-icon-edit" circle @click="openLock('wx')"></el-button>
            </el-form-item>
            <el-form-item label="手机号码:">
              <span v-show="!inputLock.telephone" class="content">{{ tableData.userPhone }}</span>
              <el-input v-model="tableData.userPhone" v-if="inputLock.telephone" ></el-input>
              <el-button class="button" type="primary" icon="el-icon-edit" circle @click="openLock('telephone')">
              </el-button>
            </el-form-item>
            <el-form-item label="密码:">
              <span v-show="!inputLock.password" class="content">{{ tableData.userPwd }}</span>
              <el-input v-model="tableData.userPwd" v-if="inputLock.password"></el-input>
              <el-button class="button" type="primary" icon="el-icon-edit" circle @click="openLock('password')">
              </el-button>
            </el-form-item>
            <el-form-item>
              <el-button class="button" type="primary" @click="updateUser">确定修改个人信息</el-button>
            </el-form-item>
          </el-form>
        </div>
      </el-main>
    </el-container>
  </el-container>
</template>

<script>
  import HeaderIndex from "../HeaderIndex";
  export default {
    name: "PersonalInfo",
    components: {HeaderIndex},
    data () {
      return {
        tableData: {
          userNick: '',
          userName: '',
          stuNum: '',
          wechatId: '',
          userPhone: '',
          userPwd: '',
        },
        inputLock: {
          name: false,
          realname: false,
          num: false,
          wx: false,
          telephone: false,
          password: false
        }
      }
    },
    created () {
      this.queryUser()
```

```
        },
        methods: {
            openLock (key) {
                this.inputLock[key] = !this.inputLock[key]
            },
            queryUser () {
                const content = this
                this.axios.get('http://localhost:8661/core/user/'+sessionStorage.getItem("userId")).then(function (res) {
                    console.log(res.data)
                    content.tableData = res.data.data
                })
            },
            updateUser(){
                const content = this
                this.axios.put('http://localhost:8661/core/user/',{
                    userId: sessionStorage.getItem("userId"),
                    userNick: this.tableData.userNick,
                    userName: this.tableData.userName,
                    stuNum: this.tableData.stuNum,
                    wechatId: this.tableData.wechatId,
                    userPhone: this.tableData.userPhone,
                    userPwd: this.tableData.userPwd,
                }).then(function (res) {
                    if(res.data.code == 200){
                        content.$message.success(res.data.msg)
                        content.queryUser();
                    }
                })
            }
        }
    }
</script>
//省略部分代码
```

7.12.2 控制器类设计

UserController 类是修改用户信息模块的控制器类，被@RestController 注解标注。在 UserController 类中，定义了一个用于更新用户信息的 updateUser()方法。updateUser()方法的代码如下：

```
@PutMapping("/user")
@ApiOperation("更新用户信息")
public Result<String> updateUser(@RequestBody UserInfo userInfo){
    int result = userService.updateUser(userInfo);
    if(result == 1){
        return Result.success("修改成功");
    }
    return Result.success("修改失败");
}
```

7.12.3 服务类设计

UserService 接口为修改用户信息模块的服务接口。在 UserController 类的 updateUser()方法中，调用了 UserService 接口的 updateUser()方法。UserService 接口的 updateUser()方法不仅有一个参数（即 UserInfo 类的对象），而且具有返回值，返回值是一个 int 类型、表示是否成功修改用户信息的变量（1 表示成功，0 表示失败）。UserService 接口的 updateUser()方法的代码如下：

```
int updateUser(UserInfo userInfo);
```

UserServiceImpl 类是 UserService 接口的实现类，被@Service 注解标注，表示修改用户信息模块的服务类。在 UserServiceImpl 类中，重写了 UserService 接口的 updateUser()方法。该方法的作用是调用 DAO 层（数据访问对象）来执行数据库操作。UserServiceImpl 类的 updateUser()方法的代码如下：

```
@Override
public int updateUser(UserInfo userInfo) {
    userInfo.setUpdateTime(new Date());
    return userMapper.updateUser(userInfo);
}
```

7.12.4 DAO 层设计

UserMapper 接口为修改用户信息模块的 DAO 层。在 UserServiceImpl 类的 updateUser()方法中，调用了 UserMapper 接口的 updateUser()方法，该方法用于更新用户信息。UserMapper 接口的 updateUser()方法不仅有一个参数（即 UserInfo 类的对象），而且具有返回值，返回值是一个 int 类型、表示是否成功修改用户信息的变量（1 表示成功，0 表示失败）。UserMapper 接口的 updateUser()方法的代码如下：

```
int updateUser(UserInfo userInfo);
```

> 在本项目中，与 UserMapper 接口中的方法绑定的 SQL 语句被写在了 UserMapper.xml 文件中。因此，读者可以在 UserMapper.xml 文件中查找与 updateUser()方法绑定的 SQL 语句。

7.13 退出登录模块设计

不论是供学生访问的导航链接，还是供管理员访问的导航链接，都有如图 7.12 所示的"退出登录"超链接。以供学生访问的导航链接为例，程序默认访问的是"寻物启事"超链接，当用户单击"退出登录"超链接时，寻物启事页面将跳转到登录页面。下面将介绍退出登录模块的设计过程。

图 7.12 "退出登录"超链接的效果

退出登录模块的相关代码被编写在 HeaderIndex.vue 文件（寻物启事网站的导航链接）中。因为"退出登录"超链接的作用只是让寻物启事页面、个人中心页面或者管理中心页面跳转到登录页面，所以退出登录模块不需要控制器类、服务类或者 DAO 层的支持。退出登录模块的代码如下：

```
<template>
  <div class="header">
    <div class="content">
      //省略部分代码
      <div class="user-info">
        <span >{{userNick}}</span>
        <el-button type="text" @click="loginOut">退出登录</el-button>
      </div>
    </div>
  </div>
</template>

<script>
  export default {
```

```
    name: "HeaderIndex",
    data(){
      return{
        userNick: sessionStorage.getItem("userNick"),
        roleName: sessionStorage.getItem("roleName")
      }
    },
    methods:{
      loginOut(){
        this.$confirm('确认退出登录？')
          .then(_ => {
            sessionStorage.removeItem("token")
            this.$router.push({
              path: '/'
            })
          }).catch(_ => {});
      }
    }
  }
</script>
//省略部分代码
```

7.14 左侧导航链接设计

管理员登录以后，程序默认打开的是管理中心页面。在管理中心页面的左侧有 3 个导航超链接，它们分别是"用户管理"超链接、"分类管理"超链接和"寻物启事审核"超链接，如图 7.13 所示。通过访问这 3 个超链接，即可打开相应的页面，并实现用户管理、分类管理和寻物启事审核的功能。

管理中心页面的左侧导航链接对应的是 ManagerIndex.vue 文件，代码如下：

图 7.13 左侧导航链接的效果图

```
<template>
  <ul class="column">
    <div class="header-logo">
      <i class="iconfont el-icon-user I"></i>
      管理中心
    </div>
    <router-link :to="{name:'UserManage'}"
                 tag="li"
                 style="cursor: pointer;">
      <p class="title">用户管理</p>
    </router-link>
    <router-link :to="{name:'KindManage'}"
                 tag="li"
                 style="cursor: pointer;">
      <p class="title">分类管理</p>
    </router-link>
    <router-link :to="{name:'SearchCheck'}"
                 tag="li"
                 style="cursor: pointer;">
      <p class="title">寻物启事审核</p>
    </router-link>
  </ul>
</template>

<script>
```

```
        export default {
            name: "ManagerIndex"
        }
</script>

<style scoped>
    .header-logo{
        height: 100px;line-height:100px;

    }
    .column{
        text-align: center;
        background-color: #ef9265;

    }
    .title{
        height: 60px;
        /*margin-top: 10px;*/
        background-color: #ef9265;
        line-height: 60px;
        list-style: none;
        color: aliceblue;
        font-family: 宋体;
        font-size: large;
        font-weight: bold;
    }
    .column li.router-link-active .title{
        color: white;
        background: #de7c49;
    }
</style>
```

7.15 用户管理模块设计

如图 7.14 所示，用户管理页面用于显示用户编号、用户名、真实姓名、学号、手机号码、微信号、身份证号码、密码、添加时间、管理员（用户身份）等信息。

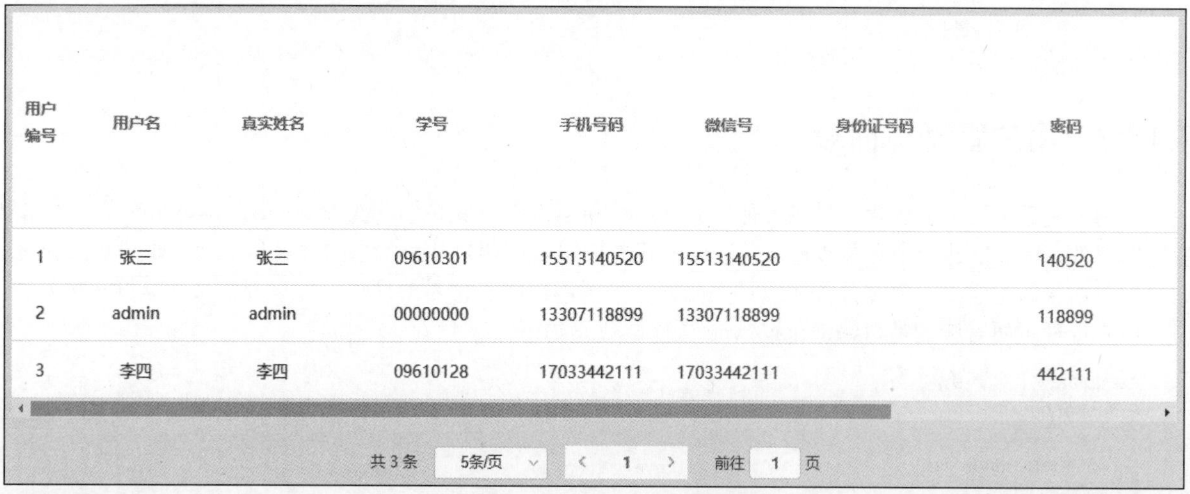

用户编号	用户名	真实姓名	学号	手机号码	微信号	身份证号码	密码
1	张三	张三	09610301	15513140520	15513140520		140520
2	admin	admin	00000000	13307118899	13307118899		118899
3	李四	李四	09610128	17033442111	17033442111		442111

共3条　5条/页　< 1 >　前往 1 页

图 7.14　用户管理页面的效果图

同时，用户管理页面也用于修改用户身份，即管理员通过 switch 开关按钮能够设置用户的身份是"学生"，还是"管理员"，如图 7.15 所示。也就是说，用户管理模块有两个功能，一个是列表查询用户信息，另一个是修改用户的身份。列表查询用户信息的设计过程与列表查询寻物启事的相同，读者可参照第 7.9 节。下面将着重介绍如何设计修改用户身份的功能。

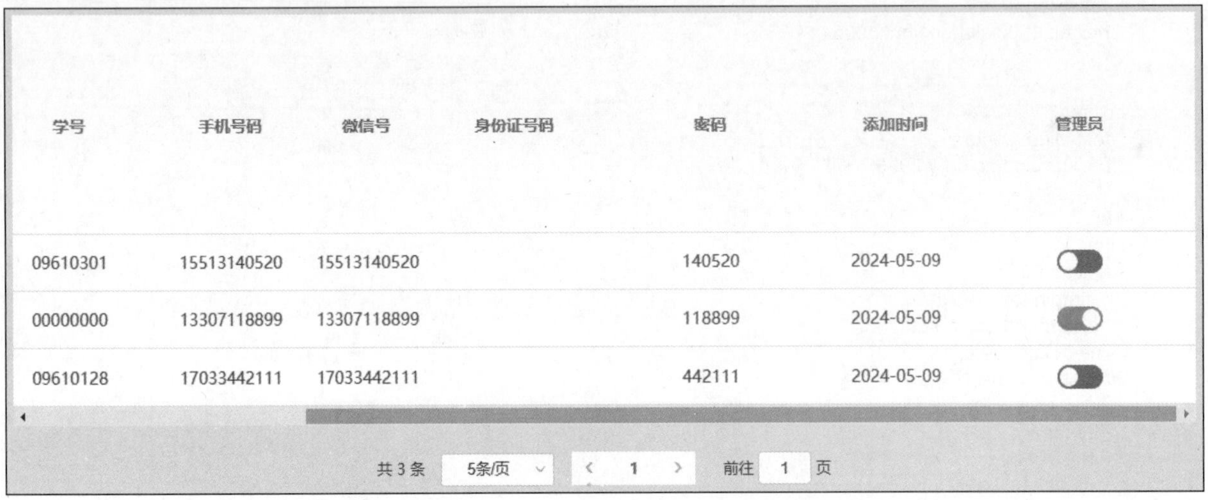

学号	手机号码	微信号	身份证号码	密码	添加时间	管理员
09610301	15513140520	15513140520		140520	2024-05-09	⚪
00000000	13307118899	13307118899		118899	2024-05-09	⚫
09610128	17033442111	17033442111		442111	2024-05-09	⚫

共 3 条   5条/页   < 1 >   前往 1 页

图 7.15 使用 switch 开关按钮修改用户身份的效果图

7.15.1 展示层对象设计

在本项目中，RoleInfoVO 类的对象是与用户管理模块中修改用户身份的功能相对应的展示层对象，其中只包含用户身份名称和用户编号这两个属性。RoleInfoVO 类中的每一个属性都与 sys_role（用户身份表）中的各个字段是一一对应的。代码如下：

```
@Data
public class RoleInfoVO {
    @ApiModelProperty("用户身份名称")
    private String roleName;

    @ApiModelProperty("用户 id")
    private String userId;
}
```

7.15.2 用户管理页面设计

在用户管理页面上，包含了表格组件、分页插件和 switch 开关按钮。在 Vue3 中，el-switch 是一个非常有用的组件，用于创建一个开关按钮，通过这个开关按钮可以切换某个状态的开启和关闭。如图 7.15 所示，如果用户的身份是学生，那么开关按钮是关闭的；如图用户的身份是管理员，那么开关按钮是开启的。

用户管理页面对应的是 UserManage.vue 文件，代码如下：

```
<template>
  <el-container>
    <el-header>
      <HeaderIndex/>
    </el-header>
    <el-container>
```

```html
            <el-aside width="200px">
                <ManagerIndex/>
            </el-aside>
            <el-main>
                <div class="main-content" style="">
                    <el-main>
                        <el-table :data="dataTable">
                            <el-table-column type="index" label="用户编号" width="50" align="center">
                            </el-table-column>
                            <el-table-column prop="userNick" label="用户名" width="120" align="center">
                            </el-table-column>
                            <el-table-column prop="userName" label="真实姓名" width="120" align="center">
                            </el-table-column>
                            <el-table-column prop="stuNum" label="学号" width="160" align="center">
                            </el-table-column>
                            <el-table-column prop="userPhone" label="手机号码" width="120" align="center">
                            </el-table-column>
                            <el-table-column prop="wechatId" label="微信号" width="120" align="center">
                            </el-table-column>
                            <el-table-column prop="numId" label="身份证号码" width="140" align="center">
                            </el-table-column>
                            <el-table-column prop="userPwd" label="密码" width="200" align="center">
                            </el-table-column>
                            <el-table-column prop="createTime" label="添加时间" width="120" align="center">
                            </el-table-column>
                            <el-table-column label="管理员" width="200" align="center" >
                                <template class="handle" #default="scope">
                                    <el-switch
                                        v-model="scope.row.roleName"
                                        active-color="#13ce66"
                                        inactive-color="#ff4949"
                                        active-value="admin"
                                        inactive-value="student"
                                        @change="changeRole(scope.row)">
                                    </el-switch>
                                </template>
                            </el-table-column>
                        </el-table>
                    </el-main>
                </div>

                <div class="page">
                    <el-pagination
                        @size-change="handleSizeChange"
                        @current-change="handleCurrentChange"
                        :current-page="pagnination.pageNum"
                        :page-sizes="[5, 10, 20, 30, 50]"
                        :page-size="pagnination.pageSize"
                        layout="total, sizes, prev, pager, next, jumper"
                        :total="pagnination.total">
                    </el-pagination>
                </div>
            </el-main>
        </el-container>
    </el-container>
</template>

<script>
    import HeaderIndex from "../HeaderIndex";
    import ManagerIndex from "./ManagerIndex";
```

```js
import {validateEMail,validatePhone} from "../../util/rule";
export default {
  name: "UserManage",
  components: {HeaderIndex,ManagerIndex},
  data(){
    return{
      dialogVisible: false,
      userName: '',
      pagination: {                          //页码信息
        pageSize: 5,                         //每页显示数据条数
        pageNum: 1,                          //当前页
        pages: 0,                            //总页数
        total: 0                             //总记录数
      },
      form: {
        userNick: '',
        userName: '',
        stuNum: '',
        userPhone: '',
        wechatId: '',
        numId: '',
      },
      dataTable: [],
    }
  },
  created() {
    this.queryUser()
  },
  methods:{
    queryUser(){
      let that = this
      this.axios.post('http://localhost:8661/core/users',{
        userName: this.userName,
        pageNum: this.pagination.pageNum,
        pageSize: this.pagination.pageSize,
      })
      .then(function (res) {
        if (res.data.code === 200) {
          console.log(res.data)
          that.pagination.pages = res.data.data.pages
          that.pagination.total = res.data.data.total
          that.dataTable = res.data.data.list
        } else {
          that.$message.error(res.data.msg)
        }
      })
    },
    handleSizeChange (val) {
      //刷新列表，刷新网页，回到第一页
      this.pagination.pageSize = val;
      this.queryUser();
    },
    handleCurrentChange (val) {
      //刷新列表，刷新页面，跳转到该页
      this.pagination.pageNum = val;
      this.queryUser();
    },
    changeRole(index){
      let that = this
      this.axios.post('http://localhost:8661/core/user/role',{
```

```
              roleName: index.roleName,
              userId: index.userId
          })
            .then(function (res) {
              if (res.data.code === 200) {
                console.log(res.data)
                that.$message.success(res.data.msg)
                that.queryUser()
              } else {
                that.$message.error(res.data.msg)
              }
            })
        }
      }
    }
</script>

<style scoped>
  .el-header, .el-footer {
      background-color: #ef9265;
      color: #333;
      text-align: center;
      line-height: 60px;
  }
  .el-aside {
      background-color: #ef9265;
      color: #333;
      height: 600px;
      text-align: center;
      line-height: 200px;
  }
  .el-main {
      background-color: #E9EEF3;
      color: #333;
      text-align: center;
      line-height: 160px;
  }
  .top-research{
      display: flex;
      justify-content: space-between;
      white-space: nowrap;
      align-items: center;
      height: 100px;
  }
</style>
```

7.15.3 控制器类设计

UserController 类是用户管理模块的控制器类。在 UserController 类中，定义了一个用于修改用户身份的 updateRole()方法。updateRole()方法的代码如下：

```
@PostMapping("/user/role")
@ApiOperation("修改用户身份")
public Result<String> updateRole(@RequestBody RoleInfoVO roleInfoVO){
    int result = userService.updateRole(roleInfoVO);
    if(result == 1){
        return Result.success("修改成功");
    }
```

```
        return Result.success("修改失败");
    }
```

7.15.4 服务类设计

UserService 接口为用户管理模块的服务接口。在 UserController 类的 updateRole()方法中，调用了 UserService 接口的 updateRole()方法。UserService 接口的 updateRole()方法不仅有一个参数（即 UserInfo 类的对象），而且具有返回值，返回值是一个 int 类型、表示是否成功修改用户身份的变量（1 表示成功，0 表示失败）。UserService 接口的 updateRole()方法的代码如下：

```
int updateRole(RoleInfoVO roleInfoVO);
```

UserServiceImpl 类是 UserService 接口的实现类，即用户管理模块的服务类。在 UserServiceImpl 类中，重写了 UserService 接口的 updateRole()方法。该方法的作用是调用 DAO 层（数据访问对象）来执行数据库操作。UserServiceImpl 类的 updateRole()方法的代码如下：

```
@Override
public int updateRole(RoleInfoVO roleInfoVO) {
    String roleId = userMapper.selectByRoleName(roleInfoVO.getRoleName());
    int result = userMapper.updateRole(roleInfoVO.getUserId(),roleId);
    return result;
}
```

7.15.5 DAO 层设计

UserMapper 接口为用户管理模块的 DAO 层。在 UserServiceImpl 类的 updateRole()方法中，分别调用了 UserMapper 接口的 selectByRoleName()方法和 updateRole()方法。selectByRoleName()方法用于获取用户身份编号，updateRole()方法用于修改用户身份。UserMapper 接口的 selectByRoleName()方法和 updateRole()方法的代码如下：

```
String selectByRoleName(String roleName);
int updateRole(@Param("userId") String userId,@Param("roleId") String roleId);
```

说明

在本项目中，与 UserMapper 接口中的方法绑定的 SQL 语句被写在了 UserMapper.xml 文件中。因此，读者可以在 UserMapper.xml 文件中查找与 selectByRoleName()方法和 updateRole()方法绑定的 SQL 语句。

7.16 分类管理模块设计

如图 7.16 所示，分类管理页面有两个功能：一个是列表查询分类，另一个是新增分类。其中，程序把列表查询分类的结果显示在表格组件中；新增分类则是通过如图 7.17 所示的对话框予以实现。下面将介绍分类管理模块的设计过程。

寻物启事网站 第7章

图 7.16 分类管理页面的效果图

图 7.17 新增分类的效果图

7.16.1 数据传输对象设计

在本项目中，KindDTO 类的对象是与分类管理模块相对应的数据传输对象，其中包含了 3 个属性，即失物分类编号、失物分类名称和失物分类的创建时间。KindDTO 类中的每一个属性都与 t_kind（失物类别表）中的各个字段是一一对应的。代码如下：

```
@Data
public class KindDTO {
    @ApiModelProperty("失物分类编号")
    private String kindId;

    @ApiModelProperty("失物分类名称")
    private String kindName;

    @ApiModelProperty("失物分类的创建时间")
    @JsonFormat(pattern = "yyyy-MM-dd")
    private Date createTime;
}
```

7.16.2 分类管理页面设计

在分类管理页面上,包含了表格组件和按钮插件。表格组件用于显示已添加的失物分类编号、失物分类名称和失物分类的创建时间;按钮组件用于新增分类,用户单击"新增分类"按钮,在弹出的对话框中输入新的分类名,单击"添加"按钮,即可完成新增分类的操作。

分类管理页面对应的是 KindManage.vue 文件,代码如下:

```
<template>
  <el-container>
    <el-header>
      <HeaderIndex/>
    </el-header>
    <el-container>
      <el-aside width="200px">
        <ManagerIndex/>
      </el-aside>
      <el-main>
        <div class="button-info" style="border: 0px solid red;height: 5vh;line-height: 5vh;">
          <el-button type="primary" @click="dialogVisible = true">新增分类</el-button>
          <el-dialog title="分类信息" :visible.sync="dialogVisible" >
            <el-form ref="form" :model="form" label-width="80px">
              <el-form-item label="分类名" prop="stuName">
                <el-input placeholder="分类名" v-model="form.kindName"></el-input>
              </el-form-item>
              <el-form-item>
                <el-button type="primary" @click="addKind">添加</el-button>
                <el-button @click="dialogVisible = false">取消</el-button>
              </el-form-item>
            </el-form>
          </el-dialog>
        </div>
        <div class="main-content" style="">
          <el-main>
            <el-table :data="dataTable">
              <el-table-column type="index" label="序号" width="50" align="center">
              </el-table-column>
              <el-table-column prop="kindName" label="分类名" width="120" align="center">
              </el-table-column>
              <el-table-column prop="createTime" label="添加时间" width="120" align="center">
              </el-table-column>
            </el-table>
          </el-main>
        </div>
      </el-main>
    </el-container>
  </el-container>
</template>

<script>
  import HeaderIndex from "../HeaderIndex";
  import ManagerIndex from "./ManagerIndex";
  export default {
    name: "KindManage",
    components: {HeaderIndex,ManagerIndex},
    data(){
      return{
        userName: '',
        dialogVisible: false,
```

```js
      dataTable: [],
      form: {
        kindName: '',
      }
    }
  },
  created() {
    this.queryKind()
  },
  methods:{
    queryKind(){
      let that = this
      this.axios.get('http://localhost:8661/core/kinds')
        .then(function (res) {
          if (res.data.code === 200) {
            console.log(res.data)
            that.dataTable = res.data.data
          } else {
            that.$message.error(res.data.msg)
          }
        })
    },
    handleSizeChange (val) {
      //刷新列表，刷新网页，回到第一页
      this.pagnination.pageSize = val;
      this.queryUser();
    },
    handleCurrentChange (val) {
      //刷新列表，刷新页面，跳转到该页
      this.pagnination.pageNum = val;
      this.queryUser();
    },
    addKind(){
      let that = this
      this.axios.post('http://localhost:8661/core/kind',{
        kindName: this.form.kindName
      })
        .then(function (res) {
          if (res.data.code === 200) {
            console.log(res.data)
            that.$message.success(res.data.msg)
            that.dialogVisible=false
            that.queryKind()
          } else {
            that.$message.error(res.data.msg)
          }
        })
    },
    deleteKind(index){
      let that = this
      this.axios.delete('http://localhost:8661/core/kind/'+index.kindId)
        .then(function (res) {
          if (res.data.code === 200) {
            console.log(res.data)
            that.$message.success(res.data.msg)
            that.queryKind()
          } else {
            that.$message.error(res.data.msg)
          }
        })
    }
```

```
        }
</script>

<style scoped>
    .el-header, .el-footer {
        background-color: #ef9265;
        color: #333;
        text-align: center;
        line-height: 60px;
    }
    .el-aside {
        background-color: #ef9265;
        color: #333;
        height: 600px;
        text-align: center;
        line-height: 200px;
    }
    .el-main {
        background-color: #E9EEF3;
        color: #333;
        text-align: center;
        line-height: 160px;
    }
    .top-research{
        display: flex;
        justify-content: space-between;
        white-space: nowrap;
        align-items: center;
        height: 100px;
    }
</style>
```

7.16.3 控制器类设计

KindController 类是分类管理模块的控制器类，被@RestController 注解标注。在 KindController 类中，定义了一个用于列表查询分类的 getList()方法和一个用于新增分类的 addKind()方法。getList()方法和 addKind()方法的代码分别如下：

```
@GetMapping(value = "/kinds")
@ApiOperation("列表查询分类")
public Result<List<KindDTO>> getList(){
    List<KindDTO> list = kindService.getList();
    if(list == null){
        return Result.fail("无分类信息");
    }
    return Result.success(list);
}


@PostMapping("/kind")
@ApiOperation("新增分类")
public Result<String> addKind(@RequestBody KindDTO kindDTO){
    int result = kindService.addKind(kindDTO);
    if(result == 1){
        return Result.success("新增分类成功");
    }
    return Result.fail("新增分类失败");
}
```

7.16.4　服务类设计

KindService 接口为分类管理模块的服务接口。在 KindController 类的 getList()方法和 addKind()方法中，分别调用了 KindService 接口的 getList()方法和 addKind()方法。KindService 接口的 getList()方法是一个无参方法；addKind()方法不仅有一个参数（即 KindDTO 类的对象），而且具有返回值，返回值是一个 int 类型、表示是否成功添加新的分类的变量（1 表示成功，0 表示失败）。KindService 接口的 getList()方法和 addKind()方法的代码分别如下：

```
List<KindDTO> getList();
int addKind(KindDTO kindDTO);
```

KindServiceImpl 类是 KindService 接口的实现类，被@Service 注解标注，表示分类管理模块的服务类。在 KindServiceImpl 类中，重写了 KindService 接口的 getList()方法和 addKind()方法。这两个方法的作用都是调用 DAO 层（数据访问对象）来执行数据库操作。KindServiceImpl 类的 getList()方法和 addKind()方法的代码分别如下：

```
@Override
public List<KindDTO> getList() {
    List<KindDTO> kindDTOS=kindMapper.selectList();
    return kindDTOS;
}

@Override
public int addKind(KindDTO kindDTO) {
    Kind kind=new Kind();
    BeanUtils.copyProperties(kindDTO,kind);
    kind.setKindId(UUID.randomUUID().toString().replace("-",""));
    kind.setCreateTime(new Date());
    kind.setUpdateTime(new Date());
    int result = kindMapper.insert(kind);
    return result;
}
```

7.16.5　DAO 层设计

KindMapper 接口为分类管理模块的 DAO 层。在 KindServiceImpl 类的 getList()方法中，调用了 KindMapper 接口的 selectList()方法，该方法用于列表查询分类；在 KindServiceImpl 类的 addKind()方法中，调用了 KindMapper 接口的 insert()方法，该方法用于新增分类。其中，selectList()方法没有参数；insert()方法不仅有一个参数（即 KindDTO 类的对象），而且具有返回值，返回值是一个 int 类型、表示是否成功添加新的分类的变量（1 表示成功，0 表示失败）。KindMapper 接口的 selectList()方法和 insert()方法的代码分别如下：

```
List<KindDTO> selectList();
int insert(Kind kind);
```

> **说明**
> 在本项目中，与 KindMapper 接口中的方法绑定的 SQL 语句被写在了 KindMapper.xml 文件中。因此，读者可以在 KindMapper.xml 文件中查找与 selectList()方法和 insert()方法绑定的 SQL 语句。

7.17 寻物启事审核模块设计

如图 7.18 所示，在寻物启事审核页面上，默认显示的是已发布但未审核的失物信息。每一条失物信息的右侧都会有"不通过"和"通过"两个按钮。如果管理员单击"通过"按钮，那么已发布的失物信息通过审核，将显示在如图 7.19 所示的寻物启事页面上；如果管理员单击"不通过"按钮，那么已发布的失物信息没有通过审核，将不会显示在寻物启事页面上。下面将介绍寻物启事审核模块的设计过程。

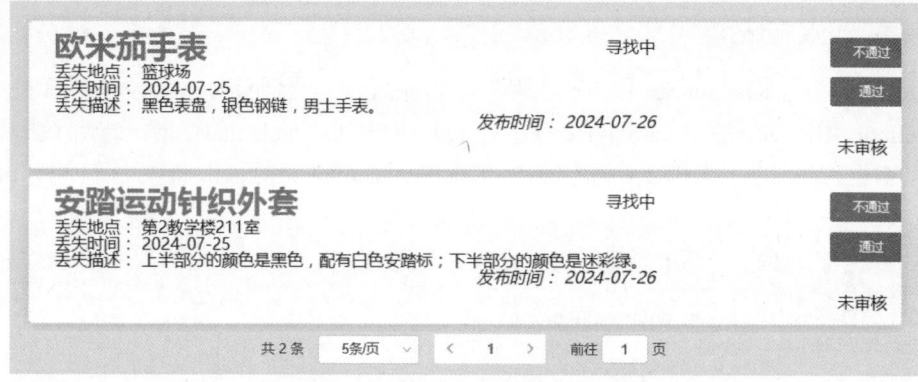

图 7.18　寻物启事审核页面的效果图

图 7.19　审核通过后寻物启事页面的效果图

7.17.1　寻物启事审核页面

在寻物启事审核页面上，包含了卡片组件、分页插件和按钮组件。卡片组件用于显示每一条已发布但未审核的失物信息，每一条失物信息的内容都是相同的，即失物名称、丢失地点、丢失时间、丢失描述和发布时间。分页插件用于把全部已发布但未审核的失物信息按照每个分页可显示的记录条数予以显示。按钮组件则允许管理员对已发布但未审核的失物信息进行审核，只有审核通过的失物信息才能显示在寻物启事页面上。

寻物启事审核页面对应的是 SearchCheck.vue 文件，代码如下：

```html
<template>
  <el-container>
    <el-header>
      <HeaderIndex/>
    </el-header>
    <el-container>
      <el-aside width="200px">
        <ManagerIndex/>
      </el-aside>
      <el-main>
        <div class="header-search" style="border: 0px solid red;width: 80%;margin:0 auto">
          <div>
            <el-radio v-model="checkStatus" :label="0">未审核</el-radio>
            <el-radio v-model="checkStatus" :label="1">审核不通过</el-radio>
            <el-radio v-model="checkStatus" :label="2">审核通过</el-radio>
          </div>
          <el-button type="primary" icon="el-icon-search" @click="querySearchGoods">搜索</el-button>
        </div>

        <div class="list" style="">
          <ul class="infinite-list" style="overflow:auto">
            <li style="border: 0px solid red;margin: 1vh auto;width: 80%;"
              v-for="(item,index) in dataTable" :key="item.lostId" :value="item" :label="item" class="infinite-list-item">
              <el-card :body-style="{ padding: '1vh',height: '20vh' }">
                <div style="border: 0px solid green;display: inline-block;width: 70%;
                  float:left;margin-left: 3vh;height: 19vh" class="card_text">
                  <div class="card_header" style="border: 0px solid red;height: 5vh;line-height: 5vh">
                    <span class="title" style="float: left;font-weight: bolder;font-size: 30px;color: #f86d0f;">
                      {{ item.lostName }}</span>
                    <span style="float: right" :class=" item.lostStatus ? 'gray' : '' ">
                      {{ item.lostStatus ? '已归还' : '寻找中' }}</span>
                  </div>
                  <div class="clearfix" style="border: 0px solid #0452d9;height: 14vh;">
                    <ul style="text-align: left">
                      <li>
                        <label>丢失地点：</label>
                        <span class="weight">{{ item.lostPlace }}</span>
                      </li>
                      <li>
                        <label>丢失时间：</label>
                        <span class="weight">{{ item.lostTime }}</span>
                      </li>
                      <li>
                        <label>丢失描述：</label>
                        <span class="weight">{{ item.lostDecp }}</span>
                      </li>
                      <li style="text-align: right">
                        <i>
                          <label>发布时间：</label>
                          <span class="weight">{{ item.lostReleaseTime }}</span>
                        </i>
                      </li>
                    </ul>
                  </div>
                </div>

                <div style="border: 0px solid #bc0aff;width: 10vh;float:right;height: 19vh;">
                  <div style="height: 15vh;border: 0px solid red;">
                    <div style="height: 6vh;line-height: 6vh">
                      <el-button v-if="item.checkStatus==0 || item.checkStatus==2" style="width: 80px;"
                        type="primary" size="mini" @click="updateCheckStatus(index,1)">不通过</el-button>
                    </div>
```

```html
                    <div style="height: 6vh;line-height: 6vh">
                      <el-button v-if="item.checkStatus==0 || item.checkStatus==1"style="width: 80px"
                        type="primary" size="mini" @click="updateCheckStatus(index,2)">通过</el-button>
                    </div>
                    <!--<el-tooltip :content="'Switch value: ' + item.checkStatus?'审核通过':'审核不通过'" placement="top">
                      <el-switch
                        v-model="item.checkStatus"
                        active-color="#13ce66"
                        inactive-color="#ff4949"
                        active-value="1"
                        inactive-value="0"
                        @change="updateCheckStatus(index)">
                      </el-switch>
                    </el-tooltip>-->
                  </div>
                  <div style=" height: 5vh;line-height: 5vh;border: 0px solid red">
                    <span v-if="item.checkStatus == 0">未审核</span>
                    <span v-if="item.checkStatus == 1">不通过</span>
                    <span v-if="item.checkStatus == 2">通过</span>
                  </div>
                </div>
              </el-card>
            </li>
          </ul>
          <el-pagination
            @size-change="handleSizeChange"
            @current-change="handleCurrentChange"
            :current-page="page.pageNum"
            :page-sizes="[5, 10, 20, 50]"
            :page-size="page.pageSize"
            layout="total, sizes, prev, pager, next, jumper"
            :total="page.total">
          </el-pagination>
        </div>
      </el-main>
    </el-container>
  </el-container>
</template>

<script>
  import HeaderIndex from "../HeaderIndex";
  import ManagerIndex from "./ManagerIndex";
  export default {
    name: "SearchCheck",
    components: {HeaderIndex,ManagerIndex},
    data () {
      return {
        date: [],
        dataTable: [],
        checkStatus: 0,
        dialogVisible: false,
        page: {
          pageSize: 5,
          pageNum: 1,
          pages: 4,
          total: 20
        },
        dialogData: null,
        kinds: {}
      }
    },
    created () {
```

```
      this.querySearchGoods()
      this.queryKind()
    },
    methods: {
      querySearchGoods() {
        const that = this
        this.axios.post('http://localhost:8661/core/losts', {
          beginTime: this.date[0],
          endTime: this.date[1],
          pageNum: this.page.pageNum,
          pageSize: this.page.pageSize,
          checkStatus: this.checkStatus,
        }).then(function (res) {
          if (res.data.code === 200) {
            that.page.pages = res.data.data.pages
            that.page.total = res.data.data.total
            that.dataTable = res.data.data.list
          } else {
            that.$message.error(res.data.msg)
          }
        })
      },
      handleSizeChange(val) {
        //刷新列表，刷新网页，回到第一页
        this.page.pageSize = val
        this.querySearchGoods()
      },
      handleCurrentChange(val) {
        //刷新列表，刷新页面，跳转到该页
        this.page.pageNum = val
        this.querySearchGoods()
      },

      queryKind(){
        let that = this
        this.axios.get('http://localhost:8661/core/kinds')
          .then(function (res) {
            if (res.data.code === 200) {
              console.log(res.data)
              that.kinds = res.data.data
            } else {
              that.$message.error(res.data.msg)
            }
          })
      },
      updateCheckStatus(index,status){

        const that = this
        this.axios.put('http://localhost:8661/core/lost/status/'+ this.dataTable[index].lostId+'/'+status)
          .then(function (res) {
            if (res.data.code === 200) {
              that.$message.success(res.data.msg)
              that.querySearchGoods()
            } else {
              that.$message.error(res.data.msg)
            }
          })
      }
    }
  }
</script>
//省略部分代码
```

7.17.2 控制器类设计

LostController 类也是寻物启事审核模块的控制器类。在 LostController 类中，定义了一个用于修改寻物启事的审核状态的 updateCheckStatus()方法。updateCheckStatus()方法的代码如下：

```java
@PutMapping("/lost/status/{lostId}/{status}")
@ApiOperation("修改寻物启事的审核状态")
public Result<String> updateCheckStatus(
        @PathVariable(name = "lostId") @ApiParam(name = "lostId",value = "失物 id") String lostId,
        @PathVariable(name = "status") @ApiParam(name = "status",value = "审核状态") String status){
    int result=lostService.updateCheckStatus(lostId,status);
    if(result==1){
        return Result.success("审核状态修改成功");
    }
    return Result.fail("审核状态修改失败");
}
```

7.17.3 服务类设计

LostService 接口也是寻物启事审核模块的服务接口。在 LostController 类的 updateCheckStatus()方法中，调用了 LostService 接口的 updateCheckStatus()方法。LostService 接口的 updateCheckStatus()方法不仅有两个参数，而且具有返回值，返回值是一个 int 类型、表示是否成功修改寻物启事的审核状态的变量（1 表示成功，0 表示失败）。LostService 接口的 updateCheckStatus()方法的代码如下：

```java
int updateCheckStatus(String lostId, String status);
```

LostServiceImpl 类是 LostService 接口的实现类，被@Service 注解标注，表示寻物启事审核模块的服务类。在 LostServiceImpl 类中，重写了 LostService 接口的 updateCheckStatus()方法。该方法的作用是调用 DAO 层（数据访问对象）来执行数据库操作。LostServiceImpl 类的 updateCheckStatus()方法的代码如下：

```java
@Override
public int updateCheckStatus(String lostId, String status) {
    return lostMapper.updateCheckStatus(lostId,status);
}
```

7.17.4 DAO 层设计

LostMapper 接口也是寻物启事审核模块的 DAO 层。在 LostServiceImpl 类的 updateCheckStatus()方法中，调用了 LostMapper 接口的 updateCheckStatus()方法，该方法用于修改寻物启事的审核状态。updateCheckStatus()方法不仅有两个参数，而且具有返回值，返回值是一个 int 类型、表示是否成功修改寻物启事的审核状态的变量（1 表示成功；0 表示失败）。LostMapper 接口的 updateCheckStatus()方法的代码如下：

```java
int updateCheckStatus(@Param("lostId")String lostId, @Param("status")String status);
```

7.18 项目运行

通过前述步骤，设计并完成了"寻物启事网站"项目的开发。"寻物启事网站"项目是一个前后端分离的项目，因此运行该项目需要两个步骤：首先启动后端，然后启动前端。下面运行本项目，以检验我们的开发成果。如图 7.20 所示，首先选择 LostApplication，单击▶快捷图标，然后选择 start，单击▶快捷图标，如图 7.21 所示，即可运行本项目。

启动项目后，寻物启事网站的登录页面将被打开，如图 7.22 所示。

图 7.20　启动后端的快捷图标

图 7.21　启动前端的快捷图标　　　　图 7.22　登录页面的效果图

如果用户的身份是学生，那么程序默认打开的是寻物启事页面。在寻物启事页面的头部有 3 个导航超链接，它们分别是"寻物启事"超链接、"个人中心"超链接和"退出登录"超链接，如图 7.23 所示。其中，在寻物启事页面上，学生既可以查看失物信息，也可以发布寻物启事，还可以联系管理员归还失物；在个人中心页面上，学生可以修改用户信息，如微信号、手机号码、密码等。

如果用户的身份是管理员，那么程序默认打开的是管理中心页面。在管理中心页面的头部有 4 个导航超链接，它们分别是"寻物启事"超链接、"个人中心"超链接、"管理中心"超链接和"退出登录"超链接，如图 7.24 所示。其中，在管理中心页面上，包含了 3 个功能，它们分别是用户管理、分类管理和寻物启事审核。

图 7.23　供学生访问的导航链接　　　　图 7.24　供管理员访问的导航链接

这样，我们就成功地检验了本项目的运行。

在 SpringBoot 项目中，可以把实体类对象称作 PO，即持久化对象（persistant object），其作用是与数据库形成映射关系。简单地说，PO 就是每一个数据库中的数据表，一个字段对应 PO 中的一个属性（PO 常被存储在项目中的 entity 包下）。此外，本章还用到了 VO 和 DTO。VO 代表值对象（value object），它是一种特殊的 Java 对象，主要用于把某个指定页面的所有数据封装起来，减少传输数据量，避免数据库中的数据外泄；DTO 代表数据传输对象（data transfer object），是一种用于在不同层之间传递数据的 Java 对象。

7.19　源　码　下　载

虽然本章详细地讲解了如何编码实现"寻物启事网站"项目的各个功能，但给出的代码都是代码片段，而非源码。为了方便读者学习，本书提供了完整的项目源码，扫描右侧二维码即可下载。

第 8 章
明日之星物业管理系统

——Spring Boot + Vue + MySQL

随着社会经济的发展，人们的生活水平不断提高，居民对物业服务质量也提出了更高的要求。物业管理系统是为了规范物业公司的日常服务行为和工作标准、减少物业公司人员的实际工作量和工作时间、建立与住户之间的有效互动、对有偿的便民服务予以透明化管理（阐明收费项目和收费金额）的一种工具。本章将使用 Java Web 开发中的 Spring Boot 框架、Vue 和 MySQL 数据库等关键技术开发一个"明日之星物业管理系统"项目。

项目微视频

本项目的核心功能及实现技术如下：

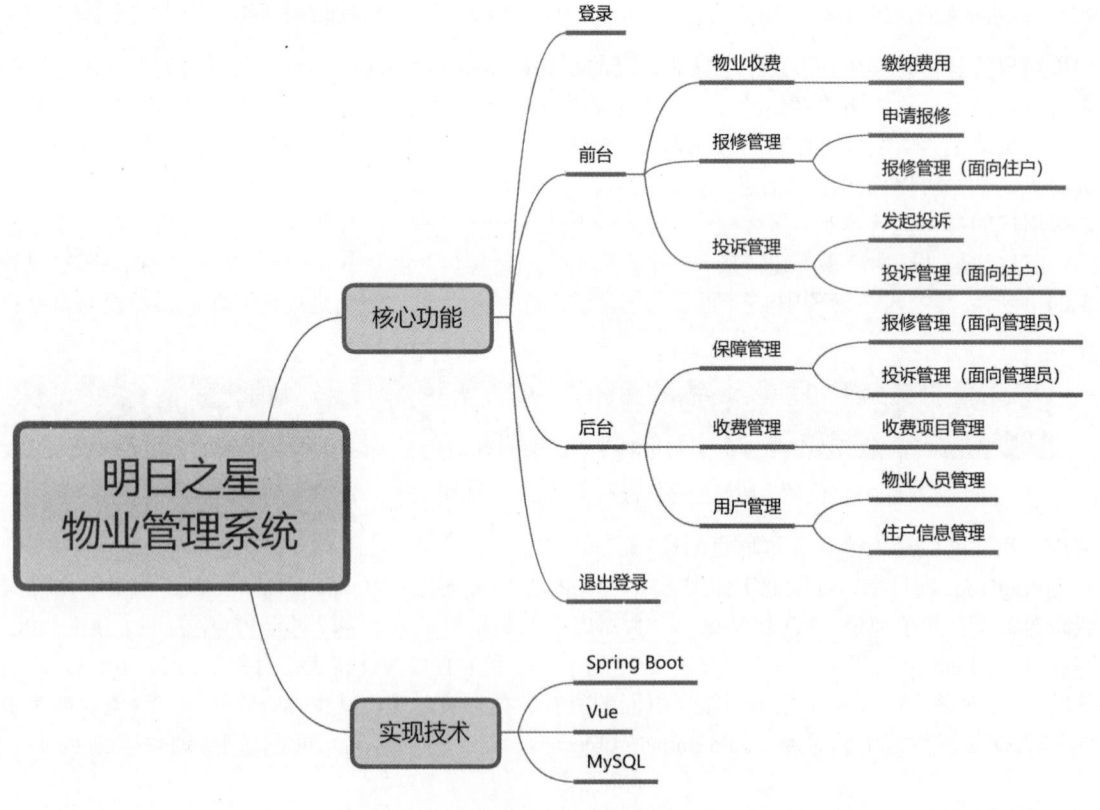

8.1 开发背景

物业公司的日常管理工作中，难免会存在一些问题，如管理效率有待提升、信息传递效率有待优化、收费有待更透明等。此外，传统物业公司的日常管理还需要投入大量的人力和物力，既增加了运营成本，又容易出现错误和延误。因此，有必要开发一个物业管理系统，对物业公司的日常管理给予智能化的技术支持。

通过该系统，物业公司可以规范日常服务行为和工作，提高工作效率，提升服务质量。

明日之星物业管理系统主要提供了以下功能：

- ☑ 对于住户：提供了查看待缴纳的费用、申请报修、报修管理、发起投诉、投诉管理等功能；
- ☑ 对于物业公司：提供了报修管理、投诉管理、收费项目管理、物业人员管理、住户信息管理等功能。

明日之星物业管理系统将实现以下目标：

- ☑ 页面简洁清晰、操作简单方便；
- ☑ 通过申请报修、发起投诉等建立与住户之间的有效互动；
- ☑ 通过投诉管理，规范物业公司的日常服务行为和工作标准；
- ☑ 通过报修管理，住户和物业公司管理员都可以跟进维修进度；
- ☑ 通过收费管理，对有偿的便民服务予以透明化处理；
- ☑ 通过用户管理，提升对物业公司人员信息和住户信息的管理效率。

8.2 系统设计

8.2.1 开发环境

本项目的开发及运行环境如下：

- ☑ 操作系统：推荐 Windows 10、11 及以上，兼容 Windows 7（SP1）。
- ☑ 开发工具：IntelliJ IDEA。
- ☑ 开发语言：Java EE。
- ☑ 数据库：MySQL 8.0。
- ☑ Web 服务器：Tomcat 9.0 及以上版本。

8.2.2 业务流程

启动项目后，打开浏览器，访问 http://localhost:8080，明日之星物业管理系统的登录页面将被打开。在登录页面上，用户输入正确的用户名和密码后，程序将对用户的身份进行判断。

如果用户的身份是住户，那么程序打开的是明日之星物业管理系统的前台页面。在页面的侧边栏上，为住户提供了3个功能模块，分别是物业收费、报修管理和投诉管理。其中，使用物业收费下的缴纳费用的功能，能够查询当月未缴纳的收费项目、收费金额和缴费的创建时间。报修管理包含申请报修和报修管理两个功能：使用申请报修功能，能够通知物业公司对报修内容进行维修；使用报修管理功能，能够跟进物业公司的维修进度。投诉管理包含发起投诉和投诉管理两个功能：使用发起投诉功能，可以阐述对物业公司服务上的不满意；使用投诉管理功能，可以跟进物业公司对投诉原因的反馈。

如果用户的身份是管理员，那么程序打开的是明日之星物业管理系统的后台页面。在页面的侧边栏上，同样为管理员提供了3个功能模块，分别是保障管理、收费管理和用户管理。其中，保障管理包含报修管理和投诉管理两个功能：使用报修管理功能，能够确认住户的报修内容是否被处理；使用投诉管理的功能，能够确认住户的投诉原因是否被反馈。收费管理包含收费项目管理功能，使用该功能可以新增收费项目，其中包含收费项目名称、收费金额、创建时间等信息。用户管理包含物业人员管理和住户信息管理两个功能：使用物业人员管理功能，能够对物业人员信息执行增、删、改、查等操作；使用住户信息管理的功能，能够对住户信息执行增、删、改、查等操作。

明日之星物业管理系统的业务流程如图 8.1 所示。

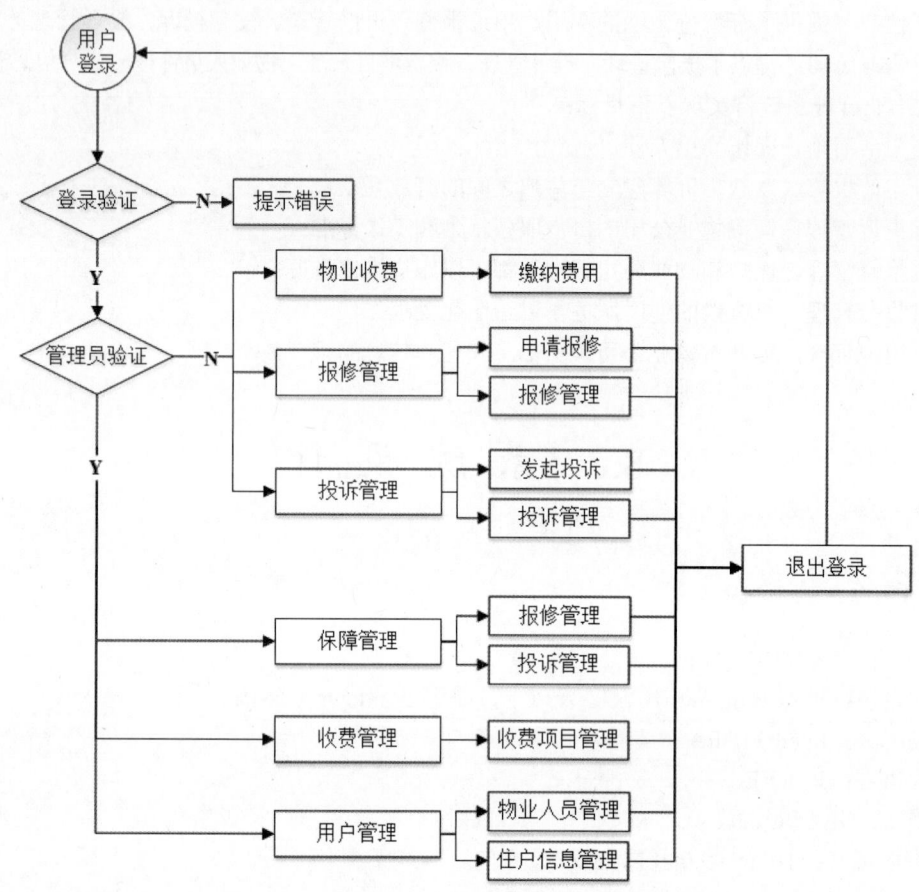

图 8.1　明日之星物业管理系统的业务流程图

8.2.3　功能结构

本项目的功能结构已经在章首页中给出。本项目实现的具体功能如下：
- ☑ 用户登录：支持住户和管理员的登录操作。
- ☑ 缴纳费用：供住户查询当月未缴纳的收费项目、收费金额和缴费的创建时间。
- ☑ 申请报修：供住户通知物业公司对报修内容进行维修。
- ☑ 报修管理（面向住户）：供住户跟进物业公司的维修进度。
- ☑ 发起投诉：供住户阐述对物业公司服务上的不满意。
- ☑ 投诉管理（面向住户）：供住户跟进物业公司对投诉原因的反馈。
- ☑ 报修管理（面向管理员）：供管理员确认住户的报修内容是否被处理。
- ☑ 投诉管理（面向管理员）：供管理员确认住户的投诉原因是否被反馈。
- ☑ 收费项目管理：供管理员新增收费项目，其中包含收费项目名称、收费金额、创建时间等信息。
- ☑ 物业人员管理：供管理员对物业人员信息执行增、删、改、查等操作。
- ☑ 住户信息管理：供管理员对住户信息执行增、删、改、查等操作。
- ☑ 退出登录：支持住户和管理员的退出登录操作。

8.3 技术准备

- Spring Boot：Spring Boot 实质上就是启动 Spring 项目的一个工具而已。简而言之，Spring Boot 是一个服务于框架的框架。也可以说，Spring Boot 是一个工具，这个工具简化了 Spring 的配置。本项目在使用 Spring Boot 实现增、删、改、查操作时的主要步骤：定义实体类→创建访问 DAO 层的接口→创建服务类→创建控制器类并使用服务类。
- Vue：Vue 是一款流行的 JavaScript 框架，用于构建交互性的 Web 应用程序。Vue 的主要目的是简化前端开发人员的工作流程，提供一种更易于理解和使用的方式来构建 Web 应用程序。本项目的缴纳费用、申请报修、发起投诉、收费项目管理、物业人员管理、住户信息管理等页面均是使用 Vue 编写的页面。
- MySQL：MySQL 是一种关系型数据库管理系统，它将数据保存在不同的表中，而不是将所有数据放在一个大仓库内，这样可以提升数据的访问速度和灵活性。本项目使用的数据库依然是 MySQL，用于连接 MySQL 的相关信息被存储在 application.properties 文件中，代码如下：

```
#数据库驱动:
spring.datasource.driver-class-name=com.mysql.cj.jdbc.Driver
#数据源名称
spring.datasource.name=defaultDataSource
#数据库连接地址
spring.datasource.url=jdbc:mysql://127.0.0.1:3306/estate_management?serverTimezone=UTC\
    &useUnicode=true&characterEncoding=UTF-8
#数据库用户名&密码:
spring.datasource.username=root
spring.datasource.password=root
```

有关 Spring Boot、Vue 和 MySQL 的知识，分别在《Spring Boot 从入门到精通》《Vue.js 从入门到精通》《Java 从入门到精通（第 7 版）》中予以详解，对这些知识不太熟悉的读者可以参考这 3 本书中的相关内容。

8.4 数据库设计

8.4.1 数据库概述

本项目采用的数据库主要包含 4 张数据表，如表 8.1 所示。

表 8.1 明日之星物业管理系统的数据库结构

表　名	表　说　明	表　名	表　说　明
sys_user	用户信息表	sys_complaint	投诉信息表
sys_repair	报修信息表	sys_charge_type	收费信息表

8.4.2 数据表设计

下面将详细介绍本项目使用的 4 张表的结构设计。

☑ sys_user（用户信息表）：主要用于存储物业人员信息和住户信息。该数据表的结构如表 8.2 所示。

表 8.2　sys_user 表结构

字 段 名 称	数 据 类 型	长　　度	是 否 主 键	说　　明
id	VARCHAR	50	主键	用户编号
user_name	VARCHAR	20		用户名称
full_name	VARCHAR	20		真实姓名
password	VARCHAR	100		用户密码
status	CHAR	1		用户状态（0 表示正常，1 表示停用）
phone	VARCHAR	11		用户手机号
login_ip	VARCHAR	50		用户最后登录 IP
login_date	DATETIME			用户最后登录时间
address	VARCHAR	100		用户住址

☑ sys_repair（报修信息表）：主要用于存储报修的相关信息。该数据表的结构如表 8.3 所示。

表 8.3　sys_repair 表结构

字 段 名 称	数 据 类 型	长　　度	是 否 主 键	说　　明
id	INT		主键	申请报修编号
user_id	VARCHAR	50		用户编号
user_name	VARCHAR	255		用户名称
title	VARCHAR	255		用户报修内容
phone	VARCHAR	12		用户手机号
date	DATETIME			申请报修的时间
text	LONGTEXT			用户报修的详细说明
address	VARCHAR	50		用户住址
is_examine	INT			管理员是否审核
examine_data	LONGTEXT			管理员予以处理的回执数据

☑ sys_complaint（投诉信息表）：主要用于存储投诉的相关信息。该数据表的结构如表 8.4 所示。

表 8.4　sys_complaint 表结构

字 段 名 称	数 据 类 型	长　　度	是 否 主 键	说　　明
id	INT		主键	投诉编号
user_id	VARCHAR	50		用户编号
user_name	VARCHAR	255		用户名称
phone	VARCHAR	12		用户手机号
title	VARCHAR	255		投诉原因
address	VARCHAR	50		用户住址
text	LONGTEXT			用户投诉的详细说明
is_examine	INT			管理员是否审核
examine_data	LONGTEXT			管理员予以处理的回执数据
date	DATETIME			发起投诉的时间

☑ sys_charge_type（收费信息表）：主要用于存储收费项目的相关信息。该数据表的结构如表 8.5 所示。

表 8.5　sys_charge_type 表结构

字段名称	数据类型	长度	是否主键	说明
id	INT		主键	收费项目编号
charge_name	VARCHAR	20		收费项目名称
charge_money	INT			收费金额
create_time	DATETIME			缴费的创建时间
update_time	DATETIME			缴费的修改时间

8.5　添加依赖和配置信息

添加依赖和配置信息是每一个 Spring Boot 项目都必不可少的环节。下面将分别介绍如何为本项目添加依赖和配置信息。

8.5.1　添加依赖

因为本项目也是使用 Maven 构建的 Spring Boot 项目，所以 pom.xml 文件是其核心配置文件，用于存储本项目所需的依赖，这些依赖会被添加到<dependencies>标签的内部。代码如下：

```xml
<?xml version="1.0" encoding="UTF-8"?>
<project xmlns="http://maven.apache.org/POM/4.0.0" xmlns:xsi="http://www.w3.org/2001/XMLSchema-instance"
    xsi:schemaLocation="http://maven.apache.org/POM/4.0.0 https://maven.apache.org/xsd/maven-4.0.0.xsd">
    <modelVersion>4.0.0</modelVersion>
    <groupId>com.kum</groupId>
    <artifactId>em_server</artifactId>
    <version>0.0.1-SNAPSHOT</version>
    <name>em_server</name>
    <description>Demo project for Spring Boot</description>

    <properties>
        <java.version>1.8</java.version>
        <project.build.sourceEncoding>UTF-8</project.build.sourceEncoding>
        <project.reporting.outputEncoding>UTF-8</project.reporting.outputEncoding>
        <spring-boot.version>2.3.7.RELEASE</spring-boot.version>
    </properties>

    <dependencies>
        <dependency>
            <groupId>org.springframework.boot</groupId>
            <artifactId>spring-boot-starter-quartz</artifactId>
        </dependency>
        <dependency>
            <groupId>org.springframework.boot</groupId>
            <artifactId>spring-boot-starter-security</artifactId>
        </dependency>
        <dependency>
            <groupId>org.springframework.boot</groupId>
            <artifactId>spring-boot-starter-web</artifactId>
        </dependency>
        <dependency>
            <groupId>com.alibaba</groupId>
            <artifactId>fastjson</artifactId>
```

```xml
        <version>1.2.72</version>
    </dependency>
    <dependency>
        <groupId>org.springframework.boot</groupId>
        <artifactId>spring-boot-devtools</artifactId>
        <scope>runtime</scope>
        <optional>true</optional>
    </dependency>
    <dependency>
        <groupId>com.baomidou</groupId>
        <artifactId>mybatis-plus-boot-starter</artifactId>
        <version>3.3.0</version>
    </dependency>
    <dependency>
        <groupId>mysql</groupId>
        <artifactId>mysql-connector-java</artifactId>
        <scope>runtime</scope>
    </dependency>
    <dependency>
        <groupId>org.projectlombok</groupId>
        <artifactId>lombok</artifactId>
        <optional>true</optional>
    </dependency>
    <dependency>
        <groupId>com.github.qcloudsms</groupId>
        <artifactId>qcloudsms</artifactId>
        <version>1.0.6</version>
    </dependency>
    <!--定时任务-->
    <dependency>
        <groupId>org.springframework.boot</groupId>
        <artifactId>spring-boot-starter-quartz</artifactId>
    </dependency>
    <!--验证码生成库-->
    <dependency>
        <groupId>com.github.penggle</groupId>
        <artifactId>kaptcha</artifactId>
        <version>2.3.2</version>
    </dependency>
    <!-- 自定义 yml-->
    <dependency>
        <groupId>org.springframework.boot</groupId>
        <artifactId>spring-boot-configuration-processor</artifactId>
        <optional>true</optional>
    </dependency>
    <dependency>
        <groupId>org.springframework.boot</groupId>
        <artifactId>spring-boot-starter-test</artifactId>
        <scope>test</scope>
        <exclusions>
            <exclusion>
                <groupId>org.junit.vintage</groupId>
                <artifactId>junit-vintage-engine</artifactId>
            </exclusion>
        </exclusions>
    </dependency>
    <dependency>
        <groupId>org.springframework.security</groupId>
        <artifactId>spring-security-test</artifactId>
        <scope>test</scope>
    </dependency>
    <dependency>
```

```xml
            <groupId>org.apache.commons</groupId>
            <artifactId>commons-lang3</artifactId>
            <version>3.10</version>
        </dependency>
    </dependencies>

    <dependencyManagement>
        <dependencies>
            <dependency>
                <groupId>org.springframework.boot</groupId>
                <artifactId>spring-boot-dependencies</artifactId>
                <version>${spring-boot.version}</version>
                <type>pom</type>
                <scope>import</scope>
            </dependency>
        </dependencies>
    </dependencyManagement>

    <build>
        <plugins>
            <plugin>
                <groupId>org.apache.maven.plugins</groupId>
                <artifactId>maven-compiler-plugin</artifactId>
                <version>3.8.1</version>
                <configuration>
                    <source>1.8</source>
                    <target>1.8</target>
                    <encoding>UTF-8</encoding>
                </configuration>
            </plugin>
            <plugin>
                <groupId>org.springframework.boot</groupId>
                <artifactId>spring-boot-maven-plugin</artifactId>
                <version>2.3.7.RELEASE</version>
                <configuration>
                    <mainClass>com.kum.EmServerApplication</mainClass>
                </configuration>
                <executions>
                    <execution>
                        <id>repackage</id>
                        <goals>
                            <goal>repackage</goal>
                        </goals>
                    </execution>
                </executions>
            </plugin>
        </plugins>
    </build>
</project>
```

8.5.2 添加配置信息

Spring Boot 支持多种格式的配置文件，最常用的是 properties 格式（默认格式），而 yml 格式是一种比较新颖的格式，也比较常用。一个 Spring Boot 项目可以加载多个配置文件，本项目加载了两个配置文件，分别是 application.properties 和 application.yml。

application.properties 包含如下的配置信息：

```
#应用名称
spring.application.name=em_server
```

```
#应用服务 WEB 访问端口
server.port=8082

#数据库驱动：
spring.datasource.driver-class-name=com.mysql.cj.jdbc.Driver
#数据源名称
spring.datasource.name=defaultDataSource
#数据库连接地址
spring.datasource.url=jdbc:mysql://127.0.0.1:3306/estate_management?serverTimezone=UTC\
      &useUnicode=true&characterEncoding=UTF-8
#数据库用户名&密码：
spring.datasource.username=root
spring.datasource.password=root
```

application.yml 包含如下的配置信息：

```
#设置 Session 过期时间
server:
  servlet:
    session:
      timeout: 1D

#MybatisPlus 开启日志
mybatis-plus:
  configuration:
    log-impl: org.apache.ibatis.logging.stdout.StdOutImpl
```

8.6 实体类设计

在 Java 语言中，实体类与数据模型存在着一一对应的关系。在 8.4.2 节中，已经介绍了本项目的 4 张数据表。下面将介绍与这 4 张数据表对应的实体类。

8.6.1 用户信息类

本项目与 sys_user（用户信息表）相对应的实体类是 SysUser 类，表示用户信息类。SysUser 类中的每一个属性都对应着 sys_user（用户信息表）中的一个字段。SysUser 类代码如下：

```
@Builder
@Data
@AllArgsConstructor
@NoArgsConstructor
public class SysUser {
    /**
     * 用户编号
     */
    @TableId(type = IdType.ASSIGN_UUID)
    private String id;
    /**
     * 用户名称
     */
    private String userName;
    /**
     * 真实姓名
     */
```

```
    private String fullName;
    /**
     * 用户密码
     */
    private String password;
    /**
     * 用户状态(0 正常 1 停用)
     */
    private String status;
    /**
     * 用户手机号
     */
    private String phone;
    /**
     * 用户最后登录 IP
     */
    private String loginIp;
    /**
     * 用户最后登录时间
     */
    private Date loginDate;

    /**
     * 用户住址
     */
    private String address;
}
```

上述代码中，@Builder、@Data、@AllArgsConstructor 和@NoArgsConstructor 都是由 Lombok 库提供的注解，具体作用如下：

- @Builder 用于实现 Builder 模式。Builder 模式是一种设计模式，即使用@Builder 注解自动生成一个内部的 Builder 类及其构建方法，从而简化代码的编写。
- @Data 用于为实体类中的所有字段自动生成 Getter 方法、Setter 方法、equals()方法、hashCode()方法、toString()方法等。
- @AllArgsConstructor 用于自动生成一个包含所有参数的构造函数。
- @NoArgsConstructor 用于自动生成一个无参的构造函数。

8.6.2 报修信息类

本项目与 sys_repair（报修信息表）相对应的实体类是 SysRepair 类，表示报修信息类。SysRepair 类中的每一个属性都对应着 sys_repair（报修信息表）中的一个字段。SysRepair 类代码如下：

```
@Data
@AllArgsConstructor
@NoArgsConstructor
public class SysRepair {
    @TableId(type = IdType.AUTO)

    private Integer id;
    /**
     * 用户编号
     */
    private String userId;
    /**
     * 用户名称
     */
    private String userName;
```

```java
/**
 * 用户报修内容
 */
private String title;
/**
 * 用户手机号
 */
private String phone;
/**
 * 用户住址
 */
private String address;
/**
 * 申请报修的时间
 */
private Date date;
/**
 * 用用户报修的详细说明
 */
private String text;
/**
 * 管理员是否审核
 */
private Integer isExamine;
/**
 * 管理员予以处理的回执数据
 */
private String examineData;
}
```

8.6.3 投诉信息类

本项目与 sys_complaint（投诉信息表）相对应的实体类是 SysComplaint 类，表示投诉信息类。SysComplaint 类中的每一个属性都对应着 sys_complaint（投诉信息表）中的一个字段。SysComplaint 类代码如下：

```java
@Data
@AllArgsConstructor
@NoArgsConstructor
public class SysComplaint {
    @TableId(type = IdType.AUTO)
    private Integer id;
    /**
     * 用户 ID
     */
    private String userId;
    /**
     * 用户名称
     */
    private String userName;
    /**
     * 投诉原因
     */
    private String title;
    /**
     * 用户手机号
     */
    private String phone;
    /**
     * 用户住址
     */
```

```
    private String address;
    /**
     * 发起投诉的时间
     */
    private Date date;
    /**
     * 用户投诉的详细说明
     */
    private String text;
    /**
     * 管理员是否审核
     */
    private Integer isExamine;
    /**
     * 管理员予以处理的回执数据
     */
    private String examineData;
}
```

8.6.4 收费信息类

本项目与 sys_charge_type（收费信息表）相对应的实体类是 SysChargeType 类，表示收费信息类。SysChargeType 类中的每一个属性都对应着 sys_charge_type（收费信息表）中的一个字段。SysChargeType 类代码如下：

```
@Data
@AllArgsConstructor
@NoArgsConstructor
public class SysChargeType {
    @TableId(type = IdType.AUTO)
    private Integer id;
    /**
     * 收费项目名称
     */
    private String chargeName;
    /**
     * 收费金额
     */
    private Integer chargeMoney;
    /**
     * 缴费的创建时间
     */
    @TableField(fill = FieldFill.INSERT)
    private Date createTime;
    /**
     * 缴费的修改时间
     */
    @TableField(fill = FieldFill.INSERT_UPDATE)
    private Date updateTime;
}
```

8.7 登录模块设计

如图 8.2 所示，明日之星物业管理系统的登录页面被打开后，用户需要先在页面上依次输入用户名、密码和验证码，再单击"登录"按钮，以完成登录操作。"登录"按钮被单击后，程序将验证当前用户输入的用户名和密码是否正确。如果用户输入的用户名和密码是正确的，则程序将以此为依据判断当前用户的身份是"管理员"还是"住户"。下面将介绍登录模块的设计过程。

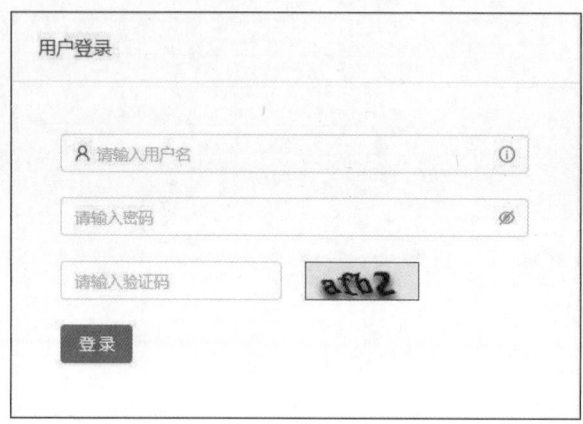

图 8.2　登录页面的效果图

8.7.1　登录页面设计

明日之星物业管理系统的登录页面是由标题、用户名输入框、密码框、验证码输入框和"登录"按钮组成的。其中，用户名输入框、密码框和验证码输入框均不能为空；"登录"按钮先判断用户输入的用户名和密码是否正确，再判断用户的身份。

明日之星物业管理系统的登录页面对应的是 home.vue 文件，代码如下：

```
<template>
  <a-layout class="content">
    <a-layout>
      <a-layout-sider width="500" style="background: #eff2f5; margin: 0 auto;">
        <a-card title="用户登录">
          <div class="login">
            <a-input v-model="login_form_data.userName" placeholder="请输入用户名">
              <a-icon slot="prefix" type="user" />
              <a-tooltip slot="suffix" title="输入用户名或手机号">
                <a-icon type="info-circle" style="color: rgba(0, 0, 0, 0.45);" />
              </a-tooltip>
            </a-input>
            <br />
            <br />
            <a-input-password v-model="login_form_data.passWord" placeholder="请输入密码" />
            <br />
            <br />
            <el-row :gutter="20">
              <el-col :span="12" :offset="0">
                <a-input v-model="login_form_data.code" placeholder="请输入验证码" />
              </el-col>
              <el-col :span="12" :offset="0">
                <img
                  @click="() => {
                    login_img_src += '?time=' + new Date().getTime()
                  }"
                  :src="login_img_src"
                  style="height: 32px; width: 100px; border: 1px solid;"
                />
```

```html
                    </el-col>
                  </el-row>
                  <br />
                  <el-row :gutter="20">
                    <el-col :span="5" :offset="0">
                      <a-button type="primary" @click="home_login()">登录</a-button>
                    </el-col>
                  </el-row>
                </div>
              </a-card>
            </a-layout-sider>
          </a-layout>
      </a-layout>
</template>
<script>
import { login, isAdmin } from '@/api/requests/rq-manage.js'
import notice_list from '@/views/home/notice_list.vue'
import rq_facilities_list from '@/views/home/rq_facilities_list.vue'
import estate_user_list from '@/views/home/estate_user_list.vue'
export default {
  name: 'home',
  components: {
    'notice-list': notice_list,
    'rq-facilities-list': rq_facilities_list,
    'estate-user-list': estate_user_list,
  },
  data () {
    return {
      login_img_src: 'http://localhost:8082/login/code',
      login_form_data: {},
      carousel_img_list: ['https://s3.ax1x.com/2023/03/13/6dz6HS.png',
                          'http://img4.imgtn.bdimg.com/it/u=290215386,829228996&fm=15&gp=0.jpg'],
      key: 'em_notice',
      noTitleKey: 'rq_facilities',
    }
  },
  created () {
    if (localStorage.getItem("isLogin")) {
      this.$store.dispatch('user/login', this.form).then(res => {
        this.$router.push('/')
      })
    }
  },
  methods: {
    home_login () {
      login(this.login_form_data).then(res => {
        if (res.code == 200) {
          this.$success({
            title: '登录成功',
          });
          localStorage.setItem("username", res.data.userName)
          this.$store.dispatch('user/login', this.form).then(res => {
            setTimeout(() => {
              window.location.href = '/'
            }, 1000)
          })
        } else {
          this.$error({
            title: ' 登录失败',
            content: res.msg,
          });
        }
```

```
        })
      },
      onTabChange (key, type) {
        console.log(key, type);
        this[type] = key;
      },
    },
  }
</script>
<style lang="scss" scoped>
.content {
    padding: 0 30px;
}
.ant-card {
    margin-top: 20px;
    .login {
        padding: 20px;
    }
}
.ant-layout-content {
    margin-left: 10px;
}
</style>
```

8.7.2 控制器类设计

SysLoginController 类是登录模块的控制器类，被@RestController 注解标注。在 SysLoginController 类中，定义了一个用于登录的 login()方法和一个用于生成登录验证码的 getKaptchaImage()方法，代码如下：

```
@PostMapping("/login")
private AjaxResult login(@RequestBody SysLogin sysLogin) {
    if (!sysLoginService.checkCode(sysLogin.getCode())) {
        return AjaxResult.error("未输入验证码或填写错误");
    }
    SysUser user = sysLoginService.login(sysLogin);
    if (user == null) {
        return AjaxResult.error("用户名或密码错误");
    }
    JSONObject jsonObject = new JSONObject();
    jsonObject.put("userName",user.getUserName());
    System.out.println(jsonObject);
    return AjaxResult.success(jsonObject);
}
/**
 * 登录验证码
 *
 * @param request
 * @param response
 * @throws Exception
 */
@GetMapping("/login/code")
public void getKaptchaImage(HttpServletRequest request, HttpServletResponse response) throws Exception {
    response.setDateHeader("Expires", 0);
    response.setHeader("Cache-Control", "no-store, no-cache, must-revalidate");
    response.addHeader("Cache-Control", "post-check=0, pre-check=0");
    response.setHeader("Pragma", "no-cache");
    response.setContentType("image/jpeg");
    String capText = producer.createText();
    //将 sessionId 或其他代表用户身份的信息和验证码文本存入 Redis
```

```
System.out.println(String.format("%s - %s", request.getSession().getId, capText));
request.getSession().setAttribute(Constants.KAPTCHA_SESSION_KEY, capText);
BufferedImage bi = producer.createImage(capText);
ServletOutputStream out = response.getOutputStream();
ImageIO.write(bi, "jpg", out);
out.flush();
out.close();
}
```

8.7.3 登录类对象设计

在 SysLoginController 类的 login()方法中有一个参数，即 SysLogin 类的对象。其中，SysLogin 类的对象就是登录类对象，其作用是把登录页面的所有数据封装起来，进而减少传输的数据量。SysLogin 类的代码如下：

```
@Data
@AllArgsConstructor
@NoArgsConstructor
public class SysLogin {
    /**
     * 用户在 session 之中存储的 key 名称
     */
    public static final String LOGIN_USER_SESSION_KEY = "user_permissions";

    /**
     * 登录用户名
     */
    private String userName;
    /**
     * 登录密码
     */
    private String passWord;
    /**
     * 验证码
     */
    private String code;
}
```

8.7.4 服务类设计

SysLoginService 类被@Service 注解标注，表示登录模块的服务类。在 SysLoginController 类的 login()方法和 getKaptchaImage()方法中，分别调用了 SysLoginService 类的用于检查输入的验证码是否正确的 checkCode()方法和 login()方法，代码如下：

```
public boolean checkCode(String code){
    HttpServletRequest request = RequestUtils.getCurrentRequest();
    System.out.println(request.getSession().getId());
    String sCode = (String) request.getSession().getAttribute(Constants.KAPTCHA_SESSION_KEY);
    if(sCode == null || !code.equals(sCode)){
        return false;
    }
    return true;
}

public SysUser login(SysLogin sysLogin){
    //用户验证
    Authentication authentication = null;
    try
```

```
        {
            //该方法会去调用 UserDetailsServiceImpl.loadUserByUsername
            authentication = authenticationManager
                .authenticate(new UsernamePasswordAuthenticationToken(sysLogin.getUserName(), sysLogin.getPassWord()));
        }
        catch (Exception e)
        {
            e.printStackTrace();
            return null;
        }
        LoginUser loginUser = (LoginUser) authentication.getPrincipal();
        List<String> userAcl = sysUserService.findUserAcl(loginUser.getUser().getId());
        loginUser.setPermissions(userAcl);
        //将登录信息存储到 Session 中
        setUserInfoToSession(loginUser);
        //删除验证码信息
        removeLoginCode();
        return loginUser.getUser();
}
```

在 SysLoginService 类的 login()方法中,调用了 SysUserService 类的 findUserAcl()方法,该方法用于获取包含用户身份的用户信息。SysUserService 类被@Service 注解标注,表示用户服务类。SysUserService 类的 findUserAcl()方法的代码如下:

```
public List<String> findUserAcl(String userId) {
    List<String> userAcl = sysUserMapper.findUserAcl(userId);
    return userAcl;
}
```

8.7.5 DAO 层设计

在 SysUserService 类的 findUserAcl()方法中,调用了 SysUserMapper 接口(DAO 层)的 findUserAcl()方法,该方法用于根据用户编号获取包含用户身份的用户信息,代码如下:

```
@Select("SELECT a.`name` FROM sys_user u   \n" +
        "JOIN sys_user_role ur ON u.id = ur.user_id\n" +
        "JOIN sys_role_acl ra ON ur.role_id = ra.role_id \n" +
        "JOIN sys_acl a ON ra.acl_id = a.id \n" +
        "WHERE u.id = #{userId}")
public List<String> findUserAcl(String userId);
```

8.8 侧边栏(面向住户)设计

如果已经登录的用户是住户,程序将在如图 8.3 所示的侧边栏中提供物业收费、报修管理和投诉管理 3 个功能模块。在物业收费功能模块下,只有缴纳费用一个功能;在报修管理功能模块下,包含申请报修和报修管理两个功能;在投诉管理功能模块下,包含发起投诉和投诉管理两个功能。每个功能都对应着一个超链接。

侧边栏(面向住户)的设计对应的是 user.js 文件,代码如下:

```
export default [
    {
```

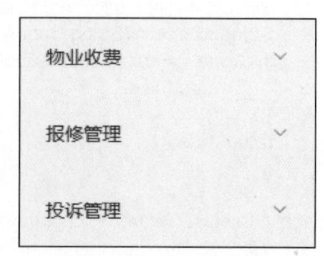

图 8.3 侧边栏(面向住户)的效果图

```js
        path: '/user/',
        component: Layout,
        redirect: '/user/pay',
        meta: {
            title: '物业收费'
        },
        children: [
            {
                path: 'pay',
                name: 'user_pay',
                component: () => import('@/views/admin/user/user_pay.vue'),
                meta: {
                    title: '缴纳费用',
                }
            },

        ]
    },
    {
        path: '/user/repair',
        component: Layout,
        redirect: '/user/repair/add',
        meta: {
            title: '报修管理'
        },
        children: [
            {
                path: 'add',
                name: 'rq_repair',
                component: () => import('@/views/admin/rq/guarantee/rq_repair_add.vue'),
                meta: {
                    title: '申请报修',
                }
            },
            {
                path: 'manager',
                name: 'rq_repair',
                component: () => import('@/views/admin/rq/guarantee/rq_repair_manager.vue'),
                meta: {
                    title: '报修管理',
                }
            },

        ]
    },
    {
        path: '/user/complaint',
        component: Layout,
        redirect: '/user/complaint/add',
        meta: {
            title: '投诉管理'
        },
        children: [
            {
                path: 'add',
                name: 'rq_complaint',
                component: () => import('@/views/admin/rq/guarantee/rq_complaint_add.vue'),
                meta: {
                    title: '发起投诉',
                }
            },
            {
```

```
            path: 'manager',
            name: 'rq_complaint',
            component: () => import('@/views/admin/rq/guarantee/rq_complaint_manager.vue'),
            meta: {
              title: '投诉管理',
            }
          },
        ]
      },
    ]
```

8.9 缴纳费用模块设计

如图 8.4 所示，住户使用缴纳费用的功能，能够查询当月未缴纳的收费项目、收费金额和缴费的创建时间。住户如果已缴纳某一项收费项目（如保安费），那么可以自行单击插槽（Slot）组件将其中蓝色的"未缴纳"修改为红色的"已缴纳"，如图 8.5 所示。下面将介绍缴纳费用模块的设计过程。

收费项目	收费金额	缴费创建时间	本月是否缴纳
保安费	20	2024-08-02 07:41:37	未缴纳
保洁费	15	2024-08-02 07:42:22	未缴纳
绿化费	10	2024-08-02 07:43:19	未缴纳

图 8.4　缴纳费用页面的效果图

收费项目	收费金额	缴费创建时间	本月是否缴纳
保安费	20	2024-08-02 07:41:37	已缴纳
保洁费	15	2024-08-02 07:42:22	未缴纳
绿化费	10	2024-08-02 07:43:19	未缴纳

图 8.5　已缴纳某一项收费项目的效果图

8.9.1 缴纳费用页面设计

缴纳费用页面是由分页插件、表格组件和插槽组件组成的。其中，表格组件的表头依次为收费项目、收费金额、缴费创建时间和本月是否缴纳。

缴纳费用页面对应的是 user_pay.vue 文件，代码如下：

```
<template>
  <page-main>
    <a-table :loading="loading" :data-source="payRecord_data_list">
      <a-table-column key="chargeName" title="收费项目" data-index="chargeName" />
      <a-table-column key="chargeMoney" title="收费金额" data-index="chargeMoney" />
      <a-table-column key="createTime" title="缴费创建时间" data-index="createTime" >
        <template slot-scope="text, record">
          {{rTime(text)}}
        </template>
      </a-table-column>
      <a-table-column key="isPayment" title="本月是否缴纳" data-index="isPayment">
        <template slot-scope="text, record">
          <a-tag v-if="text" color="red">已缴纳</a-tag>
          <a-tag v-else color="blue" @click="pay_fess(record.chargeTypeId)">未缴纳</a-tag>
        </template>
      </a-table-column>
    </a-table>
  </page-main>
</template>

<script>
import { getPayRecordOfMonth, payFess } from '@/api/requests/rq-manage.js'
import { Success, Warning } from '@/util/message.js'
import { rTime } from '@/util/time.js'

export default {
  name: 'user_pay',
  data () {
    return {
      rTime,
      loading: false,
      payRecord_data_list: [],
    }
  },
  created () {
    this.get_payRecordOfMonth()
  },
  methods: {
    get_payRecordOfMonth () {
      getPayRecordOfMonth().then(res => {
        this.payRecord_data_list = res.data
      })
    },
    pay_fess (chargeTypeId) {
      payFess(chargeTypeId).then(res => {
        this.$success({
          title: '支付物业费用回执',
          // JSX support
          content: (
            <div>
              <p>操作成功！</p>
            </div>
          ),
        });
        this.get_payRecordOfMonth()
      })
    }
  },
}
</script>
```

8.9.2 控制器类设计

SysUserController 类是缴纳费用模块的控制器类，被@RestController 注解标注。在 SysUserController 类中，定义了一个用于获取缴纳费用的 getPayRecordOfMonth()方法，代码如下：

```
@GetMapping("/pay/record/month")
public AjaxResult getPayRecordOfMonth(){
    LoginUser user = RequestUtils.getCurrentLoginUser();
    return AjaxResult.success(sysUserPlayRecordService.findByOfMonth(user.getUser().getId()));
}
```

8.9.3 服务类设计

SysUserPlayRecordService 类被@Service 注解标注，表示缴纳费用模块的服务类。在 SysUserController 类的 getPayRecordOfMonth()方法中，调用了 SysUserPlayRecordService 类的 findByOfMonth()方法，该方法用于根据用户编号获取当月需要缴纳的费用，代码如下：

```
public List<JSONObject> findByOfMonth(String userId) {
    List<JSONObject> result = new ArrayList<>();
    Calendar instance = Calendar.getInstance();
    int currentMonth = instance.get(Calendar.MONTH) + 1;
    List<SysChargeType> list = sysChargeTypeService.list();
    list.forEach(item -> {
        JSONObject jsonObject = new JSONObject();
        jsonObject.put("chargeTypeId", item.getId());
        jsonObject.put("chargeName", item.getChargeName());
        jsonObject.put("chargeMoney", item.getChargeMoney());
        jsonObject.put("createTime", item.getCreateTime());
        boolean isPayment = false;
        if (findByChargeTypeIdAndNowMonth(userId,item.getId(),currentMonth)) {
            isPayment = true;
        }
        jsonObject.put("isPayment", isPayment);
        result.add(jsonObject);
    });
    return result;
}
```

在 SysUserPlayRecordService 类的 findByOfMonth()方法中，调用了 SysChargeTypeService 类的 list()方法，该方法用于列表查询收费项目。SysChargeTypeService 类被@Service 注解标注，表示收费项目管理服务类。SysChargeTypeService 类的 list()方法的代码如下：

```
public List<SysChargeType> list(){
    return sysChargeTypeMapper.selectList(null);
}
```

8.9.4 DAO 层设计

在 SysChargeTypeService 类的 list()方法中，调用了 SysChargeTypeMapper 接口（DAO 层）的 selectList()方法，该方法用于列表查询收费项目。SysChargeTypeMapper 接口的代码如下：

```
@Mapper
public interface SysChargeTypeMapper extends BaseMapper<SysChargeType> {
```

```
    public void list();
}
```

说明

BaseMapper 是 MyBatis-Plus 提供的一个通用 Mapper 接口,它封装了常用的增、删、改、查方法,使得程序开发人员无须编写基本的数据库操作代码,从而专注于业务逻辑的实现。通过继承 BaseMapper 接口,程序开发人员可以直接使用这些方法,从而简化开发过程。

8.10 申请报修模块设计

如图 8.6 所示,住户使用申请报修的功能,能够通知物业公司对报修内容进行维修。住户首先需要按要求填写相应的内容(每一项都是必填项),然后单击"保存"按钮,以完成申请报修的操作。下面将介绍申请报修模块的设计过程。

8.10.1 申请报修页面设计

申请报修页面对应的是 rq_repair_add.vue 文件,它是由标题、标签、输入框、时间选择器、文本域、按钮等组件组成的,代码如下:

图 8.6 申请报修页面的效果图

```
<template>
  <page-main>
    <a-card title="申请报修">
      <a-form-model
        ref="current_form"
        :rules="rules"
        :model="current_form"
        style="width: 80vh;"
        :label-col="labelCol"
        :wrapper-col="wrapperCol"
      >
        <a-form-model-item label="姓名" prop="userName">
          <a-input v-model="current_form.userName" placeholder="您的姓名" />
        </a-form-model-item>
        <a-form-model-item label="联系电话" prop="phone">
          <a-input v-model="current_form.phone" placeholder="您的联系方式" />
        </a-form-model-item>
        <a-form-model-item label="住址" prop="address">
          <a-input v-model="current_form.address" placeholder="楼宇名-单元号-房间号" />
        </a-form-model-item>
        <a-form-model-item label="报修内容" prop="title">
          <a-input v-model="current_form.title" placeholder="损坏事物名称" />
        </a-form-model-item>
        <a-form-model-item label="损坏时间" prop="date1">
          <a-date-picker
            v-model="current_form.date1"
            @change="Form_date_changeHandler"
            placeholder="选择时间"
```

```html
            />
        </a-form-model-item>

        <a-form-model-item label="详细说明" prop="text">
            <a-input
                v-model="current_form.text"
                type="textarea"
                :rows="4"
                placeholder="详细说明损坏的原因，便于维修人员施工。"
            />
        </a-form-model-item>
        <a-form-model-item style="margin-left: 33.3%;">
            <a-button type="primary" @click="Save_Repair">保存</a-button>
            <a-button style="margin-left: 10px;">取消</a-button>
        </a-form-model-item>
    </a-form-model>
  </a-card>
 </page-main>
</template>
```

```javascript
<script>
import { addRepair } from '@/api/requests/rq-manage.js'
import { Success, Warning } from '@/util/message.js'

export default {
  data () {
    return {
      current_form: {},
      labelCol: { span: 8 },
      wrapperCol: { span: 14 },
      rules: {
        userName: [{ required: true, message: '此项为必填项', trigger: 'blur' }],
        phone: [{ required: true, message: '此项为必填项', trigger: 'blur' }],
        address: [{ required: true, message: '此项为必填项', trigger: 'blur' }],
        title: [{ required: true, message: '此项为必填项', trigger: 'blur' }],
        date1: [{ required: true, message: '此项为必填项', trigger: 'change' }],
        text: [{ required: true, message: '此项为必填项', trigger: 'blur' }],
      }
    }
  },
  created () {
    this.current_form.date = new Date()
  },
  methods: {
    Save_Repair () {
      this.$refs.current_form.validate(v => {
        if (v) {
          addRepair(this.current_form).then(res => {
            if (res.code == 200) {
              Success(this, '操作成功！')
            }
          })
        }
      })
    },
    Form_date_changeHandler (date) {
      this.current_form.date = date._d
    }
  },
}
</script>
```

8.10.2 控制器类设计

SysRepairController 类是申请报修模块的控制器类,被@RestController 注解标注。在 SysRepairController 类中,定义了一个用于执行添加操作的 addFacilities()方法,代码如下:

```
@PreAuthorize("@ps.hasPermi('system:repair:save')")
@PostMapping("/add")
public AjaxResult addFacilities(@RequestBody SysRepair sysRepair) {
    sysRepairService.add(sysRepair);
    return AjaxResult.success();
}
```

在 SysRepairController 类的 addFacilities()方法中,有一个参数,即 SysRepair 类的对象。其中,SysRepair 类的对象就是报修管理类对象,其作用是把如图 8.7 和图 8.8 所示的报修管理页面的所有数据封装起来,进而减少传输的数据量。

说明

因为报修管理类对象封装的数据包含申请报修页面传输的数据,所以报修管理类对象可以作为申请报修页面用于传输数据的载体。

8.10.3 服务类设计

SysRepairService 类被@Service 注解标注,表示申请报修模块的服务类。在 SysRepairController 类的 addFacilities()方法中,调用了 SysRepairService 类的 add()方法,该方法用于添加申请报修页面中的数据,代码如下:

```
public void add(SysRepair sysRepair) {
    String userId = RequestUtils.getCurrentLoginUser().getUser().getId();
    sysRepair.setUserId(userId);
    save(sysRepair);
}
```

在 SysRepairService 类的 add()方法中,调用了 SysRepairService 类的 save()方法,该方法用于保存申请报修页面中的数据,代码如下:

```
public void save(SysRepair sysRepair) {
    if (findById(sysRepair.getId()) != null) {
        sysRepairMapper.updateById(sysRepair);
        return;
    }
    sysRepairMapper.insert(sysRepair);
}
```

8.10.4 DAO 层设计

在 SysRepairService 类的 save()方法中,调用了 SysRepairMapper 接口(DAO 层)的 insert()方法,该方法用于添加申请报修页面中的数据。SysRepairMapper 接口的代码如下:

```
@Mapper
public interface SysRepairMapper extends BaseMapper<SysRepair> {
}
```

> **说明**
> 因为SysRepairMapper接口继承了BaseMapper接口，所以程序开发人员可以直接使用BaseMapper接口中常用的增、删、改、查方法，从而简化开发过程。

8.11 报修管理（面向住户）模块设计

如图8.7所示，住户使用报修管理（面向住户）的功能，能够根据"是否处理"下的状态（"未处理"或者"已处理"）跟进物业公司的维修进度。如果已申请的报修被处理，那么住户可以通过单击"已处理"插槽（Slot），在弹出的"详细信息"对话框中查看管理员的回执数据，如图8.8所示。下面将介绍报修管理（面向住户）模块的设计过程。

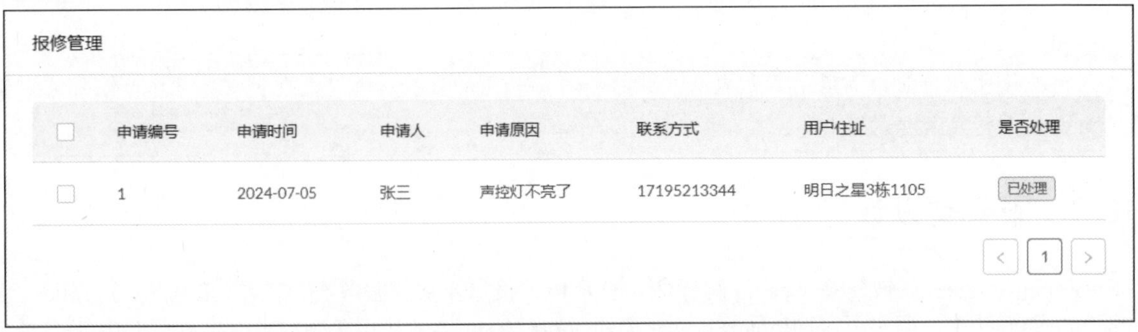

图8.7 报修管理（面向住户）页面的效果图

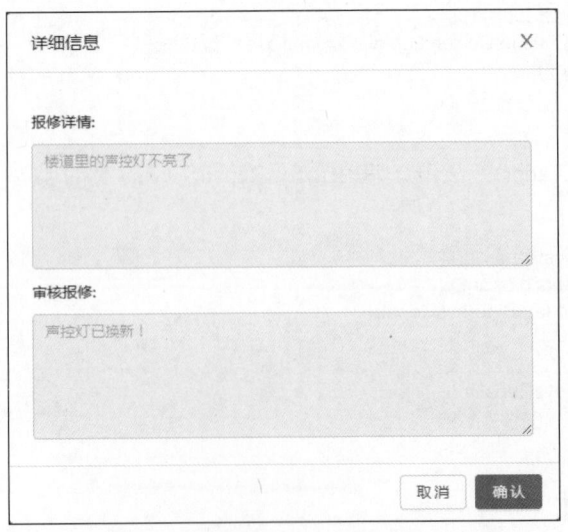

图8.8 "详细信息"对话框的效果图

8.11.1 报修管理（面向住户）页面设计

报修管理（面向住户）页面对应的是rq_repair_manager.vue文件，它是由分页插件、表格组件和插槽组件组成的。其中，表格组件的表头依次为申请编号、申请时间、申请人、申请原因、联系方式、用户住址和是否处理。rq_repair_manager.vue 文件的代码如下：

```html
<a-card :loading="loading" title="报修管理">
  <div class="head" v-if="!isUser">
    <a-button
      type="danger"
      v-if="table_selectedRowKeys.length > 0"
      style="height: 38px; margin-left: 10px;"
      @click="Del_batchData"
    >删除已选择的报修申请</a-button>
  </div>
  <a-table
    :data-source="repair_data_list"
    :row-selection="{ selectedRowKeys: table_selectedRowKeys, onChange: Table_selectChange }"
  >
    <a-table-column key="id" title="申请编号" data-index="id" />
    <a-table-column key="date" title="申请时间" data-index="date">
      <!-- rTime -->
      <template slot-scope="text, record">
        <span>{{String(record.date).substr(0,10)}}</span>
      </template>
    </a-table-column>
    <a-table-column key="userName" title="申请人" data-index="userName" />
    <a-table-column key="title" title="申请原因" data-index="title" />
    <a-table-column key="phone" title="联系方式" data-index="phone" />
    <a-table-column key="address" title="用户住址" data-index="address" />
    <a-table-column key="isExamine" title="是否处理" data-index="isExamine">
      <template slot-scope="text, record">
            <a-tooltip placement="top">
    <template slot="title">
      <span>点我查看详情</span>
    </template>
    <div @click="See_repairDateModal(record)">
                    <a-tag :color="record.isExamine == 1 ? 'red' : 'blue'" >
        {{
        record.isExamine == 1 ? '已处理' : '未处理'
        }}
      </a-tag>
      </div>
  </a-tooltip>
      </template>
    </a-table-column>
    <a-table-column v-if="!isUser" key="action" title="操作">
      <template slot-scope="text, record">
        <a-button-group>
          <a-button
            :disabled="record.isExamine == 1"
            type="primary"
            @click="Edit_repairData(record)"
          >审核</a-button>
          <a-button type="danger" @click="Del_repairData(record.id)">删除</a-button>
        </a-button-group>
      </template>
    </a-table-column>
  </a-table>
</a-card>
<!--对话框 -->
<a-modal
  v-model="repair_save_modalVisible"
  :title="repair_save_title"
  ok-text="确认"
```

```
cancel-text="取消"
:maskClosable="false"
:destroyOnClose="false"
@ok="Save_repairData"
>
// 省略部分代码
```

8.11.2 控制器类设计

SysRepairController 类是报修管理（面向住户）模块的控制器类，被@RestController 注解标注。在 SysRepairController 类中，定义了一个 getListByUserId()方法，该方法用于获取与用户编号相对应的申请报修的数据，代码如下：

```
@GetMapping("/list/user")
public AjaxResult getListByUserId() {
    String userId = RequestUtils.getCurrentLoginUser().getUser().getId();
    return AjaxResult.success(sysRepairService.findByUserId(userId));
}
```

8.11.3 服务类设计

SysRepairService 类被 @Service 注解标注，表示报修管理（面向住户）模块的服务类。在 SysRepairController 类的 getListByUserId()方法中，调用了 SysRepairService 类的 findByUserId()方法，该方法用于根据用户编号查询申请报修的数据，代码如下：

```
public List<SysRepair> findByUserId(String userId){
    QueryWrapper<SysRepair> wrapper = new QueryWrapper<>();
    wrapper.eq("user_id",userId);
    return sysRepairMapper.selectList(wrapper);
}
```

8.11.4 DAO 层设计

在 SysRepairService 类的 findByUserId()方法中，调用了 SysRepairMapper 接口（DAO 层）的 selectList() 方法，该方法用于根据用户编号查询申请报修的数据。SysRepairMapper 接口的代码如下：

```
@Mapper
public interface SysRepairMapper extends BaseMapper<SysRepair> {
}
```

> **说明**
> 因为 SysRepairMapper 接口继承了 BaseMapper 接口，所以程序开发人员可以直接使用 BaseMapper 接口中常用的增、删、改、查方法，从而简化开发过程。

8.12 发起投诉模块设计

如图 8.9 所示，住户使用发起投诉的功能，可以阐述对物业公司服务上的不满意。住户首先需要按要求填写相应的内容（每一项都是必填项），然后单击"保存"按钮，以完成发起投诉的操作。下面将介绍发起投诉模块的设计过程。

图 8.9 发起投诉页面的效果图

8.12.1 发起投诉页面设计

发起投诉页面对应的是 rq_complaint_add.vue 文件，它是由标题、标签、输入框、文本域、按钮等组件组成的，代码如下：

```
<template>
  <page-main>
    <a-card title="发起投诉">
      <a-form-model
        ref="current_form"
        :rules="rules"
        :model="current_form"
        style="width: 80vh;"
        :label-col="labelCol"
        :wrapper-col="wrapperCol"
      >
        <a-form-model-item label="姓名" prop="userName">
          <a-input v-model="current_form.userName" placeholder="您的姓名" />
        </a-form-model-item>
        <a-form-model-item label="联系电话" prop="phone">
          <a-input v-model="current_form.phone" placeholder="您的联系方式" />
        </a-form-model-item>
        <a-form-model-item label="投诉原因" prop="title">
          <a-input v-model="current_form.title" placeholder="投诉事物名称" />
        </a-form-model-item>
        <a-form-model-item label="住址" prop="address">
          <a-input v-model="current_form.address" placeholder="楼宇名-单元号-房间号" />
        </a-form-model-item>
        <a-form-model-item label="详细说明" prop="text">
          <a-input v-model="current_form.text" type="textarea" :rows="4" placeholder="详细说明投诉的具体原因" />
        </a-form-model-item>
        <a-form-model-item style="margin-left: 33.3%;">
          <a-button type="primary" @click="Save_Complaint">保存</a-button>
```

```html
          <a-button style="margin-left: 10px;">取消</a-button>
        </a-form-model-item>
      </a-form-model>
    </a-card>
  </page-main>
</template>

<script>
import { addComplaint } from '@/api/requests/rq-manage.js'
import { Success, Warning } from '@/util/message.js'

export default {
  data () {
    return {
      current_form: {},
      labelCol: { span: 8 },
      wrapperCol: { span: 14 },
      rules: {
        userName: [{ required: true, message: '此项为必填项', trigger: 'blur' }],
        phone: [{ required: true, message: '此项为必填项', trigger: 'blur' }],
        title: [{ required: true, message: '此项为必填项', trigger: 'blur' }],
        address: [{ required: true, message: '此项为必填项', trigger: 'blur' }],
        text: [{ required: true, message: '此项为必填项', trigger: 'blur' }],
      },
    }
  },
  created () {
  },
  methods: {
    Save_Complaint () {
      this.$refs.current_form.validate(v => {
        if (v) {
          addComplaint(this.current_form).then(res => {
            if (res.code == 200) {
              Success(this, '操作成功！')
            }
          })
        }
      })
    },
  },
}
</script>
```

8.12.2 控制器类设计

SysComplaintController 类是发起投诉模块的控制器类，被 @RestController 注解标注。在 SysComplaintController 类中，定义了一个用于执行添加操作的 addFacilities() 方法，代码如下：

```java
@PreAuthorize("@ps.hasPermi('system:complaint:save')")
@PostMapping("/add")
public AjaxResult addFacilities(@RequestBody SysComplaint sysComplaint) {
    sysComplaintService.add(sysComplaint);
    return AjaxResult.success();
}
```

在 SysComplaintController 类的 addFacilities() 方法中，有一个参数，即 SysComplaint 类的对象。其中，SysComplaint 类的对象就是投诉管理类对象，其作用是把如图 8.10 和图 8.11 所示的投诉管理页面的所有数据封装起来，进而减少传输的数据量。

> **说明**
> 因为投诉管理类对象封装的数据包含发起投诉页面传输的数据，所以投诉管理类对象可以作为发起投诉页面用于传输数据的载体。

8.12.3 服务类设计

SysComplaintService 类被@Service 注解标注，表示发起投诉模块的服务类。在 SysComplaintController 类的 addFacilities()方法中，调用了 SysComplaintService 类的 add()方法，该方法用于添加发起投诉页面中的数据，代码如下：

```
public void add(SysComplaint sysComplaint) {
    String userId = RequestUtils.getCurrentLoginUser().getUser().getId();
    sysComplaint.setUserId(userId);
    sysComplaint.setDate(new Date());
    save(sysComplaint);
}
```

在 SysComplaintService 类的 add()方法中，调用了 SysComplaintService 类的 save()方法，该方法用于保存发起投诉页面中的数据，代码如下：

```
public void save(SysComplaint sysComplaint) {
    if (findById(sysComplaint.getId()) != null) {
        sysComplaintMapper.updateById(sysComplaint);
        return;
    }
    sysComplaintMapper.insert(sysComplaint);
}
```

8.12.4 DAO 层设计

在 SysComplaintService 类的 save()方法中，调用了 SysComplaintMapper 接口（DAO 层）的 insert()方法，该方法用于添加发起投诉页面中的数据。SysComplaintMapper 接口的代码如下：

```
@Mapper
public interface SysComplaintMapper extends BaseMapper<SysComplaint> {
}
```

> **说明**
> 因为 SysComplaintService 接口继承了 BaseMapper 接口，所以程序开发人员可以直接使用 BaseMapper 接口中常用的增、删、改、查方法，从而简化开发过程。

8.13 投诉管理（面向住户）模块设计

如图 8.10 所示，住户使用投诉管理（面向住户）的功能，能够根据"是否处理"下的状态（"未处理"或者"已处理"）跟进物业公司对投诉原因的反馈。如果已发起的投诉被处理，那么住户可以通过单击"已处理"插槽（Slot），在弹出的"详细信息"对话框中查看管理员的回执数据，如图 8.11 所示。下面将介绍投诉管理（面向住户）模块的设计过程。

	投诉编号	发起时间	投诉人	投诉原因	联系方式	用户住址	是否处理
☐	2	2024-08-08	李四	不把业主的事当事！	15152316789	明日之星15栋901	已处理

图 8.10　投诉管理（面向住户）页面的效果图

图 8.11　发起投诉是否处理对话框的效果图

8.13.1　投诉管理（面向住户）页面设计

投诉管理（面向住户）页面对应的是 rq_complaint_manager.vue 文件，它是由分页插件、表格组件和插槽组件组成的。其中，表格组件的表头依次为投诉编号、发起时间、投诉人、投诉原因、联系方式、用户住址和是否处理。rq_complaint_manager.vue 文件的代码如下：

```
<a-card :loading="loading" title="投诉管理">
  <div class="head" v-if="!isUser">
    <a-button
      type="danger"
      v-if="table_selectedRowKeys.length > 0"
      style="height: 38px; margin-left: 10px;"
      @click="Del_batchData"
    >删除已选择的投诉</a-button>
  </div>
  <a-table
    :data-source="complaint_data_list"
    :row-selection="{ selectedRowKeys: table_selectedRowKeys, onChange: Table_selectChange }"
  >
    <a-table-column key="id" title="投诉编号" data-index="id" />
    <a-table-column key="date" title="发起时间" data-index="date">
```

```html
        <!-- rTime -->
        <template slot-scope="text, record">
          <span>{{String(record.date).substr(0,10)}}</span>
        </template>
      </a-table-column>
      <a-table-column key="userName" title="投诉人" data-index="userName" />
      <a-table-column key="title" title="投诉原因" data-index="title" />
      <a-table-column key="phone" title="联系方式" data-index="phone" />
      <a-table-column key="address" title="用户住址" data-index="address" />
      <a-table-column key="isExamine" title="是否处理" data-index="isExamine">
        <template slot-scope="text, record">
            <a-tooltip placement="top">
      <template slot="title">
          <span>点我查看详情</span>
      </template>
      <div @click="See_complaintDateModal(record)">
            <a-tag :color="record.isExamine == 1 ? 'red' : 'blue'" >
            {{
            record.isExamine == 1 ? '已处理' : '未处理'
            }}
            </a-tag>
      </div>
  </a-tooltip>
        </template>
      </a-table-column>
      <a-table-column key="action" title="操作" v-if="!isUser">
        <template slot-scope="text, record">
          <a-button-group>
            <a-button
              :disabled="record.isExamine == 1"
              type="primary"
              @click="Edit_complaintData(record)"
            >审核</a-button>
            <a-button type="danger" @click="Del_complaintData(record.id)">删除</a-button>
          </a-button-group>
        </template>
      </a-table-column>
    </a-table>
</a-card>
<!-- 提示框 -->
<a-modal
  v-model="complaint_save_modalVisible"
  :title="complaint_save_title"
  ok-text="确认"
  cancel-text="取消"
  :maskClosable="false"
  :destroyOnClose="false"
  @ok="Save_complaintData"
>
// 省略部分代码
```

8.13.2 控制器类设计

SysComplaintController 类是投诉管理（面向住户）模块的控制器类，被@RestController 注解标注。在 SysComplaintController 类中，定义了一个 getListByUserId()方法，该方法用于获取与用户编号相对应的发起投诉的数据，代码如下：

```
@GetMapping("/list/user")
public AjaxResult getListByUserId(){
    String userId = RequestUtils.getCurrentLoginUser().getUser().getId();
```

```
            return AjaxResult.success(sysComplaintService.findByUserId(userId));
}
```

8.13.3 服务类设计

SysComplaintService 类被@Service 注解标注，表示投诉管理（面向住户）模块的服务类。在 SysComplaintController 类的 getListByUserId()方法中，调用了 SysComplaintService 类的 findByUserId()方法，该方法用于根据用户编号查询发起投诉的数据，代码如下：

```
public List<SysComplaint> findByUserId(String userId){
    QueryWrapper<SysComplaint> wrapper = new QueryWrapper<>();
    wrapper.eq("user_id",userId);
    return sysComplaintMapper.selectList(wrapper);
}
```

8.13.4 DAO 层设计

在 SysComplaintService 类的 findByUserId()方法中，调用了 SysComplaintMapper 接口（DAO 层）的 selectList()方法，该方法用于根据用户编号查询发起投诉的数据。SysComplaintMapper 接口的代码如下：

```
@Mapper
public interface SysComplaintMapper extends BaseMapper<SysComplaint> {
}
```

> **说明**
>
> 因为 SysComplaintMapper 接口继承了 BaseMapper 接口，所以程序开发人员可以直接使用 BaseMapper 接口中常用的增、删、改、查方法，从而简化开发过程。

8.14 侧边栏（面向管理员）设计

如果已经登录的用户是管理员，那么程序将在如图 8.12 所示的侧边栏中为管理员提供保障管理、收费管理和用户管理 3 个功能模块。在保障管理功能模块下，包含报修管理和投诉管理两个功能；在收费管理功能模块下，只有收费项目管理一个功能；在用户管理功能模块下，包含物业人员管理和住户信息管理两个功能。每个功能都对应着一个超链接。

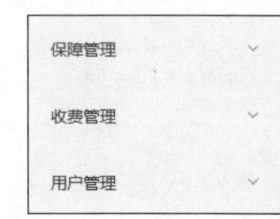

图 8.12 侧边栏（面向管理员）的效果图

侧边栏（面向管理员）的设计对应的是 admin.js 文件，代码如下：

```
export default [
  {
    path: '/guarantee-manager',
    component: Layout,
    redirect: '/guarantee-manager/repair',
    meta: {
      title: '保障管理'
    },
    children: [
      {
```

```
          path: 'repair',
          name: 'rq_repair',
          component: () => import('@/views/admin/rq/guarantee/rq_repair_manager.vue'),
          meta: {
            title: '报修管理',
          }
        },
        {
          path: 'complaint',
          name: 'rq_complaint',
          component: () => import('@/views/admin/rq/guarantee/rq_complaint_manager.vue'),
          meta: {
            title: '投诉管理',
          }
        },
      ]
    },
    {
      path: '/charge-manager',
      component: Layout,
      redirect: '/charge-manager/type',
      meta: {
        title: '收费管理'
      },
      children: [
        {
          path: 'type',
          name: 'rq_charge_type',
          component: () => import('@/views/admin/rq/charge/rq_charge_type.vue'),
          meta: {
            title: '收费项目管理',
          }
        },
      ]
    },
    {
      path: '/user-manager',
      component: Layout,
      redirect: '/user-manager/estate_user',
      meta: {
        title: '用户管理'
      },
      children: [
        {
          path: 'estate_user',
          name: 'estate_user',
          component: () => import('@/views/admin/user/estate_user_manager.vue'),
          meta: {
            title: '物业人员管理',
          }
        },
        {
          path: 'household',
          name: 'user_household',
          component: () => import('@/views/admin/user/user_household.vue'),
          meta: {
            title: '住户信息管理',
          }
        },
      ]
    }
]
```

8.15 报修管理（面向管理员）模块设计

如图 8.13 所示，管理员使用报修管理（面向管理员）的功能，能够根据"是否处理"下的状态（"未处理"或者"已处理"）确认住户的报修内容是否被处理。

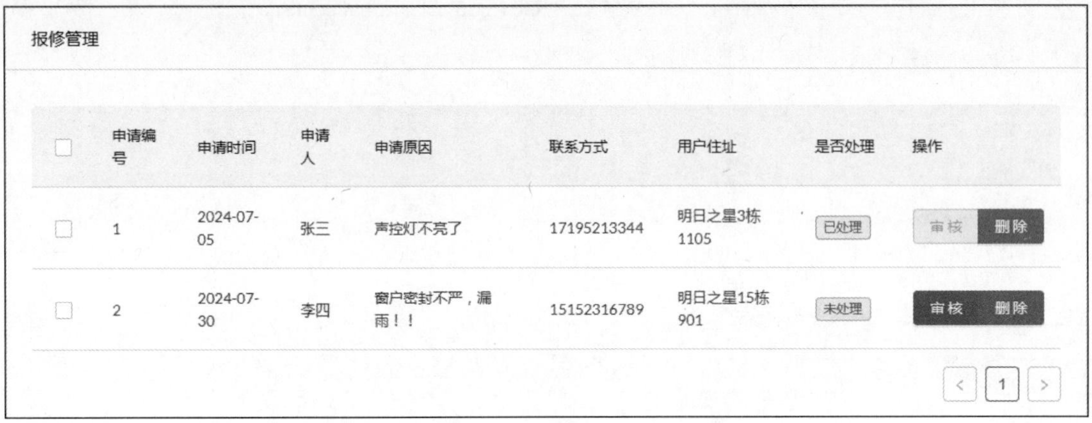

图 8.13 报修管理（面向管理员）页面的效果图

如果已申请的报修被处理，那么管理员可以通过单击"已处理"插槽（Slot），在弹出的"详细信息"对话框中查看回执数据，如图 8.14 所示。

如果已申请的报修未被处理，那么管理员可以通过单击"审核"按钮，在弹出的"申请报修审核"对话框中先输入回执数据，再单击"确认"按钮，以完成审核申请报修的操作，如图 8.15 所示。下面将介绍报修管理（面向管理员）模块的设计过程。

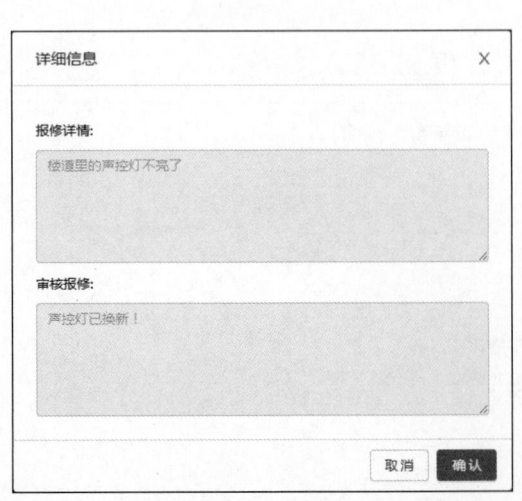

图 8.14 申请报修审核（已处理）对话框的效果图

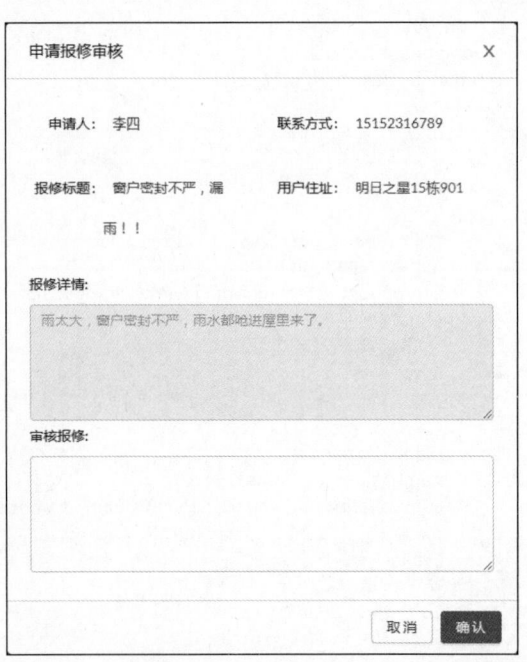

图 8.15 申请报修审核（未处理）对话框的效果图

8.15.1 报修管理（面向管理员）页面设计

报修管理（面向管理员）页面与报修管理（面向住户）页面是相同的，它们对应的都是 rq_repair_manager.vue 文件。但是，第 8.11.1 节的 rq_repair_manager.vue 文件中，只给出了报修管理（面向管理员）页面的设计代码，所以读者想要更加深入地了解报修管理（面向管理员）的页面设计，需要自行在源码中找到并查看 rq_repair_manager.vue 文件。

8.15.2 控制器类设计

SysRepairController 类是报修管理（面向管理员）模块的控制器类，被@RestController 注解标注。在 SysRepairController 类中，依次定义了 getList()方法、examineFacilities()方法和 deleteFacilities()方法。其中，getList()方法用于获取所有申请报修的数据；examineFacilities()方法用于审核未被处理的报修申请；deleteFacilities()方法用于删除一条或者多条申请报修的数据。

SysRepairController 类的 getList()方法、examineFacilities()方法和 deleteFacilities()方法的代码分别如下：

```java
@GetMapping("/list")
public AjaxResult getList() {
    return AjaxResult.success(sysRepairService.list());
}

@PreAuthorize("@ps.hasPermi('system:repair:examine')")
@PostMapping("/examine")
public AjaxResult examineFacilities(@RequestBody SysRepair sysRepair) {
    sysRepairService.examine(sysRepair);
    return AjaxResult.success();
}

@PreAuthorize("@ps.hasPermi('system:repair:delete')")
@PostMapping("/delete/{id}")
public AjaxResult deleteFacilities(@PathVariable("id")String id) {
    if (sysRepairService.delete(id)) {
        return AjaxResult.success();
    }
    return AjaxResult.error();
}
```

8.15.3 服务类设计

SysRepairService 类被@Service 注解标注，表示报修管理（面向管理员）模块的服务类。

在 SysRepairController 类的 getList()方法中，调用了 SysRepairService 类的 list()方法，用于查询所有申请报修的数据。

在 SysRepairController 类的 examineFacilities()方法中，调用了 SysRepairService 类的 examine()方法，用于审核未被处理的报修申请。

在 SysRepairController 类的 deleteFacilities()方法中，调用了 SysRepairService 类的 delete()方法，用于删除一条或者多条申请报修的数据。

SysRepairService 类的 list()方法、examine()方法和 delete()方法的代码分别如下：

```java
public List<SysRepair> list() {
    return sysRepairMapper.selectList(null);
```

```java
}
public void examine(SysRepair sysRepair){
    sysRepair.setIsExamine(1);
    save(sysRepair);
}

public void save(SysRepair sysRepair) {
    if (findById(sysRepair.getId()) != null) {
        sysRepairMapper.updateById(sysRepair);
        return;
    }
    sysRepairMapper.insert(sysRepair);
}

public boolean delete(String id) {
    return sysRepairMapper.deleteById(id) > 0;
}
```

8.15.4　DAO 层设计

在 SysRepairService 类的 list()方法中，调用了 SysRepairMapper 接口（DAO 层）的 selectList()方法，用于查询所有申请报修的数据。

在 SysRepairService 类的 examine()方法中，调用了其自身的 save()方法，而 save()方法中又调用了 SysRepairMapper 接口（DAO 层）的 updateById()方法，用于根据用户编号修改申请报修的数据。

在 SysRepairService 类的 delete()方法中，调用了 SysRepairMapper 接口（DAO 层）的 deleteById()方法，用于根据用户编号删除申请报修的数据。

SysRepairMapper 接口的代码如下：

```java
@Mapper
public interface SysRepairMapper extends BaseMapper<SysRepair> {
}
```

说明

因为 SysRepairMapper 接口继承了 BaseMapper 接口，所以程序开发人员可以直接使用 BaseMapper 接口中常用的增、删、改、查方法，从而简化开发过程。

8.16　投诉管理（面向管理员）模块设计

如图 8.16 所示，管理员使用投诉管理（面向管理员）的功能，能够根据"是否处理"下的状态（"未处理"或者"已处理"）确定住户的投诉原因是否被反馈。

如果投诉被处理，那么管理员可以通过单击"已处理"插槽（Slot），在弹出的"详细信息"对话框中查看回执数据，如图 8.17 所示。

如果投诉未被处理，那么管理员可以通过单击"审核"按钮，在弹出的"发起投诉审核"对话框中先输入回执数据，再单击"确认"按钮，以完成审核投诉的操作，如图 8.18 所示。下面将介绍投诉管理（面向管理员）模块的设计过程。

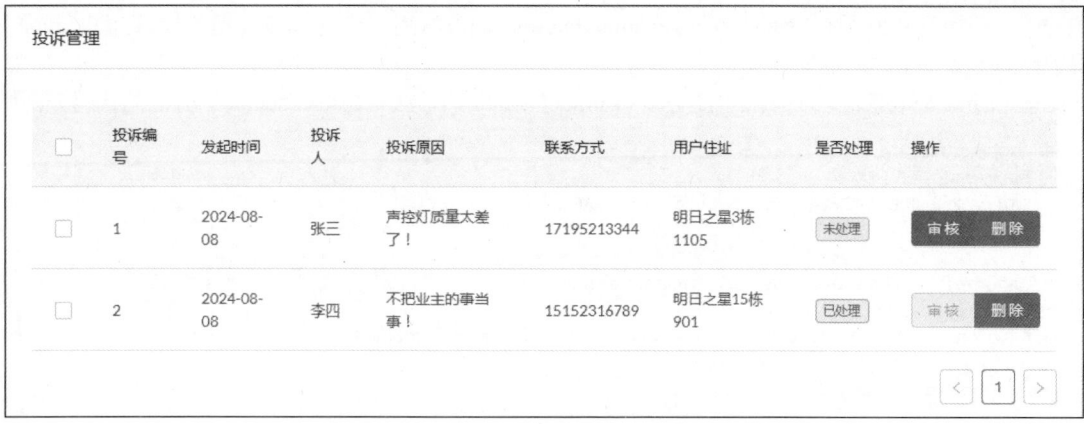

图 8.16　投诉管理（面向管理员）页面的效果图

图 8.17　发起投诉审核（已审核）对话框的效果图

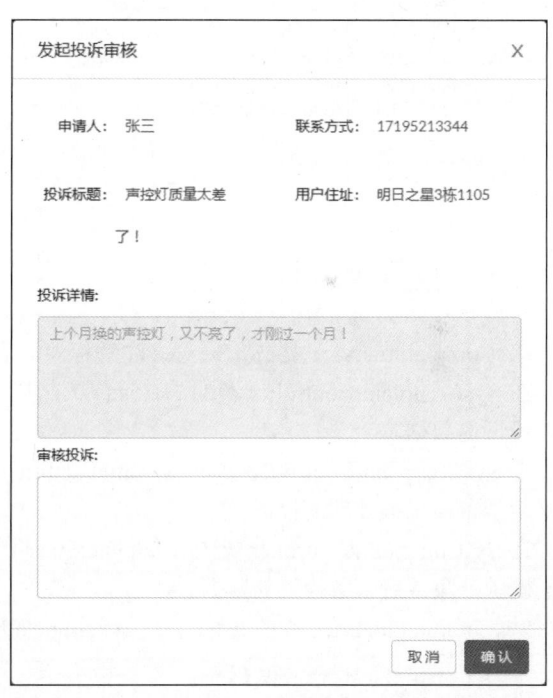

图 8.18　发起投诉审核（未审核）对话框的效果图

8.16.1　投诉管理（面向管理员）页面设计

投诉管理（面向管理员）页面与投诉管理（面向住户）页面是相同的，它们对应的都是rq_complaint_manager.vue 文件。但是，第 8.13.1 节的 rq_complaint_manager.vue 文件中，只给出了投诉管理（面向管理员）页面的设计代码，所以读者想要更加深入地了解投诉管理（面向管理员）的页面设计，需要自行在源码中找到并查看 rq_complaint_manager.vue 文件。

8.16.2　控制器类设计

SysComplaintController 类是投诉管理（面向管理员）模块的控制器类，被@RestController 注解标注。在 SysComplaintController 类中，依次定义了 getList()方法、examineFacilities()方法和 deleteFacilities()方法。其

中，getList()方法用于获取所有投诉的数据；examineFacilities()方法用于审核未被处理的投诉；deleteFacilities()方法用于删除一条或者多条投诉的数据。

SysComplaintController 类的 getList()方法、examineFacilities()方法和 deleteFacilities()方法的代码分别如下：

```java
@GetMapping("/list")
public AjaxResult getList(){
    return AjaxResult.success(sysComplaintService.list());
}

@PreAuthorize("@ps.hasPermi('system:complaint:examine')")
@PostMapping("/examine")
public AjaxResult examineFacilities(@RequestBody SysComplaint sysComplaint) {
    sysComplaintService.examine(sysComplaint);
    return AjaxResult.success();
}

@PreAuthorize("@ps.hasPermi('system:complaint:delete')")
@PostMapping("/delete/{id}")
public AjaxResult deleteFacilities(@PathVariable("id")String id) {
    if(sysComplaintService.delete(id)){
        return AjaxResult.success();
    }
    return AjaxResult.error();
}
```

8.16.3 服务类设计

SysComplaintService 类被@Service 注解标注，表示投诉管理（面向管理员）模块的服务类。

在 SysComplaintController 类的 getList()方法中，调用了 SysComplaintService 类的 list()方法，用于查询所有投诉的数据。

在 SysComplaintController 类的 examineFacilities()方法中，调用了 SysComplaintService 类的 examine()方法，用于审核未被处理的投诉。

在 SysComplaintController 类的 deleteFacilities()方法中，调用了 SysComplaintService 类的 delete()方法，用于删除一条或者多条投诉的数据。

SysComplaintService 类的 list()方法、examine()方法和 delete()方法的代码分别如下：

```java
public List<SysComplaint> list() {
    return sysComplaintMapper.selectList(null);
}

public void examine(SysComplaint sysComplaint){
    sysComplaint.setIsExamine(1);
    save(sysComplaint);
}

public void save(SysComplaint sysComplaint) {
    if (findById(sysComplaint.getId()) != null) {
        sysComplaintMapper.updateById(sysComplaint);
        return;
    }
    sysComplaintMapper.insert(sysComplaint);
}

public boolean delete(String id) {
    return sysComplaintMapper.deleteById(id) > 0;
}
```

8.16.4 DAO 层设计

在 SysComplaintService 类的 list()方法中，调用了 SysComplaintMapper 接口（DAO 层）的 selectList()方法，用于查询所有投诉的数据。

在 SysComplaintService 类的 examine()方法中，调用了其自身的 save()方法，而 save()方法中又调用了 SysComplaintMapper 接口（DAO 层）的 updateById()方法，用于根据用户编号修改投诉的数据。

在 SysComplaintService 类的 delete()方法中，调用了 SysComplaintMapper 接口（DAO 层）的 deleteById()方法，用于根据用户编号删除投诉的数据。

SysComplaintMapper 接口的代码如下：

```
@Mapper
public interface SysComplaintMapper extends BaseMapper<SysComplaint> {
}
```

说明

因为 SysComplaintMapper 接口继承了 BaseMapper 接口，所以程序开发人员可以直接使用 BaseMapper 接口中常用的增、删、改、查方法，从而简化开发过程。

8.17 收费项目管理模块设计

如图 8.19 所示，管理员使用收费项目管理的功能，能够对收费项目执行增、删、改、查的操作。

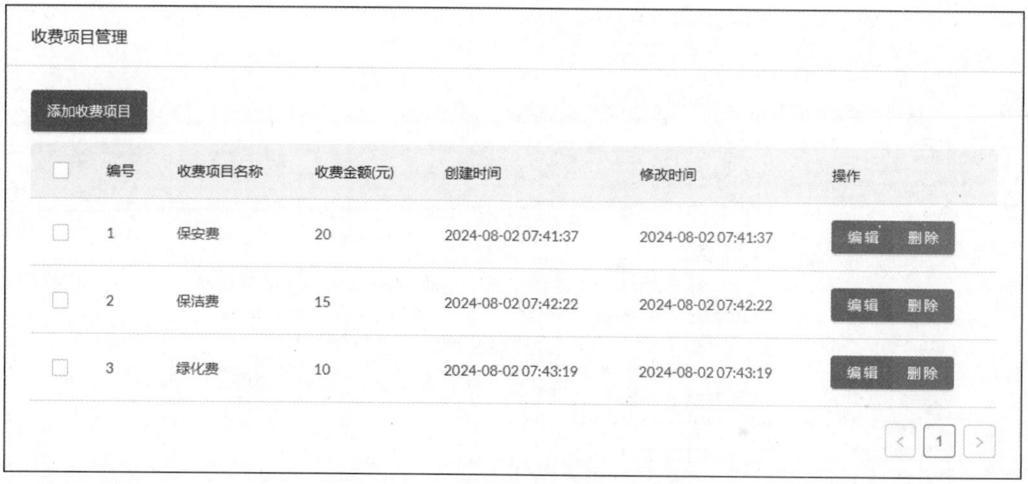

图 8.19 收费项目管理页面的效果图

管理员首先单击"添加收费项目"按钮，然后在如图 8.20 所示的"新增收费项目"对话框中输入收费项目名称和收费金额（月），最后单击"确认"按钮，以完成新增收费项目的操作。

管理员先单击"编辑"按钮，再在如图 8.21 所示的"编辑收费项目"对话框中对收费项目名称和收费金额（月）予以修改，而后单击"确认"按钮，以完成编辑收费项目的操作。下面将介绍收费项目管理（面向管理员）模块的设计过程。

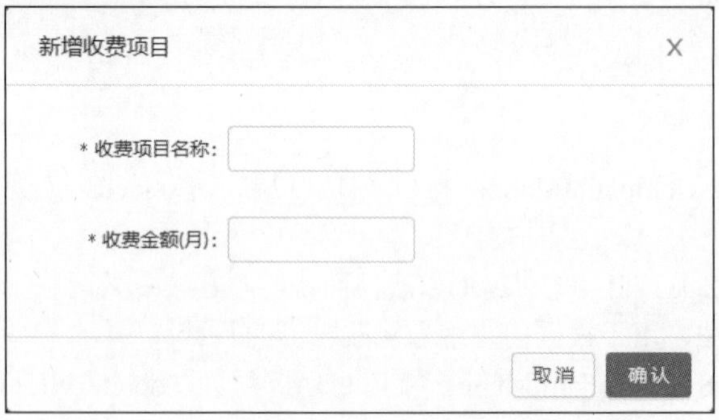

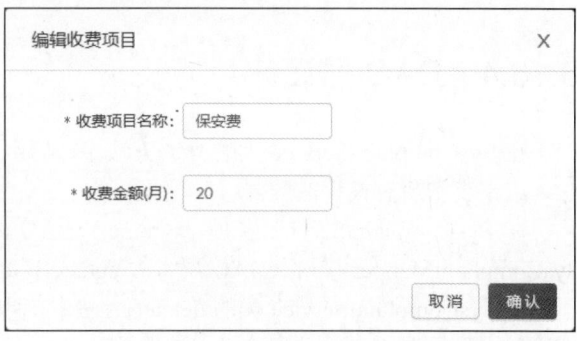

图 8.20 "新增收费项目"对话框的效果图　　图 8.21 "编辑收费项目"对话框的效果图

8.17.1 收费项目管理页面设计

收费项目管理页面对应的是 rq_charge_type.vue 文件，它是由分页插件、表格组件和按钮组件组成的。其中，表格组件的表头依次为编号、收费项目名称、收费金额(元)、创建时间、修改时间和操作。rq_charge_type.vue 文件的代码如下：

```
<template>
  <page-main>
    <a-card :loading="loading" title="收费项目管理">
      <div class="head">
        <a-button
          type="primary"
          style="height: 40px;"
          @click="chargeType_save_modalVisible = true"
        >添加收费项目</a-button>
        <a-button
          type="danger"
          v-if="table_selectedRowKeys.length > 0"
          style="height: 38px; margin-left: 10px;"
          @click="Del_batchData"
        >删除被选择的「收费项目」</a-button>
      </div>
      <a-table
        :data-source="chargeType_data_list"
        :row-selection="{ selectedRowKeys: table_selectedRowKeys, onChange: Table_selectChange }"
      >
        <a-table-column key="id" title="编号" data-index="id" />
        <a-table-column key="chargeName" title="收费项目名称" data-index="chargeName" />
        <a-table-column key="chargeMoney" title="收费金额(元)" data-index="chargeMoney" />
        <a-table-column key="createTime" title="创建时间" data-index="createTime">
          <!-- rTime -->
          <template slot-scope="text, record">
            <span>{{rTime(record.createTime)}}</span>
          </template>
        </a-table-column>
        <a-table-column key="updateTime" title="修改时间" data-index="updateTime">
          <!-- rTime -->
          <template slot-scope="text, record">
            <span>{{rTime(record.updateTime)}}</span>
          </template>
        </a-table-column>
        <a-table-column key="action" title="操作">
          <template slot-scope="text, record">
            <a-button-group>
```

```html
                <a-button type="primary" @click="Edit_chargeTypeData(record)">编辑</a-button>
                <a-button type="danger" @click="Del_chargeTypeData(record.id)">删除</a-button>
              </a-button-group>
            </template>
          </a-table-column>
        </a-table>
      </a-card>
      <!-- 新增或保存设施提示框 -->
      <a-modal
        v-model="chargeType_save_modalVisible"
        :title="chargeType_save_title"
        ok-text="确认"
        cancel-text="取消"
        :maskClosable="false"
        :destroyOnClose="false"
        @ok="Save_chargeTypeData"
      >
        <a-form-model
          :model="chargeType_form_data"
          :rules="rules"
          :label-col="labelCol"
          :wrapper-col="wrapperCol"
        >
          <a-form-model-item label="收费项目名称" prop="chargeName">
            <a-input v-model="chargeType_form_data.chargeName" />
          </a-form-model-item>
          <a-form-model-item label="收费金额(月)" prop="chargeName">
            <a-input v-model="chargeType_form_data.chargeMoney" />
          </a-form-model-item>
        </a-form-model>
      </a-modal>
    </page-main>
</template>

<script>
import { getChargeType, saveChargeType, deleteChargeType } from '@/api/requests/rq-manage.js'
import { Success, Warning } from '@/util/message.js'
import { rTime } from '@/util/time.js'

export default {
  data () {
    return {
      rTime,
      loading: false,
      labelCol: { span: 7 },
      wrapperCol: { span: 7 },
      table_selectedRowKeys: [],
      chargeType_query_type: 'chargeName',
      chargeType_query_buttonTitle: '搜索',
      chargeType_query_text: '',
      chargeType_save_title: '新增收费项目',
      chargeType_save_modalVisible: false,
      chargeType_form_data: {},
      chargeType_data_list: [],
      rules: {
        chargeName: [{ required: true, message: '此项为必填项', trigger: 'blur' }],
        chargeMoney: [{ required: true, message: '此项为必填项', trigger: 'blur' }],
      },
    }
  },
  created () {
    this.Get_chargeTypeDataList()
```

```javascript
    },
    watch: {
      chargeType_save_modalVisible (val) {
        if (!val) {
          this.chargeType_form_data = {}
        }
      }
    },
    methods: {
      Get_chargeTypeDataList () {
        getChargeType().then(res => {
          this.chargeType_query_buttonTitle = '搜索'
          this.chargeType_data_list = res.data
          this.chargeType_save_title = '新增收费项目'
        })
      },
      Query_chargeTypeDataList () {
        let text = this.chargeType_query_text
        let temp_list = []
        this.chargeType_data_list.forEach(item => {
          if (item[this.chargeType_query_type].indexOf(text) != -1) {
            temp_list.push(item)
          }
        })
        this.chargeType_query_buttonTitle = '返回'
        this.chargeType_data_list = temp_list
      },
      Edit_chargeTypeData (form) {
        this.chargeType_save_title = '编辑收费项目'
        this.chargeType_form_data = JSON.parse(JSON.stringify(form))
        this.chargeType_save_modalVisible = true
      },
      Del_chargeTypeData (id) {
        deleteChargeType(id).then(res => {
          if (res.code == 200) {
            Success(this, '操作成功')
          } else {
            Warning(this, '操作失败')
          }
          this.Get_chargeTypeDataList()
        })
      },
      Del_batchData () {
        this.table_selectedRowKeys.forEach(i => {
          this.Del_chargeTypeData(this.chargeType_data_list[i].id)
        })
        this.table_selectedRowKeys = []
      },
      Save_chargeTypeData () {
        saveChargeType(this.chargeType_form_data).then(res => {
          if (res.code == 200) {
            Success(this, '操作成功')
          } else {
            Warning(this, '操作失败')
          }
          this.chargeType_save_modalVisible = false
          this.Get_chargeTypeDataList()
        })
      },
      Table_selectChange (selectedRowKeys) {
        this.table_selectedRowKeys = selectedRowKeys;
      },
```

```
        },
    }
</script>

<style lang="scss" scoped>
.head {
    display: flex;
    justify-content: flex-start;
    margin-bottom: 10px;
    span {
        line-height: 40px;
        margin-right: 10px;
    }
    .ant-input-search {
        width: 30%;
        margin-left: 10px;
    }
}
.ant-modal-content {
    width: 24vw;
}
</style>
```

8.17.2 控制器类设计

SysChargeTypeController 类是收费项目模块的控制器类，被 @RestController 注解标注。在 SysChargeTypeController 类中，依次定义了 getList()方法、saveChargeType()方法和 deleteChargeType()方法。其中，getList()方法用于获取所有收费项目的数据；saveChargeType()方法用于保存新增的或者修改后的收费项目的数据；deleteChargeType()方法用于删除一条或者多条收费项目的数据。

SysChargeTypeController 类的 getList()方法、saveChargeType()方法和 deleteChargeType()方法的代码分别如下：

```
@GetMapping("/list")
public AjaxResult getList(){
    return AjaxResult.success(sysChargeTypeService.list());
}

@PreAuthorize("@ps.hasPermi('system:chargeType:save')")
@PostMapping("/save")
public AjaxResult saveChargeType(@RequestBody SysChargeType sysChargeType) {
    sysChargeTypeService.save(sysChargeType);
    return AjaxResult.success();
}

@PreAuthorize("@ps.hasPermi('system:chargeType:delete')")
@PostMapping("/delete")
public AjaxResult deleteChargeType(@RequestBody JSONObject jsonObject) {
    if(sysChargeTypeService.delete(jsonObject.getString("id"))){
        return AjaxResult.success();
    }
    return AjaxResult.error();
}
```

8.17.3 服务类设计

SysChargeTypeService 类被@Service 注解标注，表示收费项目模块的服务类。

在 SysChargeTypeController 类的 getList()方法中，调用了 SysChargeTypeService 类的 list()方法，用于查询所有收费项目的数据。

在 SysChargeTypeController 类的 saveChargeType()方法中，调用了 SysChargeTypeService 类的 save()方法，用于保存新增或者修改后的收费项目的数据。

在 SysChargeTypeController 类的 deleteChargeType()方法中，调用了 SysChargeTypeService 类的 delete()方法，用于删除一条或者多条收费项目的数据。

SysChargeTypeService 类的 list()方法、save()方法和 delete()方法的代码分别如下：

```java
public List<SysChargeType> list(){
    return sysChargeTypeMapper.selectList(null);
}

public void save(SysChargeType sysChargeType){
    if(findById(sysChargeType.getId()) != null){
        sysChargeTypeMapper.updateById(sysChargeType);
        return;
    }
    sysChargeTypeMapper.insert(sysChargeType);
}

public boolean delete(String id){
    return sysChargeTypeMapper.deleteById(id) > 0;
}
```

8.17.4　DAO 层设计

在 SysChargeTypeService 类的 list()方法中，调用了 SysChargeTypeMapper 接口（DAO 层）的 selectList()方法，用于查询所有收费项目的数据。

在 SysChargeTypeService 类的 save()方法中，调用了 SysChargeTypeMapper 接口（DAO 层）的 insert()方法或者 updateById()方法，用于新增收费项目的数据或者修改收费项目的数据。

在 SysChargeTypeService 类的 delete()方法中，调用了 SysChargeTypeMapper 接口（DAO 层）的 deleteById()方法，用于根据收费项目编号删除收费项目的数据。

SysChargeTypeMapper 接口的代码如下：

```java
@Mapper
public interface SysChargeTypeMapper extends BaseMapper<SysChargeType> {
    public void list();
}
```

> **说明**
>
> 因为 SysChargeTypeMapper 接口继承了 BaseMapper 接口，所以程序开发人员可以直接使用 BaseMapper 接口中常用的增、删、改、查方法，从而简化开发过程。

8.18　物业人员管理模块设计

如图 8.22 所示，管理员使用物业人员管理的功能，能够对物业人员信息执行增、删、改、查的操作。

图 8.22 物业人员管理页面的效果图

　　管理员首先单击"添加物业人员信息"按钮，然后在如图 8.23 所示的添加物业人员信息对话框中依次输入用户名、真实姓名和联系电话，最后单击"确认"按钮，以完成添加物业人员信息的操作。

　　管理员先单击"编辑"按钮，再在如图 8.24 所示的编辑物业人员信息对话框中对用户名、真实姓名或者联系电话予以修改，而后单击"确认"按钮，以完成编辑物业人员信息的操作。下面将介绍物业人员管理模块的设计过程。

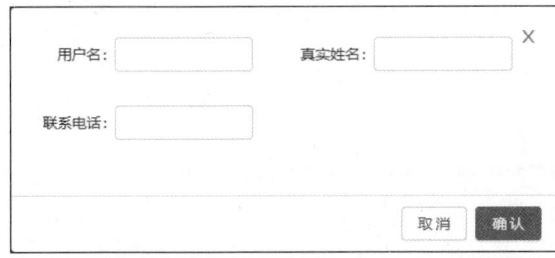

图 8.23　添加物业人员信息对话框的效果图　　　　图 8.24　编辑物业人员信息对话框的效果图

8.18.1　物业人员管理页面设计

　　物业人员管理页面对应的是 estate_user_manager.vue 文件，它是由分页插件、表格组件、switch 开关按钮和按钮组件组成的。其中，表格组件的表头依次为用户名、真实姓名、联系电话、是否启用和操作。estate_user_manager.vue 文件的代码如下：

```
<template>
  <page-main>
    <a-card :loading="loading" title="物业人员管理">
      <div class="head">
        <a-button
          type="primary"
          style="height: 40px;"
          @click="estateUser_save_modalVisible = true"
        >添加物业人员信息</a-button>
        <a-button
          type="danger"
          v-if="table_selectedRowKeys.length > 0"
          style="height: 40px; margin-left: 10px;"
          @click="Del_batchData"
        >删除已选择的物业人员</a-button>
      </div>
      <a-table
        :data-source="estateUser_data_list"
```

```html
            :row-selection="{ selectedRowKeys: table_selectedRowKeys, onChange: Table_selectChange }"
        >
            <a-table-column key="userName" title="用户名" data-index="userName" />
            <a-table-column key="fullName" title="真实姓名" data-index="fullName" />
            <a-table-column key="phone" title="联系电话" data-index="phone" />
            <a-table-column key="status" title="是否启用" data-index="status">
                <template slot-scope="text, record">
                    <a-switch v-model="record.status == 0" @change="Change_estateUserStatus(record)" />
                </template>
            </a-table-column>
            <a-table-column key="action" title="操作">
                <template slot-scope="text, record">
                    <a-button-group>
                        <a-button type="primary" @click="Edit_estateUserData(record)">编辑</a-button>
                        <a-button type="danger" @click="Del_estateUserData(record.id)">删除</a-button>
                    </a-button-group>
                </template>
            </a-table-column>
        </a-table>
    </a-card>
    <!-- 新增或保存物业管理人员提示框 -->
    <a-modal
        v-model="estateUser_save_modalVisible"
        ok-text="确认"
        cancel-text="取消"
        :maskClosable="false"
        :destroyOnClose="false"
        @ok="Save_estateUserData"
    >
        <a-form-model :model="estateUser_form_data" :label-col="labelCol" :wrapper-col="wrapperCol">
            <el-row :gutter="20">
                <el-col :span="12" :offset="0">
                    <a-form-model-item label="用户名">
                        <a-input v-model="estateUser_form_data.userName" />
                    </a-form-model-item>
                </el-col>
                <el-col :span="12" :offset="0">
                    <a-form-model-item label="真实姓名">
                        <a-input v-model="estateUser_form_data.fullName" />
                    </a-form-model-item>
                </el-col>
            </el-row>

            <el-row :gutter="20">
                <el-col :span="12" :offset="0">
                    <a-form-model-item label="联系电话">
                        <a-input v-model="estateUser_form_data.phone" />
                    </a-form-model-item>
                </el-col>
            </el-row>
        </a-form-model>
    </a-modal>
  </page-main>
</template>

<script>
import { getEstateUser, saveUser, deleteUser } from '@/api/requests/rq-manage.js'
import { Success, Warning } from '@/util/message.js'
import { rTime } from '@/util/time.js'

export default {
  data () {
```

```js
    return {
      loading: false,
      labelCol: { span: 8 },
      wrapperCol: { span: 14 },
      table_selectedRowKeys: [],
      estateUser_query_type: 'name',
      estateUser_query_buttonTitle: '搜索',
      estateUser_query_text: '',
      estateUser_save_modalVisible: false,
      estateUser_form_data: {},
      estateUser_data_list: [],
    }
  },
  created () {
    this.Get_estateUserDataList()
  },
  watch: {
    estateUser_save_modalVisible (val) {
      if (!val) {
        this.estateUser_form_data = {}
      }
    }
  },
  methods: {
    Get_estateUserDataList () {
      getEstateUser().then(res => {
        this.estateUser_query_buttonTitle = '搜索'
        this.estateUser_data_list = res.data
      })
    },
    Query_estateUserDataList () {
      let text = this.estateUser_query_text
      let temp_list = []
      this.estateUser_data_list.forEach(item => {
        if (item[this.estateUser_query_type].indexOf(text) != -1) {
          temp_list.push(item)
        }
      })
      this.estateUser_query_buttonTitle = '返回'
      this.estateUser_data_list = temp_list
    },
    Edit_estateUserData (form) {
      this.estateUser_form_data = JSON.parse(JSON.stringify(form))
      this.estateUser_save_modalVisible = true
    },
    Del_estateUserData (id) {
        deleteUser(id).then(res => {
        if (res.code == 200) {
          Success(this, '操作成功')
        } else {
          Warning(this, '操作失败')
        }
        this.Get_estateUserDataList()
      })
    },
    Del_batchData () {
      this.table_selectedRowKeys.forEach(i => {
        this.Del_estateUserData(this.estateUser_data_list[i].id)
      })
      this.table_selectedRowKeys = []
    },
    Save_estateUserData () {
```

```
          saveUser(this.estateUser_form_data).then(res => {
            if (res.code == 200) {
              Success(this, '操作成功')
            } else {
              Warning(this, '操作失败')
            }
            this.estateUser_save_modalVisible = false
            this.Get_estateUserDataList()
          })


        },
        Table_selectChange (selectedRowKeys) {
          this.table_selectedRowKeys = selectedRowKeys;
        },
        Change_estateUserStatus (r) {
          r.status = r.status == 0 ? 1 : 0
          this.estateUser_form_data = r
          this.Save_estateUserData()
        }
      },
    }
</script>

<style lang="scss" scoped>
.head {
    display: flex;
    justify-content: flex-start;
    margin-bottom: 10px;
    span {
        line-height: 40px;
        margin-right: 10px;
    }
    .ant-input-search {
        width: 30%;
        margin-left: 10px;
    }
}
</style>
```

8.18.2 控制器类设计

SysUserController 类是物业人员管理模块的控制器类,被@RestController 注解标注。在 SysUserController 类中,依次定义了 list()方法、save()方法和 delete()方法。其中,list()方法用于获取所有物业人员的信息;save()方法用于保存新添加或者修改后的物业人员的信息;delete()方法用于删除一条或者多条物业人员的信息。

SysUserController 类的 list()方法、save()方法和 delete()方法的代码分别如下:

```
@GetMapping("/list")
public AjaxResult list() {
    return AjaxResult.success(sysUserService.list());
}

@PostMapping("/save")
public AjaxResult save(@RequestBody SysUser sysUser, HttpServletRequest req) {
    sysUser.setLoginIp(IpUtils.getIpAddr());
    sysUser.setLoginDate(new Date());
    sysUserService.save(sysUser);
    return AjaxResult.success();
}
```

```
@PostMapping("/delete/{id}")
public AjaxResult delete(@PathVariable("id") String id) {
    sysUserService.deleteEstateUser(id);
    return AjaxResult.success();
}
```

8.18.3　服务类设计

SysUserService 类被@Service 注解标注，表示物业人员管理模块的服务类。

在 SysUserController 类的 list()方法中，调用了 SysUserService 类的 list()方法，用于查询所有物业人员的信息。

在 SysUserController 类的 save()方法中，调用了 SysUserService 类的 save()方法，用于保存新添加或者修改后的物业人员的信息。

在 SysUserController 类的 delete()方法中，调用了 SysUserService 类的 deleteEstateUser()方法，用于删除一条或者多条物业人员的信息。

SysUserService 类的 list()方法、save()方法和 deleteEstateUser()方法的代码分别如下：

```
public List<SysUser> list() {
    return sysUserMapper.selectList(null);
}

public void save(SysUser sysUser) {
    if (sysUser.getId() != null) {
        sysUserMapper.updateById(sysUser);
        return;
    }else{
        sysUser.setStatus("0");
        sysUser.setPassword("123456");
        sysUser.setPassword(
                bCryptPasswordEncoder.encode(sysUser.getPassword()));
        sysUserMapper.insert(sysUser);
    }
}

public void deleteEstateUser(String id) {
    sysUserMapper.deleteById(id);
}
```

8.18.4　DAO 层设计

在 SysUserService 类的 list()方法中，调用了 SysUserMapper 接口（DAO 层）的 selectList()方法，用于查询所有物业人员的信息。

在 SysUserService 类的 save()方法中，调用了 SysUserMapper 接口（DAO 层）的 insert()方法或者 updateById()方法，用于新添加或者修改物业员工的信息。

在 SysUserService 类的 deleteEstateUser()方法中，调用了 SysUserMapper 接口（DAO 层）的 deleteById()方法，用于根据编号删除物业人员的信息。

SysUserMapper 接口的代码如下：

```
@Mapper
public interface SysUserMapper extends BaseMapper<SysUser> {
}
```

> **说明**
> 因为 SysUserMapper 接口继承了 BaseMapper 接口，所以程序开发人员可以直接使用 BaseMapper 接口中常用的增、删、改、查方法，从而简化开发过程。

8.19　住户信息管理模块设计

如图 8.25 所示，管理员使用住户信息管理的功能，能够对住户信息执行增、删、改、查的操作。

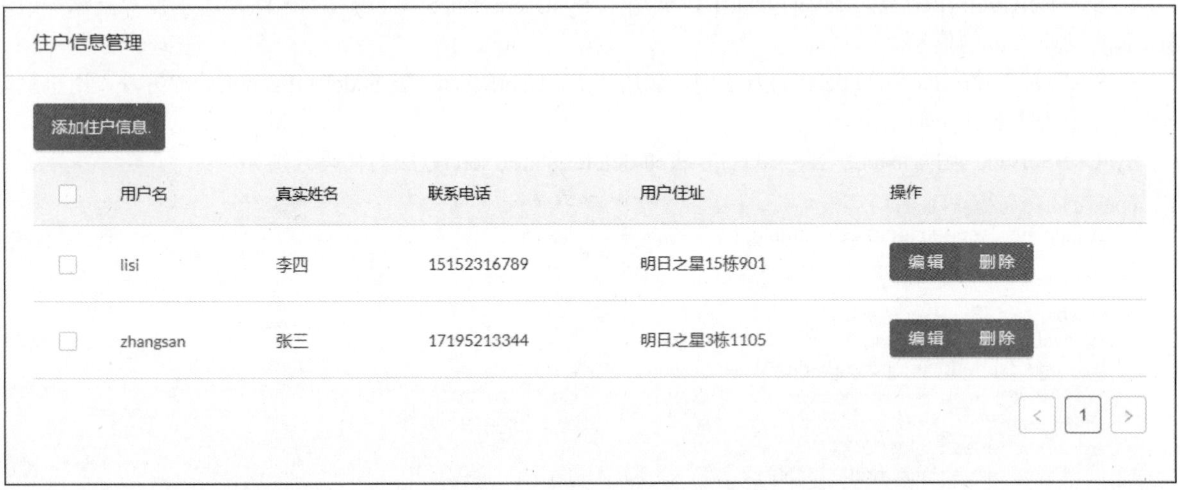

图 8.25　住户信息管理页面的效果图

管理员首先单击"添加住户信息"按钮，然后在如图 8.26 所示的添加住户信息对话框中依次输入用户名、用户住址、真实姓名和住户电话，最后单击"确认"按钮，以完成添加住户信息的操作。

管理员先单击"编辑"按钮，再在如图 8.27 所示的编辑住户信息对话框中对用户名、用户住址、真实姓名或者住户电话予以修改，而后单击"确认"按钮，以完成编辑住户信息的操作。下面将介绍住户管理模块的设计过程。

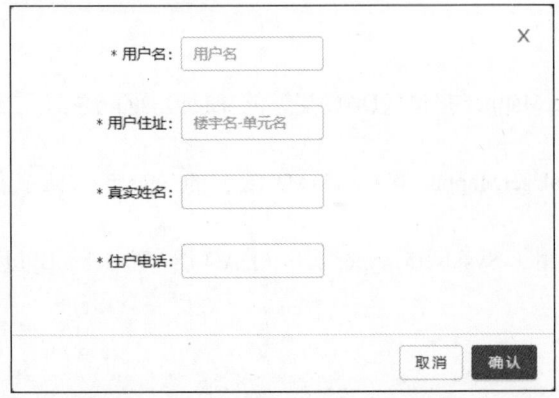

图 8.26　添加住户信息对话框的效果图

图 8.27　编辑住户信息对话框的效果图

8.19.1　住户信息管理页面设计

住户信息管理页面对应的是 user_household.vue 文件，它是由分页插件、表格组件和按钮组件组成的。其中，表格组件的表头依次为用户名、真实姓名、联系电话、用户住址和操作。user_household.vue 的代码如下：

```html
<template>
  <page-main>
    <a-card :loading="loading" title="住户信息管理">
      <div class="head">
        <a-button
          type="primary"
          style="height: 40px;"
          @click="household_save_modalVisible = true"
        >添加住户信息</a-button>
        <a-button
          type="danger"
          v-if="table_selectedRowKeys.length > 0"
          style="height: 38px; margin-left: 10px;"
          @click="Del_batchData"
        >删除已选择的住户信息</a-button>
      </div>
      <a-table
        :data-source="household_data_list"
        :row-selection="{ selectedRowKeys: table_selectedRowKeys, onChange: Table_selectChange }"
      >
        <a-table-column key="userName" title="用户名" data-index="userName" />
        <a-table-column key="fullName" title="真实姓名" data-index="fullName" />
        <a-table-column key="phone" title="联系电话" data-index="phone" />
        <a-table-column key="address" title="用户住址" data-index="address" />

        <a-table-column key="action" title="操作">
          <template slot-scope="text, record">
            <a-button-group>
              <a-button type="primary" @click="Edit_householdData(record.id)">编辑</a-button>
              <a-button type="danger" @click="Del_householdData(record.id)">删除</a-button>
            </a-button-group>
          </template>
        </a-table-column>
      </a-table>
    </a-card>
    <!-- 新增或保存设施提示框 -->
    <a-modal
      v-model="household_save_modalVisible"
      ok-text="确认"
      cancel-text="取消"
      :maskClosable="false"
      :destroyOnClose="false"
      @ok="Save_householdData"
    >
      <a-form-model
        :model="household_form_data"
        :rules="rules"
        :label-col="labelCol"
        :wrapper-col="wrapperCol"
      >
        <a-form-model-item label="用户名" prop="userName">
          <a-input v-model="household_form_data.userName" placeholder="用户名"/>
        </a-form-model-item>
        <a-form-model-item label="用户住址" prop="address">
```

```html
                    <a-input v-model="household_form_data.address" placeholder="楼宇名-单元名"/>
                </a-form-model-item>
                <a-form-model-item label="真实姓名" prop="fullName">
                    <a-input v-model="household_form_data.fullName" />
                </a-form-model-item>
                <a-form-model-item label="住户电话" prop="phone">
                    <a-input v-model="household_form_data.phone" />
                </a-form-model-item>
            </a-form-model>
        </a-modal>
    </page-main>
</template>
```

```javascript
<script>
import {
    getHousehold,
    saveHousehold,
    deleteHousehold,
    downloadHouseholds,
    getUsers,
    registerUser,
    getHouseholdById,
} from '@/api/requests/rq-manage.js'
import { Success, Warning } from '@/util/message.js'
import { download } from '@/util/download.js'
import { rTime } from '@/util/time.js'

export default {
    data () {
        return {
            rTime,
            loading: false,
            labelCol: { span: 7 },
            wrapperCol: { span: 7 },
            table_selectedRowKeys: [],
            household_query_type: 'userName',
            household_query_buttonTitle: '搜索',
            household_query_text: '',
            household_save_modalVisible: false,
            household_form_data: {},
            household_data_list: [],
            rules: {
                userName: [{ required: true, message: '此项为必填项', trigger: 'blur' }],
                fullName: [{ required: true, message: '此项为必填项', trigger: 'blur' }],
                phone: [{ required: true, message: '此项为必填项', trigger: 'blur' }],
                address: [{ required: true, message: '此项为必填项', trigger: 'blur' }],
            },
            user_data_list: [],
            household_form_search_userNames: [],
            household_form_userName_isChoice: false,
            household_user_register_from: {},
        }
    },
    created () {
        this.Get_householdDataList()
        this.Get_users()
    },
    watch: {
        household_save_modalVisible (val) {
            if (!val) {
                this.household_form_data = {}
            }
```

```js
    }
},
methods: {
    Get_householdDataList () {
        getHousehold().then(res => {
            this.household_query_buttonTitle = '搜索'
            this.household_data_list = res.data
        })
    },
    Get_users () {
        getUsers().then(res => {
            this.user_data_list = res.data
        })
    },
    Query_householdDataList () {
        let text = this.household_query_text
        let temp_list = []
        this.household_data_list.forEach(item => {
            if (item[this.household_query_type].indexOf(text) != -1) {
                temp_list.push(item)
            }
        })
        this.household_query_buttonTitle = '返回'
        this.household_data_list = temp_list
    },
    Edit_householdData (id) {
        getHouseholdById(id).then(res => {
            if (res.code == 200) {
                this.household_form_data = res.data
                this.household_save_modalVisible = true
            } else {
                Warning(this, '操作失败')
            }
        });
    },
    Del_householdData (id) {
        deleteHousehold(id).then(res => {
            if (res.code == 200) {
                Success(this, '操作成功')
            } else {
                Warning(this, '操作失败')
            }
            this.Get_householdDataList()
        })
    },
    Del_batchData () {
        this.table_selectedRowKeys.forEach(i => {
            this.Del_householdData(this.household_data_list[i].id)
        })
        this.table_selectedRowKeys = []
    },
    async Save_householdData () {
        this.household_form_data.password = '123456'
        saveHousehold(this.household_form_data).then(res => {
            if (res.code == 200) {
                Success(this, '操作成功')
            } else {
                Warning(this, '操作失败')
            }
            this.household_save_modalVisible = false
            this.Get_householdDataList()
        })
```

```
        },
        Download_householdsExcel () {
            downloadHouseholds().then(res => {
                download('社区住户信息.xlsx', res.data)
            })
        },
        Table_selectChange (selectedRowKeys) {
            this.table_selectedRowKeys = selectedRowKeys;
        },
        async Register_User (from) {
            this.$message.info('系统自动为此账户注册，默认密码为「123456」')
            return await registerUser(from).then(res => {
                return res.data
            })
        },
        Users_choice_onSelect (value) {
            let flag = false
            this.user_data_list.forEach(item => {
                if (item.userName.indexOf(value) != -1) {
                    flag = true
                    this.household_form_data.userId = item.id
                }
            })
        },
        Users_choice_handleSearch (value) {
            let flag = false
            this.household_form_search_userNames = []
            this.user_data_list.forEach(item => {
                if (item.userName.indexOf(value) != -1) {
                    flag = true
                    this.household_form_userName_isChoice = true
                    this.household_form_search_userNames.push(item.userName)
                }
            })
            if (!flag) {
                this.household_user_register_from.userName = value
                this.household_form_userName_isChoice = false
            }
        },
    },
}
</script>

<style lang="scss" scoped>
.head {
    display: flex;
    justify-content: flex-start;
    margin-bottom: 10px;
    span {
        line-height: 40px;
        margin-right: 10px;
    }
    .ant-input-search {
        width: 30%;
        margin-left: 10px;
    }
}
.ant-modal-content {
    width: 24vw;
}
</style>
```

8.19.2 控制器类设计

SysUserController 类是住户信息管理模块的控制器类,被@RestController 注解标注。在 SysUserController 类中,依次定义了 HouseholdInfoList()方法、HouseholdInfoSave()方法和 HouseholdInfoDelete()方法。其中,HouseholdInfoList()方法用于获取所有住户的信息;HouseholdInfoSave()方法用于保存新添加或者修改后的住户的信息;HouseholdInfoDelete()方法用于删除一条或者多条住户的信息。

SysUserController 类的 HouseholdInfoList()方法、HouseholdInfoSave()方法和 HouseholdInfoDelete()方法的代码分别如下:

```java
@GetMapping("/household/list")
//@PreAuthorize("@ps.hasPermi('system:user_householdInfo:list')")
public AjaxResult HouseholdInfoList() {
    return AjaxResult.success(sysUserService.HouseholdInfoList());
}

@PostMapping("/household/save")
//@PreAuthorize("@ps.hasPermi('system:user_householdInfo:save')")
public AjaxResult HouseholdInfoSave(@RequestBody SysUserInfoData sysUserInfoData) {
    sysUserService.HouseholdInfoSave(sysUserInfoData);
    return AjaxResult.success();
}

@PostMapping("/household/delete/{id}")
public AjaxResult HouseholdInfoDelete(@PathVariable("id") String id) {
    sysUserService.deleteUser(id);
    return AjaxResult.success();
}
```

8.19.3 服务类设计

SysUserService 类被@Service 注解标注,表示住户信息管理模块的服务类。

在 SysUserController 类的 HouseholdInfoList()方法中,调用了 SysUserService 类的 HouseholdInfoList()方法,用于查询所有住户的信息。

在 SysUserController 类的 HouseholdInfoSave()方法中,调用了 SysUserService 类的 HouseholdInfoSave()方法,用于保存新添加或者修改后的住户的信息。

在 SysUserController 类的 HouseholdInfoDelete()方法中,调用了 SysUserService 类的 deleteUser()方法,用于删除一条或者多条住户的信息。

SysUserService 类的 HouseholdInfoList()方法、HouseholdInfoSave()方法和 deleteUser()方法的代码分别如下:

```java
public List<SysUserInfoData> HouseholdInfoList() {
    return sysUserMapper.householdInfoList();
}

public void HouseholdInfoSave(SysUserInfoData sysUserInfoData) {
    //更新用户数据
    SysUser sysUser = findById(sysUserInfoData.getId());
    if (sysUser != null) {
        sysUser.setFullName(sysUserInfoData.getFullName());
        sysUser.setPhone(sysUserInfoData.getPhone());
        sysUser.setAddress(sysUserInfoData.getAddress());
        save(sysUser);
    }else{
        sysUser = new SysUser();
```

```java
            sysUser.setFullName(sysUserInfoData.getFullName());
            sysUser.setLoginDate(new Date());
            sysUser.setUserName(sysUserInfoData.getUserName());
            sysUser.setAddress(sysUserInfoData.getAddress());
            sysUser.setStatus("0");
            sysUser.setPhone(sysUserInfoData.getPhone());
            sysUser.setPassword(
                    bCryptPasswordEncoder.encode(sysUserInfoData.getPassword()));
            save(sysUser);
        }
    }
    public void save(SysUser sysUser) {
        if (sysUser.getId() != null) {
            sysUserMapper.updateById(sysUser);
            return;
        }else{
            sysUser.setStatus("0");
            sysUser.setPassword("123456");
            sysUser.setPassword(
                    bCryptPasswordEncoder.encode(sysUser.getPassword()));
            sysUserMapper.insert(sysUser);
        }
    }
    public void deleteUser(String id) {
        sysUserMapper.deleteById(id);
    }
```

8.19.4 DAO 层设计

在 SysUserService 类的 HouseholdInfoList()方法中，调用了 SysUserMapper 接口（DAO 层）的 householdInfoList()方法，用于查询所有住户的信息。

在 SysUserService 类的 HouseholdInfoSave()方法中，调用了其自身的 save()方法，在 save()方法中又调用了 SysUserMapper 接口（DAO 层）的 insert()方法或者 updateById()方法，用于新添加或者修改住户的信息。

在 SysUserService 类的 deleteUser()方法中，调用了 SysUserMapper 接口（DAO 层）的 deleteById()方法，用于根据用户编号删除住户的信息。

SysUserMapper 接口的代码如下：

```java
@Mapper
public interface SysUserMapper extends BaseMapper<SysUser> {
    @Select("SELECT u.* FROM sys_user u inner join sys_user_role ur on u.id=ur.user_id where ur.role_id=1")
    public List<SysUserInfoData> householdInfoList();
}
```

> **说明**
> 因为 SysUserMapper 接口继承了 BaseMapper 接口，所以程序开发人员可以直接使用 BaseMapper 接口中常用的增、删、改、查方法，从而简化开发过程。

8.20 退出登录模块设计

不论是前台页面，还是后台页面，都有"退出登录"超链接。前台页面的"退出登录"超链接如图 8.28

所示,后台页面的"退出登录"超链接如图 8.29 所示。因为"退出登录"超链接的作用只是从当前页面跳转到登录页面,所以退出登录模块不需要服务类或者 DAO 层的支持。

图 8.28　前台页面的"退出登录"超链接

图 8.29　后台页面的"退出登录"超链接

退出登录模块的相关代码被编写在 em_ui\src\layout\components\UserMenu\index.vue 文件中,代码如下:

```
<template>
  <div class="user">
    <div class="tools">
      <el-tooltip
        v-if="$store.state.settings.enableThemeSetting"
        effect="dark"
        content="主题配置"
        placement="bottom"
      >
        <span class="item" @click="$eventBus.$emit('global-theme-toggle')">
          <svg-icon name="theme" />
        </span>
      </el-tooltip>
    </div>
    <el-dropdown class="user-container" @command="handleCommand">
      <div class="user-wrapper">
        <el-avatar size="medium">
          <i class="el-icon-user-solid" />
        </el-avatar>
        {{ $store.state.user.account }}
        <i class="el-icon-caret-bottom" />
      </div>
      <el-dropdown-menu slot="dropdown" class="user-dropdown">
        <el-dropdown-item divided command="logout">退出登录</el-dropdown-item>
      </el-dropdown-menu>
    </el-dropdown>
  </div>
</template>
//省略部分代码
```

SysLoginController 类是(退出)登录模块的控制器类,被@RestController 注解标注。在 SysUserController 类中,定义了一个用于退出登录的 logout()方法,代码如下:

```
@GetMapping("/logout")
```

```
private AjaxResult logout() {
    RequestUtils.invalidate();
    return AjaxResult.success();
}
```

8.21 项目运行

通过前述步骤，设计并完成了"明日之星物业管理系统"项目的开发。该项目是一个具有前、后端的项目，因此运行本项目需要两个步骤：先启动后端，再启动前端。下面运行本项目，以检验我们的开发成果。首先选择 EmServerApplication，单击▶快捷图标，如图 8.30 所示；然后选择 serve，单击▶快捷图标，如图 8.31 所示，即可运行本项目。

启动项目后，明日之星物业管理系统的登录页面将被自动打开，如图 8.32 所示。在登录页面上，用户输入正确的用户名和密码后，程序还要对用户的身份进行判断。如果用户的身份是住户，那么程序打开的是明日之星物业管理系统的前台页面。如果用户的身份是管理员，那么程序打开的是明日之星物业管理系统的后台页面。这样，我们就成功地检验了本项目的运行。

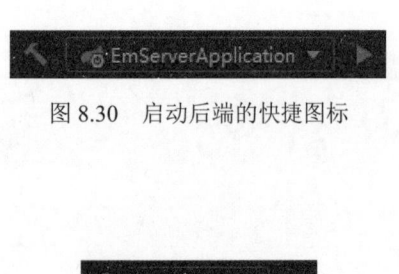

图 8.30 启动后端的快捷图标

图 8.31 启动前端的快捷图标　　　　图 8.32 明日之星物业管理系统的登录页面

Spring Boot 各层之间的交互主要遵循控制层、服务层、数据访问层以及模型层的基本架构。其中，控制层负责处理外部请求，接收输入并返回响应，它调用服务层来处理业务逻辑，并将结果返回给客户端；服务层包含业务逻辑处理，它调用数据访问层来执行数据库操作，并将处理结果返回给控制层；数据访问层负责与数据库进行交互，执行 SQL 语句，完成数据的增、删、改、查操作；模型层包含与数据库表对应的实体类，用于数据的表示和传输。

8.22 源码下载

虽然本章详细地讲解了如何编码实现"明日之星物业管理系统"项目的各个功能，但给出的代码都是代码片段，而非源码。为了方便读者学习，本书提供了完整的项目源码，扫描右侧二维码即可下载。

源码下载